THE LAWS OF SPACEFLIGHT

A GUIDEBOOK FOR NEW SPACE LAWYERS

MATTHEW J. KLEIMAN · JENIFER K. LAMIE
MARIA-VITTORIA "GIUGI" CARMINATI

FOREWORD BY MARC HOLZAPFEL, LEGAL DIRECTOR, VIRGIN GALACTIC

Cover by Kelly Book/ABA Publishing.
Image of the International Space Station courtesy of NASA.

The materials contained herein represent the views of each chapter author in his or her individual capacity and should not be construed as the views of the author's firms, employers, or clients, or of the editors or other chapter authors, or of the American Bar Association or the Section of Science & Technology Law, unless adopted pursuant to the bylaws of the Association.

Nothing contained in this book is to be considered as the rendering of legal advice for specific cases, and readers are responsible for obtaining such advice from their own legal counsel. This book is intended for educational and informational purposes only.

© 2012 American Bar Association. All rights reserved.

No part of this publication may be reproduced, stored in a retrieval system, or transmitted in any form or by any means, electronic, mechanical, photocopying, recording, or otherwise, without the prior written permission of the publisher. For permission, contact the ABA Copyrights and Contracts Department by e-mail at copyright@americanbar.org or fax at 312-988-6030, or complete the online request form at http://www.americanbar.org/utility/reprint.

Printed in the United States of America.

16 15 14 13 12   5 4 3 2 1

Library of Congress Cataloging-in-Publication Data

Kleiman, Matthew.
  The laws of spaceflight : a guidebook for new space lawyers / By Matthew Kleiman, Jenifer Lamie, and Maria-Vittoria Carminati.
      p. cm.
  Includes bibliographical references and index.
  ISBN 978-1-61438-598-1 (print : alk. paper)
  1. Space law—United States. 2. Space flight. 3. Aeronautics—Law and legislation—United States. I. Lamie, Jenifer. II. Weil, Maria-Vittoria Carminati. III. Title.
  KZD1146.U6K54 2012
  341.4'7—dc23
                                                        2012026033

Discounts are available for books ordered in bulk. Special consideration is given to state bars, CLE programs, and other bar-related organizations. Inquire at Book Publishing, ABA Publishing, American Bar Association, 321 North Clark Street, Chicago, Illinois 60654-7598.

www.ShopABA.org

Contents

About the Authors   xi
Acknowledgments   xiii
Foreword   xv
Introduction   xvii

CHAPTER 1
Where To Go and How To Get There   1

I. Spaceflight Operations   1
II. The Physics of Spaceflight   3
    A. Earth/Space Boundary   3
    B. Describing an Orbit   3
    C. Spacecraft Orbits   5
    D. Getting into and Staying in Orbit   8
III. Power and Propulsion   11
    A. Delta-v   11
    B. Spacecraft Propulsion Systems   13
    C. Internal Power Systems   15
IV. Satellite Systems   16
    A. Components   16
    B. Threats and Security   18
    C. Constellations   18
V. Hazards of the Outer Space Environment   19
    A. Sustaining Human Life   19
    B. Microgravity   20
    C. Radiation   20
    D. Micrometeoroids and Space Debris   22
VI. Practical Uses of Outer Space   24
    A. Telecommunications   24
    B. Satellite Navigation   25
    C. Earth Observation   25
    D. Space Exploration   26
    E. Space Tourism   26
Glossary   26

CHAPTER 2
The History of Spaceflight   31

   I. Theory and Early Practice   31
   II. Ready . . . Set . . . Space Race!   33
       A. Unmanned Spaceflight   33
       B. Manned Spaceflights   39
   III. Other National Space Programs   46
       A. China   46
       B. Europe   47
       C. India   47
       D. Japan   48
   IV. Commercial Spaceflight   48
       A. Suborbital   49
       B. Orbital   51
       C. Beyond LEO   54
       D. Spaceports   55

CHAPTER 3
The International Outer Space Legal Framework   57

   I. The International Lawmaking Process   57
       A. Sources of International Law   57
       B. International Dispute Resolution   58
   II. Making International Agreements in the United States   59
   III. International Sources of Space Law   60
       A. Major Outer Space Treaties   61
       B. U.N. Outer Space Resolutions   66
   IV. International Organizations Relevant to Space Operations   67
       A. United Nations Committee on Peaceful Uses of Outer Space and Office of Outer Space Affairs   67
       B. European Space Agency (ESA)   68
       C. International Telecommunications Union (ITU)   68
       D. Other International Organizations of Relevance   69

CHAPTER 4
The Development of U.S. Space Law — 71

    I. The 1950s: Creation of a Civil Space Agency — 71
    II. The 1960s: Exploring Space Not Because It Is Easy, But Because It Is Hard — 73
        A. The Wiesner Report — 73
        B. John F. Kennedy and the Apollo Program — 75
        C. Communications Satellite Act — 75
    III. The 1970s: The Space Shuttle — 75
    IV. The 1980s: Developing a Framework for Commercial Space Transportation and Remote Sensing — 76
        A. Commercial Space Launch Act — 76
        B. Land Remote-Sensing Commercialization Act — 77
        C. National Aeronautics and Space Act Amendment — 77
    V. The 1990s: Where Do We Go from Here? — 77
        A. The Augustine Report — 78
        B. Land Remote Sensing Policy Act — 79
    VI. The Early 2000s: Focusing on the Details of Commercial Space Regulation — 80
        A. Commercial Space Launch Amendments Act — 80
        B. Regulating Commercial Remote Sensing Data — 80
    VII. The Late 2000s: Codifying Space Law as an Independent Body of Law — 82

CHAPTER 5
Licensing Commercial Spaceflight — 83

    I. FAA Jurisdiction — 83
    II. Obtaining an FAA License — 86
        A. Preliminary Consultation — 87
        B. Policy Review — 88
        C. Safety Review — 88
        D. Environmental Review — 90
        E. Demonstration of Financial Responsibility — 91
        F. Postlicensing Requirements — 91
        G. Payload Reviews — 92

	III. Waivers: A Case Study	92
	IV. Experimental Permits	95
	V. Training, Medical, and Informed Consent Requirements	95
	A. SFPs	95
	B. Crew	97
	Glossary	98

Chapter 6
Liability and Insurance Issues for Private Spaceflight   103

	I. Identifying Sources of Risk	103
	II. Liability Regimes within the United States	104
	A. Liability at the National Level	105
	B. State Space Tourism Liability Laws	107
	Sample Warning Statements	111
	III. Insurance Practices	113
	A. Stages of Spaceflight to Insure	113
	B. Obtaining Insurance	114
	C. Global Insurance Market	114
	IV. Contractual Risk Allocation	115

Chapter 7
Licensing Private Telecommunication Satellites   117

	I. International Coordination of Satellite Services	118
	II. ITU Allocation and Allotment	120
	III. FCC Licensing of Private Telecommunications Satellites	122
	A. Underlying Spectrum Management Principles	123
	B. Licensing the Space Station Segment of Satellite Systems	124
	C. Licensing Earth Stations	129
	Glossary	130

Chapter 8
Licensing Private Earth Remote Sensing Satellites   133

   I. NOAA Licensing Process   133
      A. Who Must Be Licensed   134
      B. Application Process   134
      C. Licensing Conditions   135
      D. Monitoring and Compliance Requirements   137
      E. Prohibition on Collection and Release of Satellite Imagery Relating to Israel   137
   II. United Nations Resolution on Data Dissemination   138
   Glossary   139

Chapter 9
Compliance with Export Control Laws   141

   I. International Traffic in Arms Regulations   142
      A. The United States Munitions List   145
      B. ITAR Export Authorizations   147
      C. Special Export Controls for Foreign Satellite Launches   151
      D. Violations and Penalties   151
   II. Export Administration Regulations   152
   III. Export Compliance and Monitoring   155
      A. Overview of the Export Compliance Process   155
      B. The Empowered Official   156
      C. Screening for Restrictions and Diversion Risks   157
      D. Safeguarding Technical Data   160
      E. Foreign National Employees   162
      F. International Business Travel   163
      G. Requesting Clearance for Information Release   164
      H. Record Keeping   164
   IV. Export Control Reform   165
   Glossary   167

Chapter 10
The U.S. Government as Your Customer — 173

   I. U.S. Government Space Activities — 173
   II. Key Government Contracting Laws and Regulations — 176
   III. Types of Government Contracts — 179
      A. FAR-Based Contracts — 179
      B. CRADAs and Space Act Agreements — 180
   IV. Federal Contracting Process — 181
      A. Government Contracting Personnel and Agencies — 181
      B. Solicitation Process — 182
      C. Protests — 185
   V. Contract Administration — 185
      A. Accounting — 185
      B. Changes — 187
      C. Termination — 187
      D. Disputes — 188
      E. Subcontracting — 188
      F. Liability — 189
      G. Socioeconomic Obligations — 191
      H. Protecting Classified Information — 191
   VI. Ethical Obligations — 192
   VII. Intellectual Property Rights — 194
      A. Inventions — 195
      B. Data Rights — 196
   Glossary — 199

Chapter 11
Space Operations and the Environment — 205

   I. International Environmental Law — 205
   II. Contamination — 207
      A. Launch Vehicles — 208
      B. Planetary Protection — 210

CONTENTS ix

 III. Nuclear Power Sources 213
 IV. Space Debris 215
  A. Tracking 217
  B. Mitigation 218
  C. Removal 221
 Glossary 223

CHAPTER 12
Property Rights 225

 I. Tangible Property 225
 II. Intellectual Property 228
  A. Patents 228
  B. Trade Secrets 231
  C. International Treaties 232

Appendices 235

 Appendix 3-1—Treaty on Principles Governing the Activities of States in the Exploration and Use of Outer Space, including the Moon and Other Celestial Bodies 237
 Appendix 3-2—Agreement on the Rescue of Astronauts, the Return of Astronauts and the Return of Objects Launched into Outer Space 245
 Appendix 3-3—Convention on International Liability for Damage Caused by Space Objects 251
 Appendix 3-4—Convention on Registration of Objects Launched into Outer Space 263
 Appendix 3-5—Agreement Governing the Activities of States on the Moon and Other Celestial Bodies 269
 Appendix 3-6—Principles Governing the Use by States of Artificial Earth Satellites for International Direct Television Broadcasting 281
 Appendix 3-7—Principles Relating to Remote Sensing of the Earth from Outer Space 287
 Appendix 3-8—Principles Relevant to the Use of Nuclear Power Sources In Outer Space 293

Appendix 3-9—Declaration on International Cooperation in the Exploration and Use of Outer Space for the Benefit and in the Interest of All States, Taking into Particular Account the Needs of Developing Countries 303
Appendix 6—Sample Contract for Launch Services 307
Appendix 9—U.S. Munitions List Categories IV and XV, Sample Technical Assistance (or Manufacturing License) Agreement, and Sample Technology Control Plan 327
Appendix 11-1—U.S. Government Orbital Debris Mitigation Standard Practices 349
Appendix 11-2—IADC Space Debris Mitigation Guidelines 353

Table of Authorities 365

Index 369

About the Authors

MATTHEW J. KLEIMAN

Matthew Kleiman is Corporate Counsel at the Draper Laboratory in Cambridge, Massachusetts. He teaches space law at Boston University and serves as chair of the Space Law Committee of the ABA Section of Science & Technology Law. Mr. Kleiman is also a member of the International Institute of Space Law and the American Institute of Aeronautics and Astronautics Technical Committee on Legal Aspects of Aeronautics and Astronautics.

JENIFER K. LAMIE

Jenifer Lamie has an LL.M. in Space, Cyber and Telecommunications Law from the University of Nebraska College of Law. She received a J.D. cum laude from Vermont Law School and a Masters of International Studies with distinction from Otago University in New Zealand. She has accepted an offer for commission with the U.S. Army JAG Corps for July 2012.

MARIA-VITTORIA "GIUGI" CARMINATI

Maria-Vittoria "Giugi" Carminati is a complex commercial litigator in the Houston, Texas, office of Weil, Gotshal & Manges, LLP. Her practice focuses on general commercial litigation, including credit derivatives, title insurance, and shareholder disputes. Ms. Carminati is a member of the International Institute of Space Law and the founder of the ABA Young Lawyer Division's Air and Space Law Committee.

The views expressed by the authors are their own, and not those of Weil, Draper Laboratory, their respective clients and sponsors, or any other organization. The material included herein is not legal advice and should not be relied on as such.

Acknowledgments

GROUP ACKNOWLEDGMENTS

The authors wish to express their sincere gratitude to the many individuals who contributed to the development of this book. This book would not have been possible without the support of the American Bar Association, and specifically the ABA Section of Science & Technology Law (SciTech). ABA Executive Editor Sarah Forbes Orwig, Steven Brower, and the other members of the SciTech Book Committee immediately recognized the need for a practical guidebook to the law of spaceflight and trusted us with the execution of this project.

The authors would also like to thank Dennis Burnett, Bobby Cohanim, Andy Habina, Phil Hattis, Jennifer Izzo, Corinne Kaplan, Mary Luther, Steve Mirmina, Laura Montgomery, John O'Loughlin, Luke Pelican, John Sloan, and Glenn Tallia for reading early drafts of portions of this book and providing invaluable feedback. Any errors are solely the responsibility of the authors.

PERSONAL ACKNOWLEDGMENTS

I am dedicating my work in this book to my wife and best friend, Laura, whose love, inspiration, and patience has made all of this possible, and to my parents, Michael and Frayda, who provided much love and support through the years, and who had the foresight to send me to Space Camp three times when I was a kid.

I would also like to thank Melinda Brown for giving me my "dream job" and for supporting all of my extracurricular space activities; Maureen O'Rourke, Ward Farnsworth, and the rest of the Boston University Law School community for inviting me to bring space law to B.U.; Jorge Contreras, Eric Drogan, Maria Gamboa, Shawn Kaminski, and Julia Passamani of SciTech for trusting me to chair SciTech's Space Law Committee and for the outstanding support they have provided for the committee's programs; and Rachel Yates, for her service as vice chair of the Space Law Committee.

—Matthew J. Kleiman

I am dedicating my work in this book to my parents, John and Audrey Grice, for their endless love, support, and inspiration.

I would first like to thank my professors, advisors, and mentors from the University of Nebraska, particularly Matt Schaefer, Frans von der Dunk, Marvin Ammori, Gretchen Oltman, and Tim Hughes for opportunities and experiences I never imagined were within my reach. I would also like to thank some of my dearest friends and family, whose unconditional support and love reminded me this past year to see the stars in the darkest of skies: Alexandra and James Whatton; Joan Pilarczyk; Tony, Naty, and Regina Carter; Pat Lippitt; Doug and Joann Stewart; Nick Mangold; Kristen Meyerjack; Emily Greenspan; Carmela DeSomma; Luke Pelican; Rebecca Cody; and Christina Zarrella.

—Jenifer K. Lamie

I would like to dedicate my work in this book to my husband Alex Garbino, for his love of space and his desire to venture there. And to my children Lorenzo, Raphael, and Fernando, in the hope their world is made better by visionaries, adventurers, daredevils, and heroes who dare to venture beyond the unknown to improve the known. I would like to thank Doug Griffith, for accepting to be my space-law mentor based on one random starry-eyed e-mail; Milton "Skip" Smith, for taking me under his wing, supporting me, encouraging me, and all but ensuring my advancement, and who from day one believed in my future in space law; Jim Rendleman, for being a go-to source of support in my space law pursuits; Marc Holzapfel, for responding to a hand-written "message in a bottle"; Frans von der Dunk, for making me feel like I belong in the field of space law; and Joanne Gabrynowicz, for immediately taking me seriously and to whom I still owe an article.

—Maria-Vittoria "Giugi" Carminati

Foreword

Commercial space is a new and developing area both businesswise and in the law—and that is why this book is so timely. Matthew Kleiman, Jenifer Lamie, and Maria-Vittoria "Giugi" Carminati, each of whom I have had the great fortune to know over the past several years, understand that space is the future, and have the vision to provide a guidebook for this future.

Regardless of where you practice, whether in a private space company, the government, or at a law firm, you will likely come across common issues, such as where does space begin, what is an orbital vehicle, what laws govern a space vehicle, or how do the export laws affect space.

Treatises on public space law already exist and those for commercial space will eventually come. What *The Laws of Spaceflight: A Guidebook for New Space Lawyers* provides, instead, is a practical guidebook that will help attorneys quickly obtain a basic understanding of spaceflight, the industry, and legal issues their clients may confront. This book is handily broken up into concise sections. One day you may need guidance on the technical aspects of spaceflight and then you can quickly go to chapter 1. Another day you may need clarity on export compliance, and then you can simply flip to chapter 9. The book may not answer all your questions, but it will give you the basic information you need and the tools you may require to identify the issues and move forward.

So I encourage you to keep this book on your desk and use it as a handy reference guide as you develop your career. And who knows, some day you may even take it to space.

—Marc Holzapfel
Senior Vice President & Legal Director
Virgin Galactic, LLC

Introduction

I. SPACE LAW IS BORN

On October 4, 1957, the Soviet Union launched Earth's first artificial satellite, *Sputnik 1*. America's second-place finish in the first leg of the Space Race came as a great shock to the country, initiating a complete reassessment of the U.S. space program and American science and technology policy. Few people realize that *Sputnik* also had far-reaching legal implications.

Prior to *Sputnik*, the legal status of outer space was unclear. The conventional wisdom was that the rules that governed airspace would simply be extended upward to Earth orbit once humanity began operating in that domain. From as early as 1919, international air law provided that a nation's sovereignty extended vertically to the airspace over its territory. If this rule extended to outer space, the Soviet Union would have violated international law by launching *Sputnik* into an orbit that passed over many countries, including the United States, without permission. Nevertheless, President Dwight D. Eisenhower, knowing that the United States was interested in eventually overflying Soviet airspace with its own spy satellites, tacitly accepted the Soviet Union's right to operate a satellite in orbit over U.S. territory. It was thus established that the rules governing spacecraft would differ from those that governed aircraft. And the field of space law was born.

II. THE DEVELOPMENT OF SPACE LAW

The term "space law" refers to the body of international and national laws and customs that govern human activities in outer space. Immediately recognizing the legal vacuum that existed in outer space following *Sputnik*, in 1958 the United Nations formed the Committee on the Peaceful Uses of Outer Space (COPUOS). The work of the COPUOS delegates eventually resulted in the foundational instrument of the outer space legal regime, the 1967 Outer Space Treaty.

The Outer Space Treaty initiated the "golden age" of international space law development. The treaty was quickly followed by the 1968 Agreement on the Rescue of Astronauts, the Return of Astronauts, and the Return of Objects Launched into Outer Space; the 1972 Convention on International Liability for Damage Caused by Space Objects; the 1975 Convention on Registration of Objects Launched into Outer Space; and the 1979 Agreement Governing the Activities of States on the Moon and Other Celestial Bodies. This last treaty is now considered dormant because it has not been ratified by any of the major space powers.

Although no major space treaties have been adopted since the 1970s, international space law has not remained stagnant. COPUOS continues to administer the major space treaties and advises the international community on space policy matters. The U.N. General Assembly has adopted nonbinding resolutions addressing satellite television broadcasting, the remote sensing of Earth from outer space, the use of nuclear power sources in outer space, and international cooperation in the exploration and use of outer space. The International Telecommunications Union, a U.N. organization, has issued regulations concerning the operation of telecommunications satellites in geostationary orbit. There have also been numerous intergovernmental agreements concerning space-related activities, such as the 1998 Intergovernmental Agreement on Space Station Cooperation among the United States, Canada, Russia, Japan, and participating countries of the European Space Agency.

The Outer Space Treaty assigns to national governments responsibility for regulating their governmental and nongovernmental space activities. In the United States, each government agency that operates spacecraft is responsible for complying with U.S. law and international treaty obligations. The United States has also enacted laws and regulations governing various aspects of nongovernmental space operations, including the launch and reentry of private spacecraft and the operation of private telecommunication and remote sensing satellites. Many other federal and state laws, such as those relating to export control, contracts, torts, environmental protection, and intellectual property, also directly or indirectly affect U.S. space activities.

III. A PERIOD OF TRANSITION

The legal framework established by the Outer Space Treaty successfully maintained peace in outer space during the darkest days of the Cold War. The space industry, however, is now in a period of transition and

INTRODUCTION xix

The SpaceX Dragon cargo vessel is captured by the International Space Station's robotic arm on May 25, 2012, becoming the first commercial spacecraft to dock with the station.

Credit: ESA/NASA

we stand on the threshold of a new era in spaceflight. For the last half century, most space operations were conducted by government agencies. With the retirement of the Space Shuttle in 2011, private companies are preparing to assume many of the missions traditionally undertaken by governments and to open new markets in outer space.

In the midst of this transition, the spaceflight industry faces many challenges. For instance:

- Humans will soon routinely travel into outer space on spacecraft built and operated by private companies. Governments and commercial spaceflight operators will need to establish licensing criteria for private spacecraft and address questions of liability in the event of accidents.

- Many useful portions of Earth orbit are crowded with space debris. If enough debris accumulates, it will become virtually impossible to operate spacecraft safely at certain altitudes.
- Export restrictions on space technologies make it difficult for U.S. space companies to compete in the global space marketplace. A more nuanced approach to controlling the export of space technologies will be necessary for U.S. space companies to maintain their leadership in this industry.

Space lawyers will be at the forefront of helping the spaceflight community meet these and other challenges and adapt to the new commercial spaceflight paradigm.

IV. A SPACE LAW GUIDEBOOK

The primary purpose of this book is to introduce lawyers new to the space industry, either at private space companies, in government, or at law firms with space industry clients, to the laws that govern activities in outer space. This is not a treatise on space law, but a practical guidebook. It is meant to help attorneys quickly obtain a basic understanding of spaceflight, the space industry, and the legal issues that their clients confront. We hope that it will also be useful to law students, entrepreneurs, engineers, policymakers, and others who wish to learn about the laws of spaceflight.

This guidebook begins by placing space law in its larger context and then focuses on discrete legal topics. Chapter 1 describes the technical aspects of spaceflight that are most relevant to the legal issues discussed later in the book. Chapter 2 provides a high-level history of human space activities. Chapters 3 and 4 describe the development of international and U.S. space laws. The remaining chapters discuss specific issues that space lawyers routinely address: licensing private spaceflight activities, liability and insurance for spacecraft operators, the regulation of telecommunication and remote sensing satellites, export controls on space technologies, contracting with the U.S. government, environmental protection, and property rights. Being a publication of the American Bar Association that is written by three American attorneys, this guidebook naturally approaches space law from the U.S. perspective. Nevertheless, most of the issues discussed in these pages are common to all spacefaring nations.

As you begin reading this guidebook, let us be the first to welcome you to the practice of space law. This is a challenging and exciting time to be part of the spaceflight community. We hope that you will find practicing in this field as rewarding as we do. We are delighted to be your guides as you enter the legal profession's next frontier.

—Matthew J. Kleiman
Boston, Massachusetts

—Jenifer K. Lamie
Cheshire, Connecticut

—Maria-Vittoria "Giugi" Carminati
Houston, Texas

CHAPTER 1

Where To Go and How To Get There

Space Operations, Space Technology, and the Outer Space Environment

Spaceflight is a uniquely challenging human endeavor. In order to provide effective legal advice to their clients, lawyers in the space industry must have a basic understanding of the technical aspects of space operations. This chapter provides an overview of space operations, basic orbital mechanics, spacecraft power and propulsion, satellite systems, the hazards of the outer space environment, and the practical applications of space technology.[1]

I. SPACEFLIGHT OPERATIONS

There are three general types of spaceflight: suborbital, orbital, and interplanetary. On a suborbital spaceflight, the spacecraft reaches

---

1. Much of the information presented in this chapter has been synthesized from the following sources:
   JAMES R. WERTZ & WILEY J. LARSON, SPACE MISSION ANALYSIS AND DESIGN (Microcosm Press, 3d ed. 1999).
   AIR COMMAND & STAFF COLL., AU-18 SPACE PRIMER (Air Univ. Press 2009), available at http://space.maxwell.af.mil/au-18-2009/index.htm
   ALAN C. TRIBBLE, A TRIBBLE'S GUIDE TO SPACE: HOT TO GET TO SPACE AND WHAT TO DO WHEN YOU ARE THERE (Princeton Univ. Press 2002)
   LUCY ROGERS, IT'S ONLY ROCKET SCIENCE: AN INTRODUCTION IN PLAIN ENGLISH (Springer 2008)
   NASA JET PROPULSION LAB., BASICS OF SPACE FLIGHT (2011), http://www2.jpl.nasa.gov/basics/index.php
   GLENN REYNOLDS & ROBERT MERGES, OUTER SPACE: PROBLEMS OF LAW AND POLICY (Westview Press 1998)
   Wikipedia, Spaceflight, http://en.wikipedia.org/wiki/Spaceflight, and the articles linked to therein

outer space, but does not have enough energy to complete a single revolution, or orbit, around Earth. Ballistic missiles were the first suborbital spacecraft. In orbital spaceflight, the spacecraft is launched with sufficient energy to complete at least one revolution around Earth. The Space Shuttle, International Space Station (ISS), and most satellites are all orbital spacecraft. In interplanetary spaceflight, the spacecraft leaves the pull of Earth's gravity and travels to another planetary body in the solar system. The region between Earth and the Moon is not considered interplanetary space, however. It is instead referred to as cislunar space.

While science fiction often depicts interstellar space travel at faster-than-light speeds, this is not possible with current technology. The speed of light is 299,792,458 meters (186,282 miles) per second. *Voyager 1*, launched in 1977, is one of the fastest spacecraft ever sent into outer space, but is crawling to the edge of the solar system at a mere 17.26 km (10.72 miles) per second. At its current speed, it would take more than 74,000 years for *Voyager 1* to reach the Sun's closest stellar neighbor, Proxima Centauri.

Spacecraft are used for a wide variety of purposes. Common missions include telecommunications, navigation, Earth observation, space exploration, microgravity research, and, more recently, space tourism. Regardless of their objective, most space missions start the same way: with a rocket carrying the spacecraft into outer space. These rockets, or launch vehicles, launch either vertically from a fixed launch pad or horizontally from a carrier aircraft. Spacecraft that must reach a specific orbit, rendezvous with another spacecraft, or travel to another planetary body must launch within a specific period of time, or launch window, when the launch location lines up properly with the mission objective. If the spacecraft misses its launch window, its launch must be postponed until the next window opens, which may be days, weeks, or even months away.

After a suborbital or orbital spacecraft completes its mission, it is either brought back to Earth in a controlled reentry or, if Earth reentry is not practical, is left in an orbit that limits its exposure to other operational spacecraft, known as a graveyard orbit. Interplanetary spacecraft are left drifting through space or as derelicts on a planetary surface. Spacecraft reentering Earth's atmosphere generate enough aerodynamic heating to destroy the vehicle's structure and vaporize upon reentry, although some larger components may survive and fall to Earth. Spacecraft carrying people or valuable cargo are built with thermal protection systems that can withstand this intense heat and pressure. Winged spacecraft, such as the Space Shuttle and Virgin Galactic's SpaceShipTwo, then glide to a landing on a runway like an airplane. Capsule spacecraft, such as the *Apollo* Command Module, Russian *Soyuz*, and SpaceX's *Dragon*, use a combination of parachutes, rockets, and cushions to make a soft landing in the water or on the ground.

II. THE PHYSICS OF SPACEFLIGHT

A. Earth/Space Boundary

Discussions of the physics of spaceflight often begin with the following question: where does outer space begin? There is no legal definition of the demarcation between the atmosphere and outer space, nor is there a fixed point where the atmosphere ends. In fact, trace amounts of atmospheric gases can still be detected as high as 10,000 km (6,200 miles) above sea level, far above many orbiting spacecraft.

The most commonly accepted demarcation between the atmosphere and outer space is the Kármán Line, named after Dr. Theodore von Kármán, a Hungarian-American engineer and physicist. As an airplane climbs higher and higher, the thinning air provides less lift to its wings, requiring the airplane to increase speed to remain airborne. In the 1950s, Dr. Kármán calculated that above an altitude of roughly 100 km (62 miles), an airplane would need to fly so fast to generate aerodynamic lift that it would reach orbital velocity (by comparison, most airliners fly at an altitude of six or seven miles). Thus, Dr. Kármán proposed that an altitude of 100 km be designated as the boundary between air and space, and this line has become the de facto demarcation accepted by most people in the space community.

B. Describing an Orbit

The following terms are used to describe orbits (see figure 1.1) and the spacecraft that occupy them:

- *Altitude*: the height of an object above mean sea level.
- *Apogee*: the point along a spacecraft's orbit where it is furthest from Earth.
- *Attitude*: the orientation of a spacecraft in space.
- *Eccentricity*: a measure of how much an orbit deviates from a perfect circle.

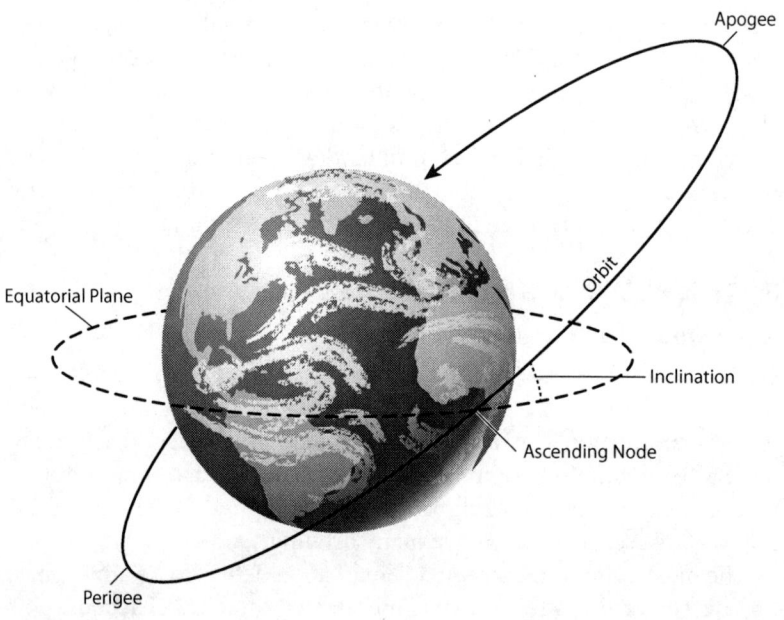

FIGURE 1.1
Elements of an orbit

- *Equatorial plane:* an imaginary plane extending from the equator on Earth to the celestial sphere.
- *Inclination:* the angle between the equatorial plane and the orbit of a spacecraft.
- *Orbital node:* one of the two points where an orbit crosses a plane of reference to which it is inclined. Around a planetary body, the ascending node is where the orbiting object moves north through the equatorial plane, and the descending node is where it moves south through the equatorial plane.
- *Orbital velocity:* the minimum speed at which a spacecraft must travel in order to remain in orbit.
- *Perigee:* the point along a spacecraft's orbit where it is closest to Earth.
- *Period:* the time it takes a spacecraft to make one full orbit around Earth.
- *Trajectory:* the path that a moving object follows through space.

II. The Physics of Spaceflight

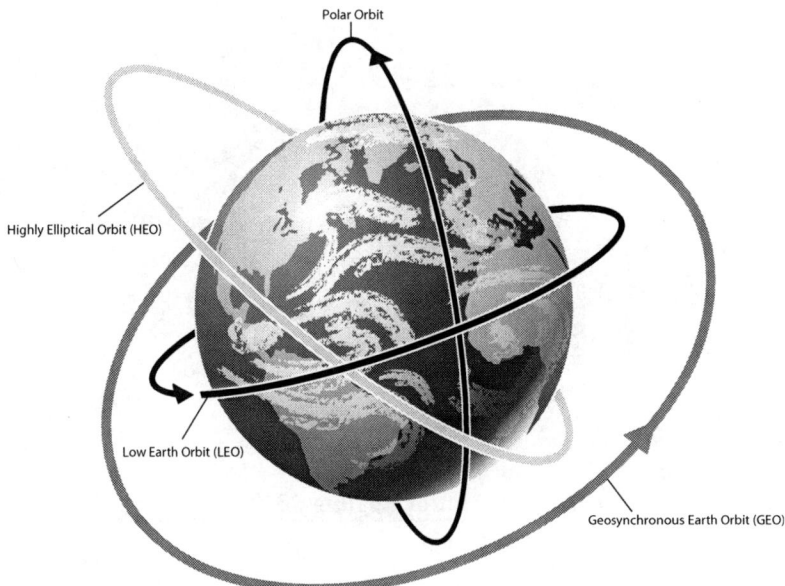

FIGURE 1.2
Common spacecraft orbits around Earth

C. Spacecraft Orbits

A spacecraft's orbit is chosen based on its mission. For instance, a satellite that will take high-resolution photographs of features on Earth will be placed in a low orbit, whereas a satellite that must view more of Earth's surface at one time will be placed in a high orbit. This section describes the orbits most commonly used by spacecraft (see figures 1.2 and 1.3 and table 1.1).

1. Low Earth Orbit

About half of all operational satellites are in low Earth orbit (LEO), at an altitude of between 160 and 1,000 km (100 to 600 miles). Spacecraft in LEO travel at about 28,000 km/h (17,500 mph) and take approximately 90 to 100 minutes to orbit Earth. These spacecraft take the least energy to reach orbit, require the least power to communicate with ground stations, and are able to take the highest resolution images of features on Earth.

A polar orbit is a type of LEO that passes above or nearly above both poles of Earth on each revolution. Polar orbits are useful for

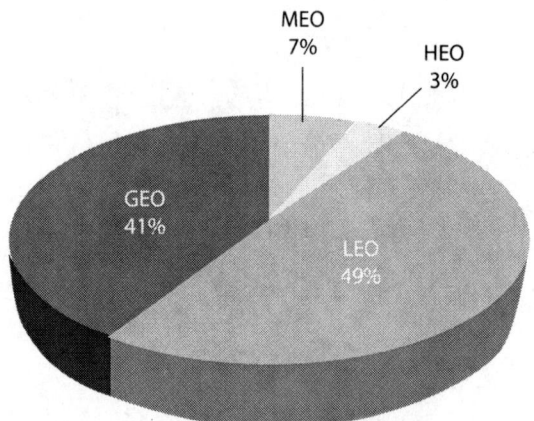

FIGURE 1.3
Operational satellites by orbit as of June 2011
Source: Satellite Industry Association

reconnaissance and other Earth observation missions since over the course of a day the satellite will pass over the entire Earth as Earth rotates beneath it. Sun-synchronous orbits are highly inclined orbits timed to pass over the same point on Earth at the same time of day, which is useful for obtaining comparable data about a particular region.

2. Medium Earth Orbit

Medium Earth orbits (MEOs) are at an altitude of between 1,000 and 35,786 km (600 and 22,236 miles). Because their orbits are higher, spacecraft in MEO orbit at a slower speed than spacecraft in LEO. MEO is therefore used to provide a longer dwell time and a larger coverage area over a given region on Earth as compared to LEO. For this reason, MEO is particularly useful for satellite navigation systems.

3. Geosynchronous Earth Orbit

Geosynchronous Earth orbit (GEO) is located at an altitude of exactly 35,786 km (22,236 miles). A satellite in GEO travels at about 11,265.40 km/h (7,000 mph) and takes about 24 hours to complete its orbit, so it remains continuously above the same longitude on Earth. Because GEO is very far from Earth, a great deal of energy is required to launch a satellite into that orbit. However, GEO has two unique benefits that make placing satellites there worth the cost. First, a spacecraft in GEO has a

TABLE 1.1
Representative Earth Orbits

Orbit Type	Representative Missions	Altitude	Period	Inclination
LEO				
• Polar sun-synchronous	Remote sensing/weather	160-900 km	98-104 min	98°
• Polar non-sun-synchronous	Earth observing, scientific	450-600 km	90-101 min	80-94°
• Inclined non-polar	ISS	340 km	91 min	51.6°
MEO	Navigation (GPS), communications, space environment	20,100 km	12 hours	55°
GEO				
• Geostationary	Communication, early warning, nuclear detection, weather	35,786 km	24 hours	0°
HEO				
• Molniya	Communications	From 495 km to 39,587 km	12 hours	63.4°

Source: Adapted from Air Command & Staff Coll., AU-18 Space Primer 90 (Air Univ. Press 2009), *available at* http://space.maxwell.af.mil/au-18-2009/index.htm

commanding view of Earth, making this orbit useful for telecommunications, early warning, and meteorological missions. Second, a satellite in GEO directly above the equator will appear to remain stationary over one point on Earth, and is therefore said to be in a geostationary orbit (GSO). Many satellite television stations are broadcast from GSO, which is why satellite television receiving dishes do not have to move and why those in the northern hemisphere always point south toward the equator.

4. Elliptical Orbits

Not all orbits are circular. For certain missions, a more elliptical orbit is desired. Highly elliptical orbits (HEO) have a perigee of about 1,060 km (660 miles) and an apogee of about 38,624 km (24,000 miles) in a single orbital period. Satellites speed up as they travel through their perigee and slow down as they near their apogee, so HEO orbits enable long dwell times and large fields of view from the apogee. These orbits are primarily used for communications, scientific research, and reconnaissance missions when GEO orbits are not available. One type of HEO is a Molniya orbit, named after the first Soviet communication satellite to use it. Two satellites in the same Molniya orbit will have a constant view of a given high-latitude area on Earth.

5. Lagrange Point Orbits

Between any two large celestial bodies, there are locations where the combined gravity of the objects interacts such that a smaller object placed at this point will stay in place relative to the larger bodies. These locations were calculated in 1772 by Italian-French mathematician Joseph Louis Lagrange, and are hence referred to as Lagrange Points. Orbits around these points are ideal for certain types of space missions, such as scientific observation missions that must stare at a particular point in space, such as stars and galaxies, for long periods of time.

6. Other Specialized Orbits

A parking orbit is a temporary orbit used to store a satellite that is waiting to move on to the next stage of its mission. For example, a Mars probe might have to wait in a parking orbit for Earth and Mars to be in proper alignment before it can leave Earth orbit and continue on its journey to the Red Planet.

A graveyard orbit, as discussed above, is an orbit several hundred miles above GSO to which GSO satellites are moved at the end of their operational life in order to make room for other satellites. Satellites in GSO are placed in graveyard orbits instead of de-orbited and vaporized in Earth's atmosphere because it is very energy intensive to move a satellite from GSO back down to LEO. Unfortunately, spacecraft operators do not always move their spacecraft to graveyard orbits at the end of their useful lives. In 2009, for example, only 11 of 21 satellites in GSO that reached their end of life were moved to proper graveyard orbits.[2] Satellites that are not properly disposed of remain in orbit as uncontrolled space debris and are a hazard to other spacecraft.

D. Getting into and Staying in Orbit

The physics that enable spacecraft to reach and maintain orbit were first described in the second half of the seventeenth century by Dr. Isaac Newton. Newton's law of universal gravitation states that every particle in the universe attracts every other particle with a predictable, measurable force. Newton's three laws of motion state:

1. Every body continues in a state of rest, or of uniform motion in a straight line, unless it is compelled to change that state

---

2. European Space Agency, Classification of Geosynchronous Objects 126 (Feb. 2010), http://lfvn.astronomer.ru/files/COGO-issue12.pdf.

II. The Physics of Spaceflight

by a force imposed upon it. This is also known as the law of inertia.
2. When a force is applied to a body, the change of momentum is proportional to, and in the direction of, the applied force.
3. For every action there is a reaction that is equal in magnitude but opposite in direction to the action.

Getting into orbit requires accelerating a spacecraft to a velocity sufficient to circle Earth faster than gravity can pull the spacecraft back down to Earth. The mechanics of achieving orbit can be understood by visualizing a cannon sitting on top of Earth, as shown in figure 1.4. If a cannonball is simply dropped while standing still, Newton's law of universal gravitation says that the cannonball will fall straight to the ground. However, if the cannonball is fired out of the cannon, Newton's first law of motion says that inertia will cause the cannonball to travel in a straight line until some other force acts upon it. That other force is gravity, which continues to pull on the cannonball at the same rate as when the cannonball was motionless. Accordingly, the cannonball's inertia will carry it forward and gravity will simultaneously pull it toward Earth, resulting in an arc-like trajectory (Scenario A). Since the force of gravity remains constant, increasing the speed at which the cannonball is launched will allow the cannonball to travel a greater distance before it is pulled to the ground (Scenario B).

The cannonball will achieve a circular orbit only when its launch velocity is fast enough that the curve of Earth falls away from the cannonball at the same rate that gravity pulls it toward Earth, leaving the cannonball in a state of perpetual free-fall (Scenario C). In other words, a spacecraft will remain in orbit when its inertia (tendency to move forward at a constant speed and trajectory) and the pull of gravity are balanced. Because gravity's pull on a spacecraft weakens as the spacecraft moves further away from Earth, a spacecraft must slow down as it climbs into a higher orbit in order to keep inertia and gravity in balance. Thus, orbital velocity for LEO is approximately 28,163.52 km/h (17,500 mph), whereas orbital velocity for GEO is only about 1,126.54 km/h (7,000 mph).

If a spacecraft maintains the balance of inertia and gravity, it will theoretically remain in orbit indefinitely. However, traces of atmospheric gas in low orbits create drag that slows the spacecraft and gradually decreases its altitude. Left unchecked, the spacecraft will eventually reenter the atmosphere and be vaporized. A spacecraft in a low orbit

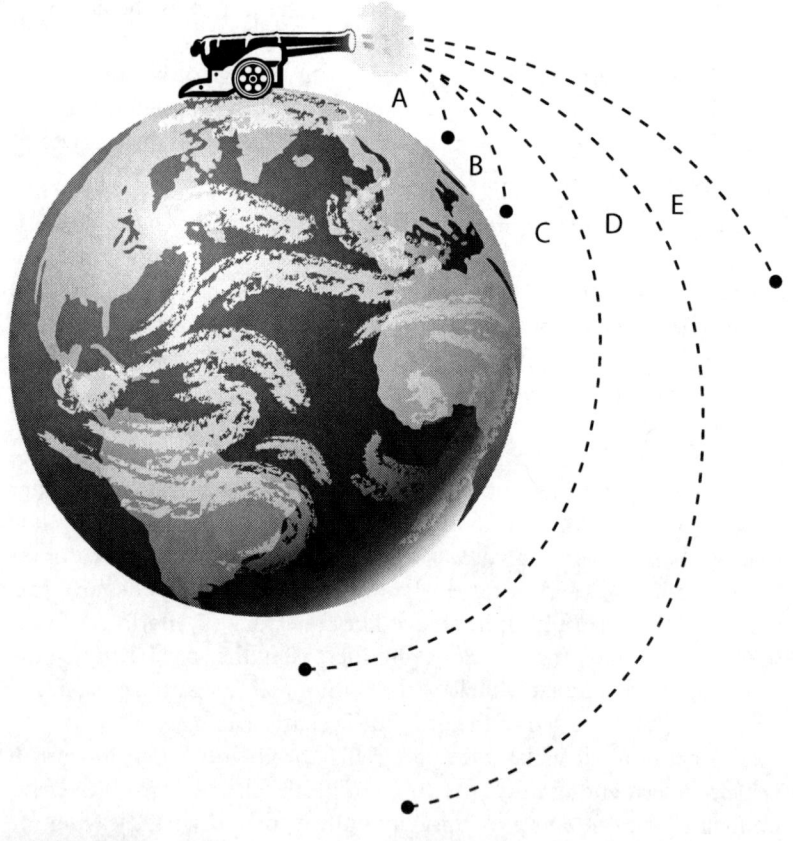

FIGURE 1.4
Newton's cannonball

counters this effect by regularly firing its engines to boost itself back to its desired orbital velocity and altitude.

If the velocity of the cannonball is increased beyond the velocity required to maintain a circular orbit, the cannonball will travel further from Earth before gravity pulls it back, resulting in an increasingly elliptical orbit (Scenario D). Spacecraft operators use this effect to enter into elliptical orbits and to move a spacecraft from a lower circular orbit to a higher circular orbit. To change orbits, the spacecraft leaves its lower orbit by firing its engines to increase its velocity and enter into an elliptical transfer orbit, with the apogee at the desired altitude of the new orbit. Once the spacecraft reaches the apogee, it fires its engines a second time to bring its inertia and gravity back into balance in the higher orbit. This maneuver is known as a Hohmann transfer orbit, in recognition of Walter Hohmann, the German scientist who published a description of it in 1925. A Hohmann transfer orbit used to reach GEO or GSO is referred to as a geosynchronous transfer orbit or geostationary transfer orbit (GTO).

If the cannonball's launch velocity is further increased to the point where it travels so far from Earth that Earth's gravity is no longer strong enough to pull the cannonball back, the cannonball will escape Earth's gravity (Scenario E). Escape velocity varies according to the amount of gravitational force generated by a body, which depends on the body's size. Earth's escape velocity is 40,295 km/h (25,038 mph).

III. POWER AND PROPULSION

A. Delta-v

Each space mission will require the spacecraft to perform a variety of orbital maneuvers. As discussed above, each orbital maneuver will require a change in velocity, which in turn will require the spacecraft to expend energy. The measure of the energy required to perform an orbital maneuver is referred to as delta-v (Δv, denoting "change in velocity"). The total delta-v that will be required to perform all of the orbital maneuvers that a spacecraft must accomplish during the course of its mission is called the mission's delta-v budget.

Changing velocity usually requires the burning of propellant carried by a spacecraft. Propellant is a limited resource, so calculating the delta-v budget is one of the most critical elements of space mission planning. Underestimating a mission's delta-v budget will result in mission failure because the spacecraft will run out of fuel before it can accomplish all of its mission objectives.

Table 1.2 lists the typical delta-v required to perform various orbital maneuvers. The exact delta-v necessary to perform an orbital maneuver varies depending on the characteristics of each spacecraft.

Comparing the delta-v required to perform different tasks reveals how much more or less energy intensive those tasks are to accomplish. For instance, the difference between the delta-v required for a suborbital launch (1.4 km/s) and for a launch to LEO (9.5 km/s) explains why a spacecraft must accelerate to about 28,163.52 km/h (17,500

TABLE 1.2
Typical Delta-v Required for Various Orbital Maneuvers

Maneuver	Typical Delta-v
Launch to 100 km (suborbital)	1.4 km/s
Launch to LEO	9.5 km/s
LEO to GEO	3.9 km/s
LEO to Moon surface	5.9 km/s
LEO to near-Earth asteroid	4.5 km/s
LEO to Mars surface	10.7 km/s
LEO to Solar System escape	8.7 km/s
Moon to low lunar orbit	1.9 km/s
Mars to low Mars orbit	4.1 km/s
Orbit control:	
Station-keeping (GEO)	50–55 m/s/year
Drag compensation:	
• alt.: 400–500 km	< 100 m/s/year
• alt.: 500–600 km	< 25 m/s/year
• alt.: >600 km	< 7.5 m/s/year
Attitude control: three-axis control	2–6 m/s/year

mph) to reach LEO, but need only reach 2,500–3,000 mph for a suborbital flight. Put another way, the amount of energy required to lift an object the size of a fighter jet to a typical LEO altitude would enable a car to drive about two-thirds of the way around the globe. The amount of energy required to then accelerate that object to orbital velocity—to allow it to remain in orbit and not fall back to Earth—would take the same car around the globe six times.[3]

Table 1.2 also shows that launching to LEO is one of the most energy-intensive tasks that spacecraft perform. This lends credence to the expression common in the space industry, "Once you get to Earth orbit, you're halfway to anywhere in the solar system."[4] It also explains why launch is usually the riskiest and most expensive part of a space mission.

---

3. Gen. William L. Shelton, Commander, Air Force Space Command, Keynote Speech at the 15th Annual FAA Commercial Space Transportation Conference (Feb. 16, 2012), http://www.parabolicarc.com/2012/02/20/video-gen-shelton-of-air-force-space-command-at-faa-conference.

4. Attributed to science fiction author Robert A. Heinlein.

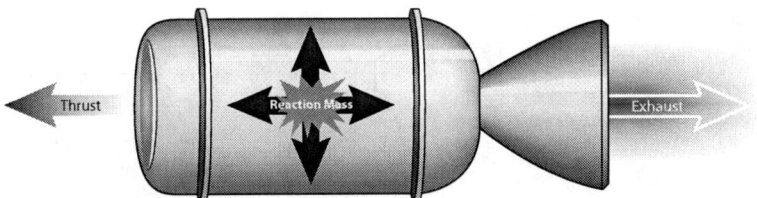

FIGURE 1.5
Reaction engine

B. Spacecraft Propulsion Systems

Delta-v is generated by spacecraft propulsion systems. The most commonly used propulsive system is a reaction engine (see figure 1.5). Reaction engines produce thrust by expelling reaction mass. Due to Newton's third law of motion (for every action there is an equal and opposite reaction), the engine will move in the direction opposite of the direction the reaction mass is expelled. Because of Newton's second law of motion (when a force is applied to a body, the change of momentum is proportional to, and in the direction of, the applied force), the faster the reaction mass is expelled, the faster the spacecraft will accelerate in the opposite direction.

There are several types of reaction engines used on spacecraft (see table 1.3). Cold gas propulsion generates thrust by expelling a controlled, pressurized gas source through a nozzle. Chemical combustion engines heat the reaction mass to a high temperature by combusting a solid, liquid, or gaseous fuel with an oxidizer. The hot reaction mass is then allowed to escape through a bell-shaped nozzle, which accelerates the mass to high speeds. Chemical combustion engines produce the most thrust of any reaction engine and are therefore used for high delta-v maneuvers. A third type, the electromagnetic propulsion engine, uses electricity and magnetic fields to accelerate the reaction mass, such as plasma or ions, to a high speed. These engines have the potential to generate much higher thrust velocity and weigh much less than combustion engines, but power constraints on current spacecraft result in electric drives delivering much less thrust than chemical rockets.

Launch vehicles are usually multistage rockets, with each stage containing its own engine(s) and propellant. Stages can be either parallel stages mounted alongside one another or tandem or serial stages

TABLE 1.3
Comparison of Various Types of Reaction Engines

Type	Advantages	Disadvantages	Common Uses
Cold gas	Extremely simple, reliable, very low cost	Very low performance, heaviest of all systems for given performance level	Orbit maintenance and maneuvering, attitude control
Solid	Simple, reliable, relatively low cost	Limited performance, higher thrust leads to safety issues, performance not adjustable, cannot stop/restart	Launch, orbit insertion
Liquid	Storable, high performance, throttleable, restartable	Complicated, toxic	Launch, orbit insertion, orbit maintenance and maneuvering, attitude control
Hybrid (solid & liquid)	Nonexplosive, nontoxic, throttleable, restartable	Requires oxidizer fuel system, bulkier than solids	Orbit insertion, orbit maintenance and maneuvering
Electromagnetic	Very high performance	Very high power, low thrust, complicated, not well developed	Orbit insertion at apogee, orbit maintenance and maneuvering, interplanetary travel

Source: Adapted from James R. Wertz & Wiley J. Larson, Space Mission Analysis and Design 688, 693 (tbls.) (Microcosm Press, 3d ed. 1999).

that are mounted one on top of the other. In lieu of a first stage, some spacecraft, such as Virgin Galactic's SpaceShipTwo, are carried to a high altitude by a carrier aircraft.

For staged rockets, the largest and most powerful engines are in the first stage since the most thrust is required to break through the thick atmosphere and high gravity at low altitudes, and the engine must lift everything else being carried by the launch vehicle. The first stage is discarded after its propellant has been expended (figure 1.6), which reduces the overall weight of the launch vehicle so that later stages are not required to generate as much thrust or carry as much fuel as would otherwise be necessary.

Not all space propulsion requires an internal reaction mass. Field propulsion uses solar wind, solar radiation, magnetic fields, or gravitational fields to propel spacecraft. A solar sail uses the pressure of solar radiation against a large collection surface to propel a spacecraft, simi-

FIGURE 1.6
The expended first stage of the Saturn V rocket separating from the second and third stages during the launch of Apollo 11
Credit: NASA

lar to a sail on a sailboat. A tether propulsion system uses a long tether that interacts with Earth's magnetic field to change a spacecraft's orbit. A gravity assist uses a large body's gravity like a slingshot to accelerate an interplanetary spacecraft toward its destination.

C. Internal Power Systems

In addition to propulsion systems for generating delta-v, spacecraft require power to operate sensors, communications equipment, and other internal systems. The preferred power source for most space systems is solar arrays mounted on the side of a spacecraft or deployed on large, steerable wings (figure 1.7). Solar power is not always available, however, such as in Earth's shadow or in deep space where the Sun's energy is too weak to be effective. Rechargeable storage batteries, fuel cells, and nuclear power are therefore used as additional power sources.

Nuclear power sources are particularly useful for long-duration missions. There are three basic types of nuclear power generators for spacecraft: radioisotope thermoelectric generators (RTGs), radioisotope heater units (RHUs), and nuclear reactors. RTGs and RHUs use

FIGURE 1.7
Solar array on the International Space Station
Credit: NASA

heat generated by the decay of radioactive materials, such as plutonium-238, to produce electricity (RTGs) or to warm sensitive instruments (RHUs). RTGs and RHUs have been used on many long-duration, deep-exploration missions, including *Pioneer 10*, *Pioneer 11*, the Voyager and Viking programs, *Galileo*, *Cassini*, *New Horizons*, and the Mars Science Laboratory. Traditional nuclear reactors can also be used to power spacecraft, although they have been used much less frequently than RTGs and RHUs due to environmental and safety concerns.

IV. SATELLITE SYSTEMS

A. Components

While "artificial satellite" refers broadly to any object that has been placed into orbit by human activity, colloquially the term is used to mean an unmanned spacecraft that performs a practical function in orbit with respect to an activity on Earth, such as telecommunications, remote sensing, or navigation. As shown in figure 1.8, most satellite

IV. Satellite Systems

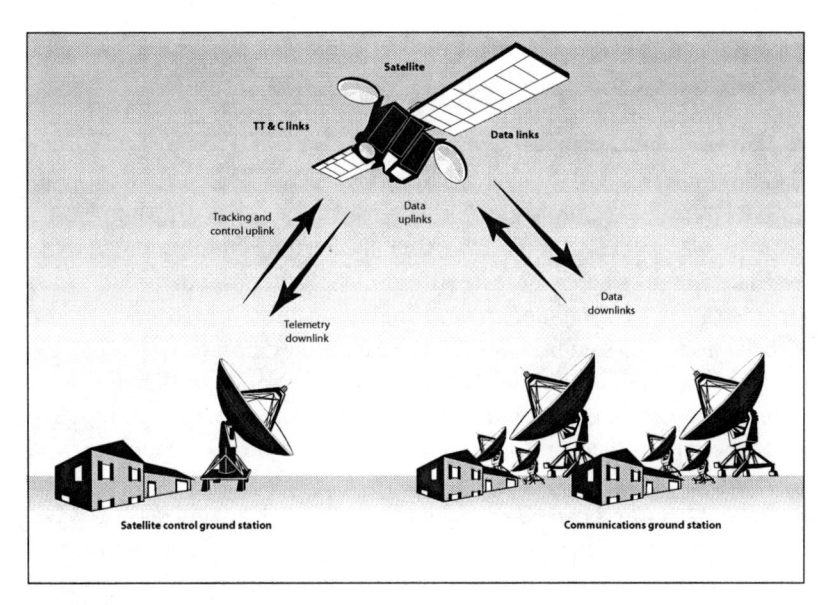

FIGURE 1.8
Components of a satellite system

Credit: U.S. Gen. Accounting Office, GAO-02-781, Critical Infrastructure Protection: Commercial Satellite Security Should Be More Fully Addressed (Aug. 2002), http://www.gao.gov/new.items/d02781.pdf

systems consist of (1) ground stations; (2) tracking, telemetry, and control (TT&C) links; (3) data links; and (4) one or more satellites.

There are two types of ground stations: control stations and communications stations. Control stations track and control satellites to ensure that they remain in the proper orbits and monitor their performance. Communications stations process imagery and other data and often provide a link to ground-based networks, including the Internet.

There are two types of links between satellites and ground stations. TT&C links exchange commands and status information between control stations and satellites. Data links exchange communications, navigation, and imaging data between communications stations and satellites. Uplinks are communications that go from Earth to space, downlinks are communications that go from space to Earth, and crosslinks are communications between satellites in orbit.

The final component of the system is the satellite itself. Every satellite has a payload, which is the equipment needed to perform the mission, such as cameras or transponders, and a bus that carries the payload and provides electrical power, computer processers, and propulsion to the entire spacecraft.

B. Threats and Security

Satellite systems are subject to a variety of natural and man-made threats. Ground stations can be damaged by natural disasters, suffer from power outages, or be intentionally harmed by criminal activity or state action during an armed conflict. Satellites can be damaged or destroyed by solar and cosmic radiation, micrometeoroids, space debris, and anti-satellite weapons. Satellite systems can also be disabled by unintentional interference from overlapping radio signals or intentional interference from signal jamming and cyber-attacks, such as malicious software viruses, denial of service attacks, spoofing, and data interception.

Satellite systems can be protected from these threats in several ways. The satellite itself may be hardened against radiation, micrometeoroids, and orbital debris. Ground stations can be physically reinforced to protect against natural disasters and attacks, equipped with high-power radio uplinks that are difficult to jam or overpower, and use other physical and cyber security controls. Communications links may be encrypted and employ unique satellite-specific digital interfaces. To minimize disruption in the event a single satellite or ground station is disabled, satellite networks may use multiple satellites and ground stations to provide redundancy.

C. Constellations

Satellites often work together in a constellation, which is "a number of satellites with coordinated ground coverage, operating together under shared control, synchronized so that they overlap well in coverage and complement rather than interfere with other satellites' important coverage."[5] Examples of satellite systems that operate in constellations include the Global Positioning System (GPS), the Sirius and XM radio systems, and the Iridium and Globalstar satellite telephone services. Figure 1.9 is an artist's rendering of the GPS satellite constellation.

---

5. Wikipedia, Satellite Constellation, http://en.wikipedia.org/wiki/Satellite_constellation.

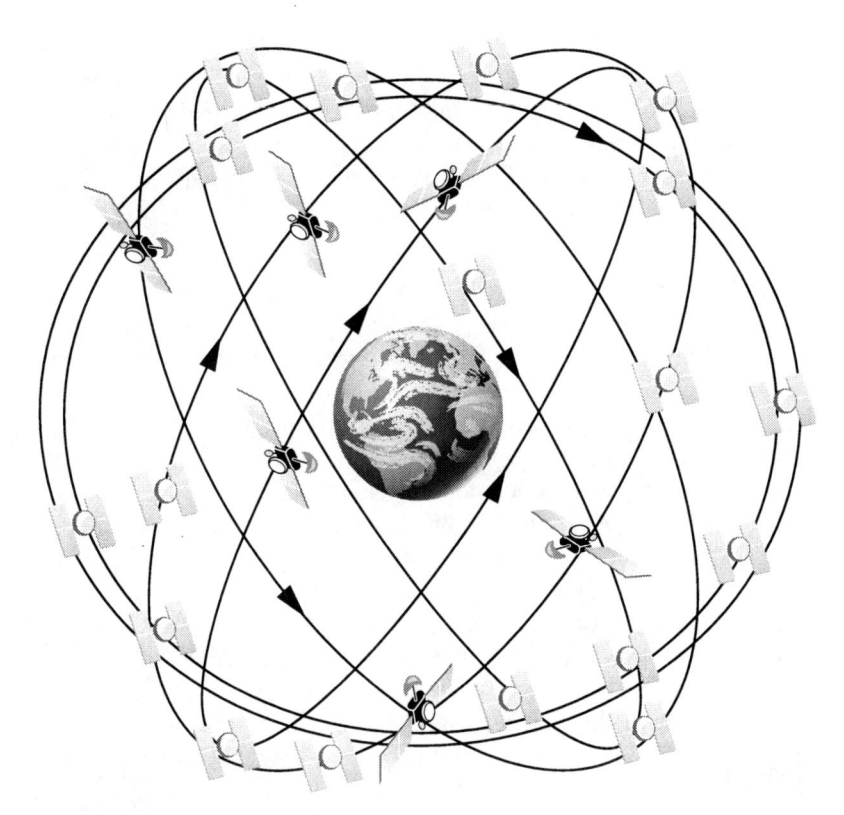

FIGURE 1.9
Artist's rendering of the GPS satellite constellation

Credit: GPS.gov

V. HAZARDS OF THE OUTER SPACE ENVIRONMENT

Outer space is a hazardous environment, both for people and equipment. Sustaining human life and providing protection from radiation, microgravity, micrometeoroids, and space debris are often the most challenging aspects of space operations. This section discusses some of the most significant hazards in outer space and the technologies used to mitigate them.

A. Sustaining Human Life

Outer space is a cold, airless vacuum. An astronaut would not survive in outer space without basic life support systems, often referred to as environmental control and life support systems (ECLSS). Astronauts must bring with them all of the food, water, and air they will need to stay alive. Spacecraft and spacesuits must be pressurized to maintain the correct air pressure on the astronauts' bodies and a comfortable body temperature. An ECLSS system must also provide for the collection and removal of human body waste.

B. Microgravity

When people think of traveling to outer space, they often think of the feeling of weightlessness, or microgravity. It is a common misconception that the weightlessness is because there is no gravity in outer space. It is true that gravity decreases with distance from massive celestial bodies, such as stars, planets, moons, and so traveling far enough into deep space would reduce the effects of gravity to near zero. But gravity maintains a strong hold on a spacecraft and its inhabitants in orbit. In fact, astronauts in orbit weigh only about 10 percent less than they do on Earth. The microgravity environment in orbit is caused not by a lack of gravity but by the perpetual free-fall of orbital motion. Because a spacecraft and its inhabitants are both falling at the same rate, the inhabitants are not pressed against the spacecraft by gravity or acceleration as they would be if standing still. Accordingly, astronauts in orbit are indeed weightless—but only in relation to their immediate environment.

The microgravity environment of outer space has many benefits, including permitting research that would be impossible on Earth and allowing astronauts and space tourists to experience the thrill of "flying." Nevertheless, microgravity has serious health effects. In the short term, disruption of the vestibular system causes nausea, known as space adaptation syndrome. Astronauts on long-duration spaceflights also suffer bone loss and weakening of the muscular and cardiovascular systems. These effects can be slowed, but not entirely prevented, while in space by exercising regularly (figure 1.10) and wearing specialized clothing that increases resistance on body movements.

C. Radiation

Outer space is saturated with electromagnetic radiation. There are generally three sources of space radiation: particles trapped in Earth's magnetic field (known as the Van Allen radiation belts), particles ejected from the Sun during solar storms, and galactic cosmic rays (heavy,

V. Hazards of the Outer Space Environment   21

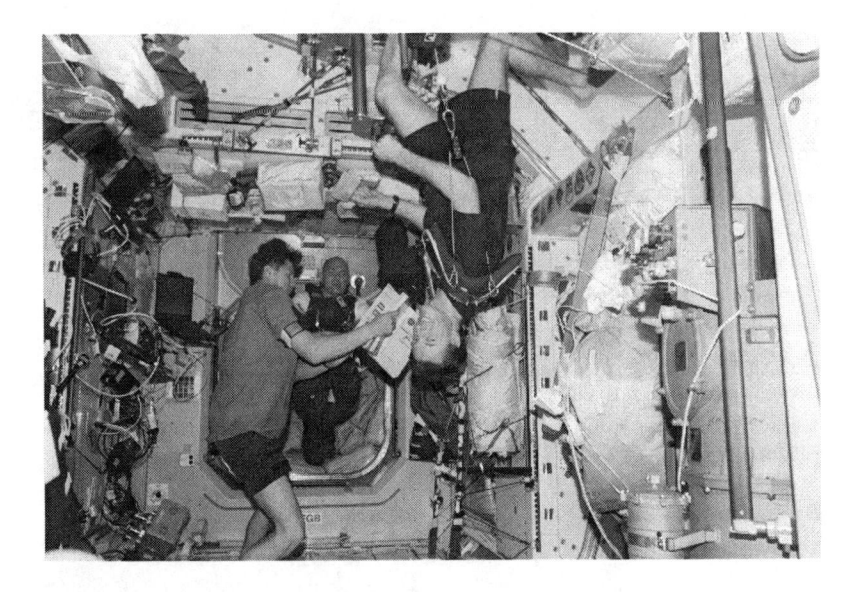

FIGURE 1.10
Astronauts exercising on the ISS
Credit: NASA

high-energy ions that originate outside of the solar system). Excessive exposure to radiation can cause severe health problems, such as cancer and even death, and wreak havoc on electronic systems by altering the states of integrated circuits, which can result in data loss and performance errors. A single solar flare can deliver a deadly dose of radiation or destroy an unprotected spacecraft's electronic systems in a matter of minutes.

On Earth, we are protected from space radiation by Earth's atmosphere and magnetic field. Nevertheless, we are still exposed to the equivalent of four to five chest X-rays each year from ambient environmental radiation (table 1.4). In Earth orbit, however, there is no atmospheric protection, and the protection of the magnetic field weakens as altitude increases. Astronauts on the ISS receive the equivalent of about eight chest X-rays *each day*. For this reason, astronauts are subject to per-mission and total career radiation exposure limits. Developing effective radiation protection is one of the greatest technological challenges to enabling long-duration, deep-space human space exploration missions.

TABLE 1.4
Comparison of Radiation Doses*

Description	Radiation Exposure (milliSievert)
Single chest X-ray	0.06
One year of normal radiation on Earth	3.0
8 days on the Space Shuttle	5.59
9-day round trip mission to the Moon	11.4
6 months on the ISS	80–160
Lowest dose received during 1945 Hiroshima bomb	200
3-year round trip to Mars	1,200

*Actual doses vary based on the solar cycle.

Source: NASA, SPACE FARING: THE RADIATION CHALLENGE (2008), http://www.nasa.gov/pdf/284277main_Radiation_MS.pdf

D. Micrometeoroids and Space Debris

Micrometeoroids, particles of dust the size of a grain of sand that date back to the creation of the solar system, are very common in outer space. In fact, most "shooting stars" seen at night are micrometeoroids burning up as they enter the atmosphere. Despite their small size, repeated high-velocity impacts of micrometeoroids on spacecraft act like a sandblaster, and long-term exposure can degrade critical exposed systems, such as camera lenses, antennas, and solar arrays.

The micrometeoroid problem has been exacerbated by human space activity over the last half century. The U.S. government is currently able to track about 22,000 man-made objects in Earth orbit. However, it is estimated that more than 500,000 man-made objects larger than a centimeter, and millions of objects smaller than a centimeter, are currently circling Earth in commonly used orbits. Of these objects, only about 1,000 are operational spacecraft.[6] The rest are dead satellites, discarded equipment, spent rocket boosters, fragments from collisions and explosions, paint chips, and other byproducts of human space activities, collectively known as space debris (figures 1.11 and 1.12). Space debris in low orbits will eventually succumb to atmospheric drag and burn up, but debris in higher orbits will remain in space for hundreds or thousands of years, or even forever.

Critical systems on spacecraft are protected from micrometeoroids and small space debris by thin foils of film placed a short distance

---

6. NASA, Frequently Asked Questions: Orbital Debris, http://www.nasa.gov/news/debris_faq.html.

V. Hazards of the Outer Space Environment   23

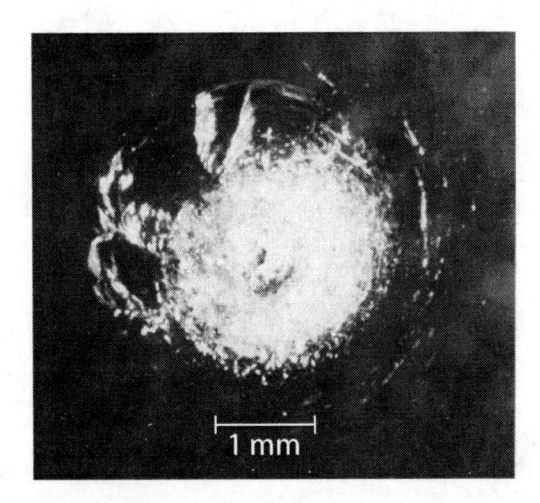

FIGURE 1.11
Window pit on the Space Shuttle *Challenger* caused by orbital debris in 1983

Credit: NASA

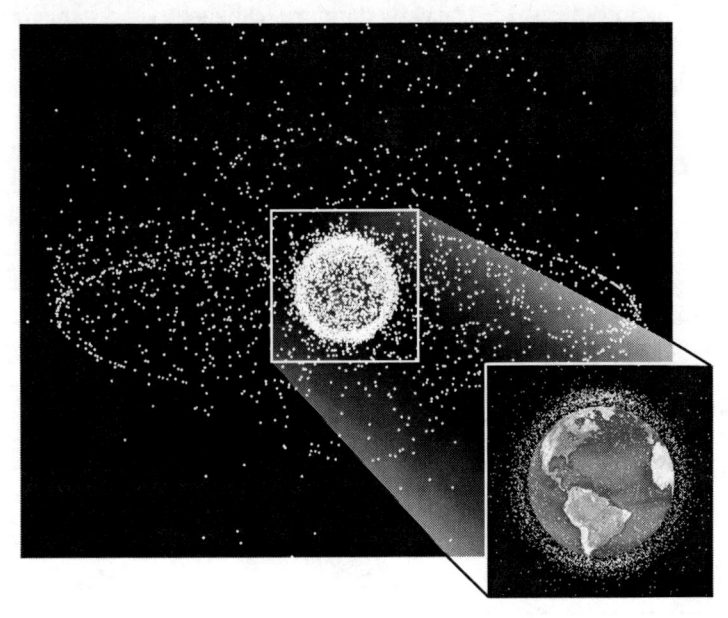

FIGURE 1.12
Computer-generated images of tracked space debris in Earth orbit

Credit: NASA

outside of the outer wall of a spacecraft, known as Whipple shields (designed by the astronomer Fred Whipple). Larger debris, however, must be actively avoided. Spacecraft routinely maneuver to avoid collisions, but this is not always possible, either because not all debris is tracked or the operator does not receive sufficient notice to perform an avoidance maneuver. In 2009, an Iridium satellite was destroyed in a collision with a defunct Russian military satellite, the first major collision between an operational satellite and space debris.

The Iridium collision generated thousands of new pieces of trackable space debris. Experts fear that if the amount of debris in orbit reaches a critical mass, one collision could set off a chain reaction of other collisions that would render portions of Earth orbit virtually unusable. This collisional cascading is known as the Kessler syndrome, named after NASA scientist Donald Kessler, who first proposed this possibility in 1978. Efforts underway to track, mitigate, and remove space debris are discussed in greater detail in chapter 11.

VI. PRACTICAL USES OF OUTER SPACE

Operating in outer space is clearly complex, expensive, and dangerous. Yet there are many practical applications of spaceflight, including telecommunications, satellite navigation, Earth observation, space exploration, and space tourism, that make it worth the effort (see figure 1.13).

A. Telecommunications

The most widespread use of outer space is for commercial, military, and civil telecommunications. The use of satellites for telecommunications was first proposed by Arthur C. Clarke in 1945. He postulated that three satellites arranged equidistantly around Earth in geostationary orbit would provide worldwide radio coverage. INTELSAT launched the first commercial communications satellite in 1965 and achieved the global coverage that Clarke envisioned by 1969. Today, the commercial telecommunications satellite industry is a nearly $180 billion industry, which represents 61 percent of total space industry revenues. The largest segment of the telecommunications sector is satellite television broadcasting, which in 2010 represented almost 80 percent of all satellite service revenues.[7]

---

7. FUTRON CORP. & SATELLITE INDUS. ASS'N, STATE OF THE SATELLITE INDUSTRY REPORT (Aug. 2011), http://www.sia.org/PDF/2011_State_of_Satellite_Industry_Report_(August2011).pdf.

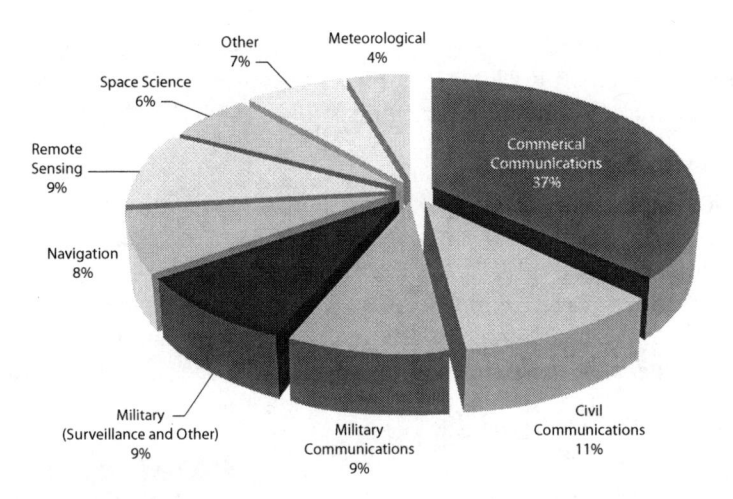

FIGURE 1.13
Operational satellites by function as of June 2011
Source: Satellite Industry Association

B. Satellite Navigation

Satellite navigation systems allow users with small receivers to determine their exact location by triangulating the time signals transmitted by a constellation of orbiting satellites. The U.S. NAVSTAR Global Positioning System and the Russian GLONASS are the only functioning global navigation satellite systems (GNSS). China has a regional navigation system, Beidou, which it is expanding to be a GNSS, and the European Union is in the initial phase of deploying its Galileo GNSS.

The U.S. GPS system is operated by the U.S. Air Force and is freely available to anyone with a compatible receiver. The first experimental GPS satellite was launched in 1978, and the system became fully operational in 1994. The GPS satellite constellation consists of 32 satellites in MEO dispersed in six different orbital planes (figure 1.9). At least four GPS satellites must be "visible" to the receiver at any one time in order for the receiver to accurately determine its position.

C. Earth Observation

Earth observation satellites, also referred to as remote sensing satellites, are essentially telescopes, cameras, and other sensors that point toward Earth instead of into space. Military remote sensing satellites are mostly used for reconnaissance and early warning purposes. Civil and commercial remote sensing satellites are used for mapping, meteorology, and environmental monitoring purposes. Commercial remote sensing was a $1 billion industry in 2010.

D. Space Exploration

Space exploration is conducted by both crewed and autonomous spacecraft. Except for the Apollo missions, human space exploration has been confined to LEO. Robotic spacecraft, on the other hand, have visited every planet in the solar system, many of their moons, and other celestial bodies such as asteroids and comets. Space exploration also includes the exploration of the cosmos with specialized observatories in outer space, such as the Hubble Space Telescope.

E. Space Tourism

Space tourism is space travel for recreational purposes. To date, seven space tourists have each paid the Russian government between $20 million and $40 million for week-long visits to the ISS. No tourist has paid for a suborbital spaceflight, although companies such as Virgin Galactic and XCOR Aerospace have presold hundreds of seats for flights that are expected to begin in the near future. Tickets on Virgin Galactic and XCOR Aerospace flights will initially cost $200,000 and $95,000 per ticket, respectively. Other space tourism companies are developing orbital crewed vehicles, private orbital space habitats, and even circumlunar missions to the Moon.

The unique challenges of operating in outer space permeate every facet of a space lawyer's practice. With these challenges in mind, the remainder of this book explores the laws, regulations, and customs that govern human activities in outer space.

GLOSSARY

Altitude: The height of an object above mean sea level.

Artificial satellite: Any object that has been placed into orbit by human activity.

Apogee: The point along a spacecraft's orbit where it is furthest from Earth.

Ascending node: The point along an orbit where the orbiting object moves north through the Equatorial Plane.

Glossary

Attitude: The orientation of a spacecraft in space.

Bus: The part of a satellite that carries the payload and provides electrical power, computer processors, and propulsion to the entire spacecraft.

Chemical combustion engine: Reaction engine that heats the reaction mass to a high temperature by combusting a solid, liquid, or gaseous fuel with an oxidizer.

Cislunar space: The region between Earth's atmosphere and the Moon.

Cold gas propulsion: A reaction engine that generates thrust by expelling a controlled, pressurized gas source through a nozzle.

Delta-v (Δv): The measure of the energy required to perform an orbital maneuver.

Delta-v budget: The total delta-v that will be required to perform all of the orbital maneuvers that a spacecraft must accomplish during the course of its mission.

Descending node: The point along an orbit where the orbiting object moves south through the Equatorial Plane.

Eccentricity: A measure of how much an orbit deviates from a perfect circle.

Electromagnetic propulsion system: A reaction engine that uses electricity and magnetic fields to accelerate the reaction mass, such as plasma or ions, to a high speed.

Equatorial Plane: An imaginary plane extending from the equator on Earth to the celestial sphere.

Escape velocity: The velocity at which an object escapes a body's gravitational pull and will move away from that body indefinitely.

Field propulsion: A propulsion method that uses solar wind, solar radiation, magnetic fields, and gravitational fields to propel spacecraft.

Geostationary Orbit (GSO): A geosynchronous orbit directly above the equator that appears to remain stationary over one point on Earth.

Geosynchronous Earth Orbit (GEO): Orbit located at an altitude of exactly 35,786 kilometers (22,236 miles).

Geosynchronous Transfer Orbit/Geostationary Transfer Orbit (GTO): A Hohmann transfer orbit used to reach GEO or GSO, respectively.

Global navigation satellite systems (GNSS): A satellite navigation system with worldwide coverage.

Graveyard orbit: An orbit several hundred miles above GSO to which GSO satellites are moved at the end of their operational life in order to make room for other satellites.

Gravity assist: A field propulsion method that uses a large body's gravity like a slingshot to accelerate an interplanetary probe towards its destination.

Highly Elliptical Orbits (HEO): An elliptical orbit that has a perigee of about 1,060 kilometers (660 miles) and an apogee of about 38,624 kilometers (24,000 miles) in a single orbital period.

Hohmann transfer orbit: An elliptical orbit used to move from a lower circular orbit to a higher circular orbit.

Inclination: The angle between the Equatorial plane and the orbit of a spacecraft.

Interplanetary spaceflight: Spaceflight where the spacecraft leaves the pull of Earth's gravity and travels to another planetary body in the Solar System.

Kármán Line: Altitude 100 km above sea level that is commonly used to define the boundary between Earth's atmosphere and outer space.

Kessler Syndrome: A scenario where the amount of space debris in Earth orbit reaches a critical mass where one collision sets off a chain reaction of other collisions that renders Earth orbit virtually unusable. Also referred to as "collisional cascading."

Lagrange points: Orbital locations between two large celestial bodies where the combined gravity of the objects interacts such that a smaller object placed at this point will stay in place relative to the larger bodies.

Launch vehicle: Rocket used to launch a spacecraft into outer space.

Launch window: The period of time when a launch vehicle must be launched in order to reach its target in outer space.

Law of Inertia: The second Law of Motion, which states that every body continues in a state of rest, or of uniform motion in a straight line, unless it is compelled to change that state by a force imposed upon it.

Law of Universal Gravitation: Every particle in the universe attracts every other particle with a predictable, measurable force.

Laws of Motion: (1) Every body continues in a state of rest, or of uniform motion in a straight line, unless it is compelled to change that state by a force imposed upon it; (2) when a force is applied to a body, the change of momentum is proportional to, and in the direction of, the applied force; and (3) for every action there is a reaction that is equal in magnitude but opposite in direction to the action.

Low Earth Orbit (LEO): Orbit located at an altitude of between 160 and 1,000 kilometers (100 to 600 miles)

Medium Earth Orbit (MEO): Orbit located at an altitude of between 1,000 and 35,786 kilometers (600 and 22,236 miles).

Microgravity: Weightlessness, where the force of gravity is just above zero.

Micrometeoroids: Particles of dust in outer space the size of a grain of sand that date back to the creation of the Solar System.

Multistage rockets: A launch vehicle with multiple stages, each containing its own engines and propellant. Stages can be either parallel stages mounted alongside one another or tandem or serial stages that are mounted one on top of the other.

Orbit: The gravitationally curved path of an object around a point in space.

Orbital node: One of the two points where an orbit crosses a plane of reference to which it is inclined.

Orbital spaceflight: Spaceflight where the spacecraft is launched with sufficient energy to complete at least one revolution around Earth.

Orbital velocity: The minimum speed at which a spacecraft must travel in order to remain in orbit.

Parking orbit: A temporary orbit used to store a satellite that is waiting to move on to the next stage of its mission.

Payload: The equipment a satellite needs to perform its mission, such as cameras or transponders.

Perigee: The point along a spacecraft's orbit where it is closest to Earth.

Period: The time it takes a spacecraft to make one full orbit about Earth.

Perpetual free-fall: A state where an object in orbit above Earth is travelling fast enough that the curve of Earth falls away from the object at the same rate that gravity pulls the object toward Earth.

Polar Orbit: A type of LEO that passes above or nearly above both poles of the Earth.

Radioisotope heater unit (RHU): A small device that uses the heat generated from the decay of a small amount of plutonium-238 or other radioactive material to keep instruments warm in outer space.

Radioisotope thermoelectric generator (RTG): A spacecraft power system that produces electricity from the decay of radioactive materials, such as plutonium-238.

Reaction engine: An engine that produces thrust by expelling reaction mass.

Reaction mass: Mass against which a system operates in order to produce acceleration.

Satellite constellation: A number of satellites with coordinated ground coverage, operating together under shared control, synchronized so that they overlap well in coverage and complement rather than interfere with other satellites' important coverage.

Solar sail: A field propulsion device that uses the pressure of solar radiation against a large collection surface to propel a spacecraft.

Space Adaptation Syndrome: Disruption of the vestibular system that causes nausea and usually affects astronauts during their first several days in outer space.

Space debris: Dead satellites, discarded equipment, spent rocket boosters, fragments from collisions and explosions, paint chips, and other byproducts of human space activities.

Suborbital spaceflight: Spaceflight where the spacecraft reaches outer space, but does not have sufficient energy to complete a full revolution around the Earth before reentering the atmosphere.

Sun-synchronous orbit: An orbit timed to pass over the same point on Earth at the same time of day.

Tether propulsion: A field propulsion system that uses a long tether that interacts with Earth's magnetic field to change a spacecraft's orbit.

Trajectory: The path that a moving object follows through space.

Van Allen radiation belts: Radioactive particles trapped in Earth's magnetic field.

Whipple Shields: Thin foils of film placed a short distance outside of the outer wall of a spacecraft to protect it from micrometeoroids and small space debris.

CHAPTER 2

The History of Spaceflight

Law is not made in a vacuum—not even space law. Understanding the laws that apply to human space activities requires an understanding of the historical context in which they were written. This chapter provides a brief history of human activities in outer space and an overview of current governmental and nongovernmental space activities. These descriptions are not exhaustive, but provide a sense of the state of the space industry today and how it came to be.

I. THEORY AND EARLY PRACTICE

The first realistic method of spaceflight was described by Russian mathematics teacher Konstantin Tsiolkovsky, who published his ideas in 1903 in *The Exploration of Cosmic Space by Means of Reaction Devices*. At the time Tsiolkovsky's ideas were not widely known outside of Russia and he was never able to put his theory into practice.

Dr. Robert Goddard, commonly referred to as the father of modern rocketry, developed in the early twentieth century the basic technology for space transportation that is still used today. Incidentally, Goddard developed this technology largely independently and without government funding. In 1912, he proved that rockets could work in a vacuum, which meant they could work in outer space. In 1914, Goddard received U.S. patents for solid- and liquid-fueled rockets using two or three stages. In 1919, Goddard published *A Method of Attaining Extreme Altitude*, which provided the "theoretical foundation for future American rocket developments such as staging that would be critical for the quest to land on the moon."[1]

---

1. Maj. Burton "Ernie" Catledge, USAF & LCDR Jeremy Powell, USN, *History*, in EARLY DEVELOPMENTS IN ROCKETRY, at 1, *available at* http://space.au.af.mil/au-18-2009/au-18_chap01.pdf (citing David O. Woodbury, OUTWARD BOUND FOR SPACE 21–22 (Little, Brown & Co. 1961)).

FIGURE 2.1
Dr. Robert H. Goddard and a liquid oxygen-gasoline rocket in the frame from which it was fired on March 16, 1926, at Auburn, Massachusetts

Credit: NASA

Goddard launched the first liquid-fueled rocket on March 16, 1926, in Auburn, Massachusetts (figure 2.1). The 12.5-m (41-foot) tall rocket traveled 56 meters (184 feet) in 2.5 seconds, reaching 96.56 km/ph (60 mph). In 1929 Goddard repeated the feat, but added weather instruments. The readings from the weather instruments "provided some of the earlier weather readings from 'on-board' sensors."[2]

In 1930, Goddard obtained financial backing from Charles Lindbergh and the Guggenheim Foundation. With this funding, he moved his work to New Mexico, where he remained until his death in 1945. Goddard launched a total of 34 rockets, reaching altitudes as high as 2.6 km (1.6 mi) and speeds of up to 885 km/h (550 mph). His later work largely focused on improving his early rockets, which led to the development of a gyro-controlled guidance system, gimbaled noz-

---

2. *Id.*

zles, small high-speed centrifugal pumps, and variable-thrust rocket engines, technologies that are still used today.

The next major developments in rocket science took place in Germany leading up to and during World War II. In 1942, Germany launched the A-4 rocket, which became the first man-made object to reach the edge of space. In 1944, Germany successfully launched its successor, the infamous V-2 rocket, which reached an altitude of 175 km (108.74 miles). Germany's (destructive) successful use of space technology led to a post-World War II race between the Soviet Union and the United States to recover German technology and scientists. Although the Soviet Union ended up with a majority of the hardware, most of the scientists and technicians wound up in the United States. The race to space—as an extension of the Cold War—had officially begun.

II. READY . . . SET . . . SPACE RACE!

Despite being left largely only with hardware from the German rocket program, the Soviet Union made great strides in the development of space technology. In October 1957 the Soviet Union launched Earth's first artificial satellite, *Sputnik 1*. A month later, the Soviet Union launched *Sputnik 2* carrying a research dog named Layka. The Soviet Union had achieved two firsts in less than 40 days: launch of a satellite and launch of a biological life form. The launch of Sputniks 1 and 2 caused an uproar in the United States, which until then had been unable to develop a coordinated space program. The controversy helped spur the development of the National Aeronautics and Space Administration (NASA), the U.S. manned and unmanned space programs, and the subsequent American landings on the Moon.

A. Unmanned Spaceflight

The 1960s were marked by tremendous developments in satellite programs. Unmanned satellites can be divided into three general categories: communications, remote sensing, and exploration. The development of each is addressed below.

1. Communications

In 1958, President Eisenhower transmitted a Christmas message to the world sent from an orbiting satellite. This message marked the first voice transmission beamed to Earth from outer space. In 1960, NASA launched *Echo*, a 100-foot diameter plastic balloon with an aluminum coating. *Echo* was designed as a passive reflector of radio signals emitted from Earth. But it proved difficult to use, and was soon eclipsed by nongovernment communications satellites. In 1962, AT&T financed the development and launch of *Telstar*, the first commercial communication satellite. Unlike modern telecommunications satellites, which are mostly in geosynchronous orbit, *Telstar* orbited in a low Earth orbit, so it did not provide continuous coverage over a single geographic area. Nevertheless, when in sight of its ground station, *Telstar* provided communications between the United States, the United Kingdom, and France. In 1963, NASA placed the first satellite in geosynchronous orbit, *Syncom*, which transmitted the Tokyo Olympic Games. The Soviet Union launched its first communication satellite system, Molniya, in 1965. These satellites provided continuous coverage over the Soviet Union by using highly elliptical orbits.

In 1964, delegates from the United States, Western Europe, Australia, Canada, and Japan met to discuss the development of a global telecommunications market. Among them, the delegations represented 90 percent of the world's international telephone traffic. The meeting resulted in the creation of the Interim International Telecommunications Satellite (INTELSAT) organization. Interim INTELSAT was an intergovernmental consortium formed for the purpose of allowing countries to cooperate and contribute to the development of a global satellite telecommunications network. In 1973, Interim INTELSAT lost its "interim" status and became INTELSAT, an intergovernmental organization in its own right.

INTELSAT's satellite network quickly grew in coverage and capability. INTELSAT launched its first satellite, *Early Bird* (INTELSAT I), in 1965 with 240 operational telephone circuits. INTELSAT achieved global telecommunications coverage in 1969 with the INTELSAT III series of satellites. In 1980, INTELSAT launched the INTELSAT V series, with 12,000 circuits and two television channels. On July 18, 2001, after 37 years as an intergovernmental organization, INTELSAT became a private corporation, Intelsat, Ltd.

2. Remote Sensing

Remote sensing of Earth's surface is performed for a variety of purposes, both civilian and military. The first photograph of Earth from outer space was taken in 1946 by a captured V-2 rocket launched from the White Sands Missile Range in New Mexico. The first U.S. military

FIGURE 2.2
First military reconnaissance imagery taken by Corona, showing the Mys Shmidta Air Field, USSR, on August 18, 1960
Credit: NRO

reconnaissance satellite, Corona, was launched in 1960 (figure 2.2). That year the United States also took the first television picture from outer space using NASA's Television Infrared Operational Satellite (TIROS) (figure 2.3). Although TIROS-1 was only in operation for 78 days, it proved that collection of weather observations from space was possible.

Encouraged by the success of TIROS, the United States launched a pair of Environmental Science Service Administration satellites, collectively called the TIROS Operational System (TOS). TOS provided worldwide weather observations without interruptions. Several improved versions of TIROS were subsequently launched and operated through 1985. From 1978 to 1986 NASA, the U.S. Weather Bureau, the National Oceanographic and Atmospheric Administration (NOAA), and RCA AstroElectronics launched 12 TIROS-N/NOAA satellites. Two of them have since failed, but the remaining 10 continue to provide daily observations of global weather systems.

The 1970s also saw the development of civil data-collection satellites called the land satellite series (Landsat). The Landsat program is

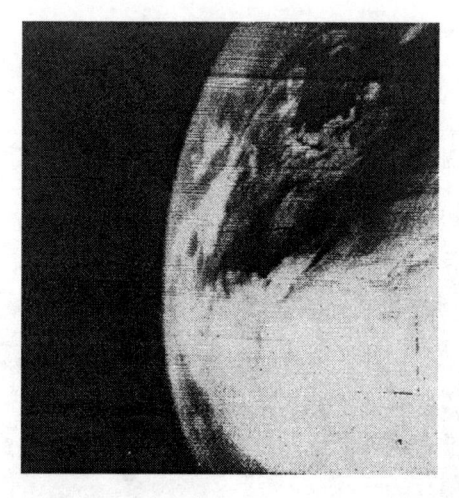

FIGURE 2.3
First television picture of Earth taken from space on April 1, 1960, by TIROS-1
Credit: NASA

composed of a series of Earth-observing satellite missions jointly managed by NASA and the U.S. Geological Survey. *Landsat 1* was the first Earth-observing satellite to be launched with the express intent to study and monitor Earth's landmass, and to make the data available to the public. The Landsat series uses infrared, microwave, and other advanced imaging capabilities, which made it, a first of its kind, available for civilian purposes. *Landsat 1,* then known as the Earth Resources Technology Satellite, was launched in 1972. NASA launched six subsequent Landsat satellites (one of which was destroyed in a launch failure), each carrying a variety of multispectral and optical imagers. The information collected is now freely available to the public.

The commercial remote sensing industry came of age in the 1990s after the passage of the Land Remote Sensing Policy Act of 1992 and the issuance of Presidential Decision Directive 23 by President Clinton in 1994, which laid the groundwork for private companies to launch competitive, high-resolution imaging satellites. The first commercial high-resolution remote sensing satellite, *Ikonos*, was launched by Geo-Eye, Inc., in 1999. Google, Inc.'s incorporation of high-resolution satellite images into its Google Maps and Google Earth programs in 2005 greatly increased the public's use and awareness of satellite imagery.

FIGURE 2.4
Artist rendering of the Cassini probe entering orbit around Saturn
Credit: NASA/JPL

3. Space Exploration

The exploration of other bodies in the solar system beyond the Moon has been conducted solely by unmanned probes. The United States has launched more than 25 robotic planetary probes. Probes have been sent to explore every planet in the solar system, many of their moons, and other celestial bodies, such as comets and asteroids.

Among the most notable feats are the two Voyager probes that the United States launched in 1977, both of which are still operational and sending data back to Earth (figure 2.5). *Voyager 1* flew by Jupiter in 1980 and Saturn in 1981 and is currently approaching the edge of the solar system approximately 18 billion kilometers from Earth. *Voyager 2* launched several weeks before *Voyager 1*, passing Jupiter in 1979 and Saturn in 1981. *Voyager 2* took a slightly slower trajectory than *Voyager 1* in order to visit Uranus and Neptune, which it did in 1986 and 1989, respectively. *Voyager 2* is now approximately 14.5 billion km (9 billion miles) from Earth.[3] *Voyager 1* and *Voyager 2* both carry a disc containing Earth sounds, music, pictures, and greetings in 60 languages.

---

3. For the current location of the Voyager probes, see NASA Jet Propulsion Lab., Where Are the Voyagers?, http://voyager.jpl.nasa.gov/where/index.html.

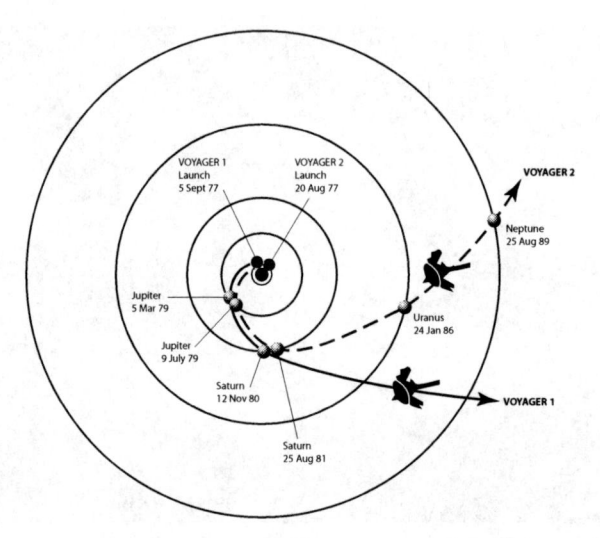

FIGURE 2.5
A diagram of the trajectories that enabled NASA's twin *Voyager* spacecraft to tour the four gas giant planets and achieve velocity to escape the solar system

Credit: NASA

More recently, the United States has launched and operated a series of Mars orbiters and rovers. The Mars series consists of the Mars Climate Orbiter, Mars Global Surveyor, Mars Odyssey, the Mars Pathfinder, the Mars Exploration Rovers, and, launched in November 2011, the Mars Science Laboratory (MSL). The MSL rover, Curiosity, is twice as long and three times as heavy as the Mars Exploration Rovers, Spirit and Opportunity, and carries instruments contributed by the Russian, Spanish, and Canadian space agencies (figure 2.6). The MSL will analyze Martian soil and rock samples for evidence that Mars could have once supported life.

Lastly, NASA has launched a series of space-borne observatories, known as the "Great Observatories," to conduct astronomical studies over many different electromagnetic wavelengths (visible, gamma rays, X-rays, and infrared). The most famous is the Hubble Space Telescope (HST). The HST is a cooperative effort between the European Space Agency (ESA) and NASA. It was launched from the Space Shuttle *Discovery* on April 25, 1990. Since then, NASA has conducted five servic-

FIGURE 2.6
Artist rendering of the Mars Science Laboratory rover, Curiosity, at work on Mars

Credit: NASA/JPL

ing missions to correct a debilitating flaw in the HST's optical system and to upgrade the telescope's instruments and operational systems (figure 2.7). The HST is an extraordinary instrument orbiting at an altitude of about 600 km (375 miles). It is the largest on-orbit observatory built to date and can image objects up to 14 billion light years away with resolutions seven to 10 times greater than any Earth-based telescope. The next space observatory, the James Webb Space Telescope, is currently expected to launch in 2018.

B. Manned Spaceflights

Just as the Soviet Union initially bested the United States in the unmanned space race, it also initially bested the United States in the manned space race. On April 12, 1961, the Soviet Union sent Yuri Gagarin into space, making him the first person to travel to space and to orbit Earth. Two years later the Soviet Union launched the first woman into space, Valentine Tereshkova.

The United States' first manned spaceflight program was Mercury. The Mercury vehicle was a capsule capable of carrying one astronaut. The Redstone launch vehicle was used for the initial two suborbital

FIGURE 2.7
The Hubble Space Telescope secured in the Space Shuttle *Discovery's* remote manipulator arm for servicing in 1997
Credit: NASA

flights, and the Atlas launch vehicle was used to lift the Mercury capsule for the four subsequent orbital flights. Each astronaut named his own capsule and added the number "7" to recognize the contribution of all seven astronauts in the Mercury team.[4] The Mercury capsule landed in the ocean upon reentry. This was different from the Soviet design, where the cosmonauts ejected from their capsules following reentry and were retrieved on the ground. On January 31, 1961, after a series of successful suborbital test flights carrying small mammals, a chimpanzee named Ham flew on a Mercury capsule. On May 5, 1961, Alan Shepard followed suit, completing a suborbital flight, and on February 20, 1961, John Glenn (figure 2.8) became the first American to orbit Earth.

Mercury was followed by the Gemini program. The Gemini capsule was larger than Mercury, capable of carrying two astronauts. The

---

4. NASA, Mercury: America's First Astronauts, http://www.nasa.gov/mission_pages/mercury/missions/spacecraft.html.

FIGURE 2.8
John Glenn entering *Friendship 7*

Credit: NASA

Gemini program allowed the United States to practice and perfect skills that would be necessary for the Apollo missions, including rendezvous, docking, extravehicular activities (figure 2.9), and long-duration spaceflight. The Gemini program consisted of 10 manned missions and two unmanned missions in 1965 and 1966, with the following program objectives:

- to subject two men and supporting equipment to long-duration flights
- to effect rendezvous and docking with other orbiting vehicles, and to maneuver the docked vehicles in space, using the propulsion system of the target vehicle for such maneuvers
- to perfect methods of reentry and landing the spacecraft at a preselected land-landing point
- to gain additional information concerning the effects of weightlessness on crew members and to record the physiological reactions of crew members during long-duration flights.[5]

---

5. NASA, Gemini, http://www-pao.ksc.nasa.gov/kscpao/history/gemini/gemini-overview.htm (last visited Dec. 29, 2011).

FIGURE 2.9
Astronaut Ed White becomes the first U.S. spacewalker on Gemini 4
Credit: NASA

The Gemini program placed the United States solidly ahead of the Soviet Union in the race to the Moon. Nonetheless, the Soviet Union forged ahead with the Voshkod capsule, which was the Soviet Union's equivalent of Gemini. It carried three astronauts (a first of its kind) and led to the first spacewalk and the first emergency manual reentry of a spacecraft.

The Apollo program allowed the United States to reach its much-anticipated goal of landing an astronaut on the surface of the Moon and returning him safely to Earth. Unlike the Mercury and Gemini spacecraft, the Apollo capsule was shaped like a teardrop rather than a bell. In order to lift both the capsule and lunar lander in one launch vehicle and accelerate them to Earth escape velocity, NASA developed a new rocket, the enormous Saturn V (figure 2.10). On July 20, 1969, Apollo 11 astronauts Neil Armstrong and Buzz Aldrin landed on the Moon, followed by the landings of Apollos 12, 14, 15, 16, and 17. Apollo 13 was not able to land on the Moon due to the accidental explosion of an oxygen tank. An anxious world watched NASA's feverish—and ultimately successful—effort to bring the crew home safely. The Apollo program ended in 1972. Parallel to the U.S. Apollo program, the Soviet Union developed the *Soyuz* spacecraft, which is still

FIGURE 2.10
Tower view of Apollo 11 launching on top of a Saturn V rocket

Credit: NASA

in use today. Although the Soviet Union never did land cosmonauts on the Moon, this reliable spacecraft has proven invaluable in transporting crewmembers to and from the International Space Station (ISS).

The Space Shuttle era began with the launch of the *Columbia* on April 12, 1981. The term "Space Shuttle" is the colloquial name given to the Space Transportation System (STS). STS consisted of the orbiter vehicle, twin solid rocket boosters, a giant external liquid fuel tank, and three Space Shuttle main engines. The STS was a first of its kind, because the orbiter was a reusable spacecraft that glided to a runway landing on Earth. The orbiter was in fact the only part of the Space Shuttle "stack" to make it into orbit and back to Earth.

The Space Shuttle fleet consisted of five individual orbiters: *Columbia*, *Challenger*, *Discovery*, *Atlantis*, and *Endeavour*, and one atmospheric

FIGURE 2.11
Launch of the Space Shuttle *Atlantis* (STS-43) on August 2, 1991
Credit: NASA

test vehicle, *Enterprise*. *Challenger* and *Columbia* were tragically lost in accidents that took the lives of all their crew members in 1986 and 2003, respectively. The Space Shuttle fleet (excluding *Enterprise*, which never flew in space) completed 135 missions from April 12, 1981, to July 21, 2011, flying a total of 864,401,200 km (537,114,016 miles) and completing 20,952 orbits around Earth. The Space Shuttle fleet transported cargo, experiments, labs, satellites, and astronauts into space, to the Soviet *Mir* space station, and to the ISS.

The ISS is about the length and width of an American football field and is the largest and most expensive space station ever built. The ISS is a joint project of the American, the Russian, the Japanese, the European, and the Canadian space agencies. The first ISS module was launched in 1998. It has been continuously inhabited since November 2, 2000, and is scheduled to remain in orbit until at least 2020. As of August 2011, there have been 135 launches to the ISS: 74 Russian crew and cargo vehicles, 37 Space Shuttles, two European cargo vehicles, and

II. Ready . . . Set . . . Space Race!　　45

FIGURE 2.12
The Space Shuttle *Endeavour* docked with the ISS on May 23, 2011
Credit: NASA

two Japanese cargo vehicles.[6] Since the retirement of the Space Shuttle in July 2011, NASA has relied on Russian *Soyuz* spacecraft to continue transporting astronauts to and from the ISS.

NASA initially planned to succeed the Space Shuttle with the Constellation program, which would have developed a family of vehicles to take people and cargo to low Earth orbit (LEO) and eventually to the Moon and to Mars. Due to schedule delays, technical difficulties, and cost overruns, the Constellation program was cancelled in 2010. Currently, as discussed below, NASA is working with private companies to develop commercial cargo and crew transportation services that will provide transportation to and from the ISS. NASA is also developing the Space Launch System and the Orion Multi-Purpose Crew Vehicle to take astronauts to destinations beyond LEO.

---

6. NASA, International Space Station Facts and Figures, http://www.nasa.gov/mission_pages/station/main/onthestation/facts_and_figures.html.

III. OTHER NATIONAL SPACE PROGRAMS

While the Cold War space race was primarily a competition between the United States and Russia, a number of other countries have also developed active space programs. Among the most capable of these programs are those of China, Europe, India, and Japan.

A. China

China has had a limited space program since the early 1960s. In the last decades, however, China's space capabilities have grown exponentially in parallel with its rapid economic expansion. In 2003, China became the third country to launch a human into outer space when it launched taikonaut Yang Liwei aboard *Shenzhou 5*. China completed its first spacewalk in 2008 and has indicated that it plans to land a person on the Moon by 2020. It also plans to build a manned space station and collect samples from the Moon by 2016.[7] China has three active launch sites and in 2010 the country conducted 15 successful satellite launches.

China relies mainly on a family of rockets called Chang Zheng ("Long March"). China is attempting to develop a new generation of rockets, called Kaituozhe ("Pioneer"), but the program has been in development since 2000 and has been subject to significant failures. Its future remains "uncertain."[8] China also controls approximately 70 satellites. Despite its successes, the Chinese civil space endeavors face numerous internal hurdles, "including substantial bureaucratic and organizational inefficiencies."[9]

China has also pursued military space developments, focused on two areas: the use of space assets and other advanced sensors for guided weapons applications, called "reconnaissance-strike complexes," and disrupting, degrading, denying, and destroying adversary space assets, called "counterspace" weapons. Despite these efforts, Chinese military capabilities remain limited by China's few communication and weather satellites available for military use, as well as the lack of a comprehensive global satellite navigation system. In addition, satellite and launch failures over the past several years have led to program delays. All in

---

7. Edward Wong & Kenneth Chang, *Space Plan from China Broadens Challenge to U.S.*, N.Y. TIMES, Dec. 29, 2011, http://www.nytimes.com/2011/12/30/world/asia/china-unveils-ambitious-plan-to-explore-space.html.

8. U.S.-CHINA ECON. & SEC. REVIEW COMM., 2011 REPORT TO CONGRESS 200 (Nov. 2011), http://www.uscc.gov/annual_report/2011/annual_report_full_11.pdf.

9. *Id.* at 207.

III. *Other National Space Programs*   47

all, "[w]hile a leader in the business of launching satellites, China is still years behind the United States in space. Its human spaceflight accomplishments to date put it roughly where the United States and the Soviet Union were in the mid-1960s."[10]

B. Europe

Europe, through ESA, continues to be one of the world's leading space programs. ESA operates a family of launch vehicles called Ariane. The Ariane 5 is ESA's heavy launcher, and although it is launched six to seven times a year, only one or two of those launches are for institutional purposes; the remainder are for commercial purposes. Ariane 5 is actually a fleet consisting of two launchers, Ariane 5 ECA and Ariane 5 ES. Ariane 5 ECA is the GTO "workhorse," delivering communication satellites to geostationary orbit. Ariane 5 ES is used in low and medium Earth orbits.[11] ESA has also launched the Russian *Soyuz* from French Guiana, which was the first time the *Soyuz* was launched from anywhere other than Kazakhstan or Russia. ESA is also currently working on developing a small launch vehicle, the Vega, which will be capable of transporting satellites ranging from 300 to 2000 kg (661 to 4,409 lbs.). ESA also has an active space science program.

C. India

India's space agency is the Indian Space Research Organization (ISRO). ISRO's main focus is the development of satellites, launch vehicles, and sounding rockets for unmanned purposes, such as telecommunications and weather monitoring. India's fleet consists of two operational launch vehicles and one more under development: the Polar Satellite Launch Vehicle (PSLV), the Geosynchronous Satellite Launch Vehicle (GSLV), and the GSLV Mark III. The PSLV has completed 19 successful flights, and ISRO refers to it as its "workhorse launch vehicle."[12] The PSLV is capable of launching 1,600 kg (3,527 lbs.) satellites to a sun-synchronous orbit or a 1050 kg (2,314 lbs.) satellite to geosynchronous orbit. The GSLV launch vehicles can place satellites weighing 2,000–2,500 kg (4,409–5,511 lbs.) in geosynchronous

---

10. Wong & Change, *supra* note 7.
11. European Space Agency, Europe's Launchers 4, http://esamultimedia.esa.int/multimedia/publications/europe-launchers/pageflip.html.
12. Indian Space Research Org., PSLV, http://www.isro.org/Launchvehicles/PSLV/pslv.aspx.

transfer orbit. As of the end of 2011, GSLV Mark I & II have completed seven flights, all for the purpose of placing satellites into space. The GSLV Mark III is currently under development and is destined to transport satellites weighing 4,500–5,000 kg (9,920–11,023 lbs.). ISRO touts the GSLV Mark III as enabling India to be competitive in the commercial launch market.

D. Japan

Japan's space agency, Japan Aerospace Exploration Agency (JAXA), is a significant player in space research and exploration. JAXA has several indigenous spacecraft: two launch vehicles (H-IIB Launch Vehicle and H-IIA Launch Vehicle); one transfer vehicle (H-II Transfer Vehicle), which is used to resupply the ISS; and the S-210/S-520 and SS-20 sounding rockets. Between 2005 and 2011, JAXA completed 18 launches. JAXA has been flying astronauts into space, aboard both the Space Shuttle and the *Soyuz*, since 1992. JAXA also developed and operates an ISS module, *Kibo*, which is the largest laboratory module on the ISS. In addition, JAXA currently has 15 satellites in operation and eight more under development. JAXA's various satellites are for lunar and planetary exploration, astronomical observation, telecommunications, and Earth observation missions.

IV. COMMERCIAL SPACEFLIGHT

In 1995, Dr. Peter Diamandis proposed a $10 million "X Prize" for the first private (non-government-funded) team that could demonstrate its ability to launch three humans to an altitude of 100 km (62 miles) above Earth's surface twice within two weeks. This prize eventually became known as the Ansari X Prize, following a multimillion donation from entrepreneurs Anousheh Ansari and Amir Ansari. Scaled Composites, the company operated by legendary aerospace designer Burt Rutan, with funding provided by Microsoft cofounder Paul Allen, won the Ansari X Prize on October 4, 2004, with their vehicle, SpaceShipOne. SpaceShipOne flew past the 100 km (62 miles) mark twice within 10 days, carrying a civilian pilot and a payload equivalent to two other passengers. Since then, numerous other entrepreneurs have raised and invested money in the development of private commercial spaceflight, both suborbital and orbital.

A. Suborbital

The Federal Aviation Administration (FAA) has identified six reusable suborbital launch service providers: Armadillo Aerospace, Blue Origin, Masten Space Systems, UP Aerospace, Virgin Galactic, and XCOR Aerospace.[13] A reusable launch vehicle (RLV) is a vehicle that accesses outer space, operates within the space environment, and returns safely to Earth, where it can be reused. A suborbital RLV (SRLV) is a vehicle that does not reach sufficient speed to actually enter an orbit around Earth. Rather, it enters space for a few minutes—approximately five—and reenters Earth's atmosphere, also to be reused. Those few moments in space give its passengers or payload a few minutes of weightlessness. SRLVs are to be used for both tourism and nontourism purposes.

1. Tourism

Four companies are currently developing suborbital spacecraft to fly tourists into outer space. One of the most visible commercial suborbital efforts has been Sir Richard Branson's Virgin Galactic. Virgin Galactic partnered with Burt Rutan's Scaled Composites immediately after Scaled Composites won the Ansari X Prize and over time developed the Virgin Galactic system, consisting of a spacecraft, SpaceShipTwo, and its carrier aircraft, WhiteKnightTwo. SpaceShipTwo is designed to carry six passengers and two pilots to an altitude of about 110 km (68.35 miles) at a top speed of Mach 3, or three times the speed of sound. SpaceShipTwo is an air-launched glider, with a hybrid rocket motor and thrusters. After spending about six minutes in space, SpaceShipTwo uses a unique feathering system to fold its wings to reenter the atmosphere like a shuttlecock. The spacecraft then unfolds its wings and glides to a landing on a runway. The first two ships are named *Enterprise* and *Voyager* in honor of the science fiction series *Star Trek*.

In order to escape Earth's dense lower atmosphere, SpaceShipTwo is carried aloft to an altitude of about 15,240 meters (50,000 feet) by the WhiteKnightTwo mothership. WhiteKnightTwo looks like a single long wing holding two fuselages. SpaceShipTwo sits between WhiteKnightTwo's twin fuselages. Virgin Galactic describes the WhiteKnightTwo as "the largest carbon composite aviation vehicle ever built and the

---

13. FAA, U.S. Dep't of Transp., The U.S. Commercial Suborbital Industry: A Space Renaissance in the Making 2 (2011).

most fuel efficient of its size."[14] A ticket on SpaceShipTwo currently costs $200,000.

Another aspiring suborbital tourism operator is XCOR Aerospace. XCOR was founded in 1999 and is based in California. XCOR focuses on the development of RLVs as well as rocket engines. XCOR is currently developing the Lynx. Similar to SpaceShipTwo and the Space Shuttle orbiter, the Lynx is a winged vehicle capable of landing on a runway. Unlike SpaceShipTwo and the orbiters, however, it does not use an independent launcher. Instead, it takes off from a runway under its own power. The suborbital trip aboard the Lynx is designed to be an "individualized" experience, where each passenger experiences a suborbital flight sitting alone alongside the pilot. A Lynx suborbital ticket currently costs $95,000. XCOR has announced $40 million in wet lease agreements for the Lynx system. Pursuant to the wet lease agreement, XCOR, as the lessor, will provide another operator, the lessee, a complete space craft package, which includes the craft, a crew, aircraft maintenance, and insurance.

Armadillo Aerospace was founded in 2000 by John Carmack, a successful video game programmer, founder of the software company id Software, and "self-taught aerospace engineer."[15] Armadillo Aerospace is developing a suborbital vehicle called the Hyperion. First flight tests are planned for 2014. The Hyperion is designed to seat two people, reach an altitude of 100 km (62 miles), carry a 20 km (44 lbs) payload, and uses liquid oxygen/alcohol propulsion. The ticket price per seat is $102,000. Armadillo Aerospace has also started development on a suborbital transport inertially guided tube vehicle, which can reach over 100 km (62 miles) under an FAA amateur classification or waiver. An amateur rocket has a total impulse of 200,000 lbs or less, and cannot reach an altitude of 150 km above sea level. Once a rocket exceeds amateur rocket criteria, it is considered "licensed," which means it requires either an FAA license or experimental permit in order to fly.

Blue Origin was founded in 2000 by Amazon.com founder Jeff Bezos. Perhaps the most secretive of the current crop of commercial space companies, Blue Origin has or is developing three vehicles, all of which are capsule-style spacecraft: a suborbital vehicle called New Shepard and two test vehicles: Goddard, which first flew in 2006,

---

14. Virgin Galactic, Spaceships, http://www.virgingalactic.com/overview/spaceships.
15. FAA, *supra* note 13, at 7.

and the PM-2, which suffered a test failure in August 2011. The New Shepard will have three or more seats, reach an altitude of 100 km (62 miles), carry three people or 200 kg (440 lbs), as well as 120 kg (254 lbs) on the outside of the vehicle. Blue Origin has not announced the expected ticket price.

2. Nontourism

Suborbital environments offer opportunities for scientific research by providing small windows of microgravity in which to perform experiments. The advantage of the suborbital model is the repeatability of microgravity experiments at relatively low cost. Suborbital flights offer an ideal platform for experiments that need only a few minutes of microgravity, but those few minutes must be repeated multiple times in short order. Suborbital research is therefore an important market for the burgeoning commercial space sector.

Virgin Galactic, XCOR, Armadillo, and Blue Origin are pursuing the suborbital research market. Three other companies, Masten Space Systems, UP Aerospace, and Whittinghill Aerospace, are also currently testing or operating suborbital research platforms. All seven companies are participating in NASA's Flight Opportunities Program, which provides suborbital flight and payload integration services for research and scientific missions. Additional companies are developing reusable suborbital launch vehicles, but flight testing has not yet begun and operations are slated for 2015 or later. Such companies include Copenhagen Suborbitals, Dassault Aviation, EADS Astrium, Rocketplane Global, and Sierra Nevada Corporation. More recently, Space Exploration Technologies Corp. (SpaceX) applied for an FAA experimental permit to operate a SRLV consisting of a Falcon 9 First Stage tank, a Merlin 1D engine, four steel landing legs, and a steel support structure. In recognition of its appearance, this vehicle is called the Grasshopper.[16]

B. Orbital

An orbital spacecraft is a spacecraft that is launched with enough energy to complete at least one revolution around Earth. As discussed in chapter 1, achieving orbital spaceflight is many times more energy-intensive, and therefore much more difficult and expensive, than achieving suborbital spaceflight. However, unlike suborbital spaceflight, an orbital

---

16. *Id.* at 21.

spacecraft can remain in space for days, weeks, months, or years, depending on a variety of factors, such as atmospheric drag. Developing a launch system that can provide safe, reliable, and cost-effective access to LEO is the holy grail of commercial spaceflight. Several companies have taken on this challenge.

1. Tourism

Orbital space tourism has so far been the province of a few very wealthy individuals. From 2001 to 2009, seven "space tourists" paid between $20 and $40 million each to reach the ISS aboard a Russian *Soyuz* spacecraft. These space tourists are Dennis Tito (U.S.), Mark Shuttleworth (South Africa), Gregory Olsen (U.S.), Anousheh Ansari (Iran/U.S.), Charles Simonyi (Hungary/U.S.), Richard Garriott (U.S./U.K.), and Guy Laliberté (Canada).

In the future, the ISS will no longer be the only orbital destination for space tourists. Bigelow Aerospace was founded by real estate investor and Budget Suites hotel chain founder Robert Bigelow in 1998 to develop flexible, configurable space habitats using expandable module technology licensed from NASA. Two prototype modules, *Genesis I* and *II*, were launched in 2006 and 2007, respectively. Although the media often refers to Bigelow's habitats as "space hotels" because of its founder's hotel industry background, Bigelow's initial market is actually governments and industry that want to perform space research but do not have access to the ISS. The primary obstacle to Bigelow launching an operational habitat is the lack of an affordable commercial launch vehicle. This will hopefully be overcome when at least one of the companies developing commercial cargo and crew transportation for NASA is ready to begin commercial operations.

2. Nontourism

Private companies in the United States have been providing orbital launch services to nongovernment customers since U.S. space law was changed to permit such services in the 1980s (see chapter 4). For example, Lockheed Martin Commercial Launch Services and Boeing Launch Services provide launch services for commercial customers aboard the Atlas and Delta rockets, respectively, and Orbital Science Corporation launches commercial payloads on its Minotaur, Pegasus, and Taurus launch vehicles.

New launch systems are currently being developed by both entrepreneurial and more established launch service providers. Much of this work is being stimulated by NASA's Commercial Crew and Cargo Program (CCCP), which is coordinating and at least partially funding the development of commercial space transportation services that will be able to transfer crew and cargo to and from the ISS. The three major CCCP programs are the Commercial Orbital Transportation Service (COTS), the Commercial Resupply Service (CRS), and the Commercial Crew Development program (CCDev).

The COTS program, first announced by NASA on January 18, 2006, is subsidizing the development of commercial spacecraft to ferry cargo to the ISS. Commercial spacecraft developers are funded through a series of milestone-based Space Act Agreements (see chapter 10). The first two companies to receive COTS contracts were SpaceX and Rocketplane Kistler. NASA terminated its contract with Rocketplane in 2007 after Rocketplane was unable to raise sufficient private funding for its vehicle. Rocketplane was replaced by Orbital Sciences Corporation in 2008.

Under COTS, SpaceX developed the Dragon spacecraft and Falcon 9 launch vehicle, which is launched from Cape Canaveral Air Force Station in Florida. SpaceX successfully completed its first COTS demonstration flight in December 2010, when it became the first private company to successfully launch an orbital spacecraft and return it safely to Earth. On its second COTS demonstration flight in May 2012, the Dragon successfully docked with the ISS, becoming the first commercial spacecraft to do so. Orbital Sciences is developing the Cygnus spacecraft and Antares (formerly Taurus II) launch vehicle, which will launch from the Mid-Atlantic Regional Spaceport on Wallops Island, Virginia.

CRS is the program to actually procure the commercial resupply services for the ISS, using the vehicles developed under COTS. On December 22, 2008, NASA announced that it had awarded SpaceX and Orbital Sciences contracts to fly a minimum of 12 and eight CRS missions, respectively, at a projected combined cost of $3.5 billion.

The CCDev program is subsidizing the development of commercial vehicles to transport astronauts to and from the ISS, also through a series of milestone-based Space Act Agreements. The first round of $50 million was announced in 2010 and contracts were awarded to five companies: Blue Origin, Boeing, Paragon Space Development Corporation, Sierra Nevada Corporation, and United Launch Alliance. On April 18, 2011, NASA awarded a second round of CCDev contracts, totaling nearly $270 million, to four companies: Blue Origin, Boeing,

Sierra Nevada, and SpaceX. With this funding, Blue Origin is developing a capsule spacecraft based on its suborbital New Shepard spacecraft, Boeing is developing a capsule spacecraft known as the CST-100, SpaceX is developing a human-rated version of its Dragon capsule, and Sierra Nevada is developing a lifting body spacecraft called the Dream Chaser that will launch on an Atlas 5 launch vehicle and land on a runway like the Space Shuttle. As of this writing, NASA plans to award a third round of commercial crew program funding, to be known as the Commercial Crew integrated Capability (CCiCap) initiative, in mid-2012. The CCiCap initiative is expected to result in the significant maturation of the commercial crew transportation systems.

C. Beyond LEO

Most commercial activity is focused on suborbital and orbital spaceflight. Nevertheless, there are at least two commercial ventures planning near-term, beyond-LEO operations. Space Adventures, the company that arranged the orbital spaceflights for the first seven space tourists aboard the Russian *Soyuz* to the ISS, recently announced a circumlunar trip to the Moon using a modified *Soyuz* spacecraft. This spacecraft will take one crewmember and two passengers on a trip to circle the Moon and return to Earth. Tickets are reportedly selling for $150 million each, and as of this writing, one ticket has already been sold and Space Adventures is in contract negotiations for the second.[17]

The second beyond-LEO venture is the Google Lunar X-Prize competition sponsored by Google, Inc., and the X-Prize Foundation. This X-Prize will be awarded to the team that lands the first privately funded, unmanned spacecraft on the Moon, travels at least 500 meters (1,640 feet) across the lunar surface, and transmits high-resolution video back to Earth, all before the end of 2015. The grand prize is $20 million, with a $5 million second-place prize, up to $4 million in bonus prizes awarded for accomplishing certain mission objectives, such as finding evidence of water and visiting an Apollo site or other area of historical interest on the lunar surface, and "a $1 million prize that will go to the team that demonstrates the greatest attempts to promote diver-

---

17. Denise Chow, *Space Tourist Trips Around the Moon Get Roomier Spaceship*, SPACE.COM., May 5, 2011, http://www.space.com/11584-space-tourism-private-moon-flights-details.html.

sity in the field of space exploration."[18] As of this writing, 26 teams are actively participating in the competition.

D. Spaceports

Launching a commercial spacecraft requires a launching site. The development of commercial launch service providers has led to the development of airports for commercial space operations, known as spaceports. There are currently seven nonfederal government spaceports in the United States: California Spaceport (CA), Cape Canaveral Spaceport (FL), Cecil Field Spaceport (FL), Kodiak Launch Complex (AL), Mid-Atlantic Regional Spaceport (VA), Mojave Air and Spaceport (CA), Oklahoma Air and Space Port (OK), and Spaceport America (NM).[19] California Spaceport caters only to orbital flights. Cecil Field Spaceport, Mojave Air and Space Port, Oklahoma Air and Space Port, and Spaceport America cater only to suborbital flights. Cape Canaveral Spaceport, Kodiak Launch Complex, and Mid-Atlantic Regional Spaceport cater to both orbital and suborbital flights.

The spaceflight industry has come a long way since Robert Goddard's first tentative rocket launches in the 1920s. The industry is now in a period of transition. A relatively stable group of government space agencies and their contractors are being joined in outer space by emerging space powers and a growing number of entrepreneurial space companies that are assuming functions traditionally undertaken by governments, opening new commercial markets in outer space, and enabling an exponentially greater number of people to experience the thrill of spaceflight.

---

18. Google Lunar X-Prize Rules Overview, http://www.googlelunarxprize.org/prize-details/rules-overview.

19. FAA, *supra* note 13, at 22.

CHAPTER 3

The International Outer Space Legal Framework

International law provides the overarching framework for the outer space legal regime. For those who have never practiced or been exposed to international law, Section I of this chapter is a synopsis of the international lawmaking process. Section II describes how international agreements are made in the United States. Section III discusses the sources of space law in particular, including overviews of the pertinent United Nations treaties and resolutions relating to space activities. Section IV describes some of the more important international organizations that focus on space law.

I. THE INTERNATIONAL LAWMAKING PROCESS

A. Sources of International Law

International law is created by international agreements and customary international law. This section briefly describes each.

1. Treaties

Article 2 of the Vienna Convention on Law of Treaties defines a "treaty" as "an international agreement concluded between States in written form and governed by international law."[1] Treaties are legally binding, unlike political commitments. The Vienna Convention also provides rules of treaty interpretation under Articles 31–33. Article 31 states that treaties should be "interpreted in good faith in accordance with the ordinary meaning to be given to the terms of the treaty in their context and in the light of its object and purpose." The *travaux préparatoires*, or "preparatory works," produced during the drafting

---

1. Vienna Convention on the Law of Treaties, May 23, 1969, 1155 U.N.T.S. 331, *reprinted in* 8 I.L.M. 679 (entered into force Jan. 27, 1980).

of a treaty are also taken into consideration when interpreting treaty language if it is ambiguous, obscure, or "leads to a result which is manifestly absurd or unreasonable."[2] This approach resembles the fundamental canons of statutory interpretation in the United States.

2. Customary International Law

Customary international law requires general State practice and *opinio juris*. *Opinio juris*, "an opinion of law," means simply that a State acted in a certain way under a belief that such action was a legal obligation. Evidence of customary international law is often found in, for example, U.N. General Assembly resolutions and domestic legislation. A State may not be bound by a customary international law if able to show it consistently objected to the rule *before* the rule rose to the level of international law. Additionally, while treaties can bind only the parties, the treaty itself may serve as evidence of customary international law.[3] This means that a treaty widely accepted in the international community could potentially rise to the level of customary international law and become applicable even to nonparties.

The U.S. Supreme Court has long held that customary international law is indeed part of U.S. law. In *The Paquete Habana* case, the Court held that international law "must be ascertained and administered by the courts of justice of appropriate jurisdiction."[4]

B. International Dispute Resolution

The International Court of Justice (ICJ) is the primary judicial organ of the United Nations. It hears disputes between States that have acceded to the ICJ's jurisdiction and may issue advisory opinions on legal questions. The United States has withdrawn from compulsory jurisdiction of ICJ and accepts the ICJ's jurisdiction only on a case-by-case basis.[5]

Arbitration is an alternative form of dispute resolution that has become popular with businesses entering bilateral investment treaties (BITs). BITs often have arbitration clauses to avoid court resolution when a dispute arises. Private international commercial arbitration helps businesses resolve disputes through methods that are more effi-

---

2. Vienna Convention, at art. 32.
3. Vienna Convention, at art. 37.
4. *The Paquete Habana*, 175 U.S. 677 (1900).
5. Case Concerning Military and Paramilitary Activities in and Against Nicaragua (Nicaragua v. United States) 1984 I.C.J. (Judgment of Nov. 26).

cient and less costly than litigation. Parties consent to arbitration usually through a written agreement. Arbitration decisions are typically binding on the parties, who often rely heavily on the United Nations Convention on the Recognition and Enforcement of Foreign Arbitral Awards for enforcement because it limits the ways parties can challenge an award from arbitration.[6]

On December 6, 2011, the Permanent Court of Arbitration, an intergovernmental organization formed for the purpose of facilitating dispute resolution between States, adopted Optional Rules for Arbitration of Disputes Relating to Outer Space Activities. The Rules are based on the 2010 arbitration rules of the United Nations Commission on International Trade Law, but are modified to address the particular needs of disputes involving outer space activities.[7]

II. MAKING INTERNATIONAL AGREEMENTS IN THE UNITED STATES

The United States has four methods of making international agreements: treaties, congressional executive agreements, presidential executive agreements, and treaty-authorized executive agreements. Under Article II of the U.S. Constitution, the president can consent to a treaty when he has the support of two-thirds of the Senate. Article VI of the Constitution establishes that courts are bound by treaties, and that treaties are above state law. However, the federal government's constitutionally designated right to enter into treaties does not abrogate state rights under the Tenth Amendment.[8]

The procedure for approving executive agreements is covered in Department of State Circular 175. A congressional executive agreement requires the consent of the president and a majority of both houses of Congress. For presidential executive agreements, the president's power must derive from the Constitution because the president is acting without congressional consent.[9] Finally, treaty-authorized executive agreements, which are far less common, only arise when a

---

6. U.N. Convention on the Recognition and Enforcement of Foreign Arbitral Awards, done at New York, June 10, 1958, 21 U.S.T. 2517, T.I.A.S. No. 6997, 330 U.N.T.S. 38, *available at* http://www.uncitral.org/uncitral/en/uncitral_texts/arbitration/NYConvention.html.

7. The Rules can be found at http://www.pca-cpa.org/upload/files/Outer%20Space%20Rules.pdf.

8. Missouri v. Holland, 252 U.S. 416 (1920).

9. United States v. Pink, 315 U.S. 203 (1942).

treaty already in effect gives the president authority to enter into additional agreements.

Whether international laws are automatically part of domestic law depends on whether a country has either a monist or dualist system. In monist countries, international laws are automatically incorporated into domestic law. In dualist countries, international laws are never automatically part of domestic laws. Dualist countries require an additional act that specifically incorporates the international agreement into domestic law.

The United States has a hybrid of the monist and dualist approaches. Treaties can either be self-executing or non-self-executing. For self-executing treaties, no further action needs to be taken at the domestic level to implement the treaty. A treaty is non-self-executing if it achieves the equivalent of lawmaking, a power that lies exclusively with Congress.[10] Courts will also consider the intent of the United States in signing the treaty to determine whether it is non-self-executing, including committee reports, the Resolution of Consent, and presidential transmittal letters. If there are no documents available to help determine intent, then the post ratification understanding and drafting history may help to glean intent.[11]

U.S. law has several important canons of construction applicable to treaty interpretation. The Charming Betsy doctrine provides that a court must not construe an act of Congress in a way that violates the law of nations if another possible interpretation exists.[12] The Last in Time rule states that when a federal statute conflicts with a treaty, the one enacted last will prevail. Finally, *lex specialis* states that when two laws are in conflict, the law covering a more specific matter will trump a more general law.

III. INTERNATIONAL SOURCES OF SPACE LAW

Space law is no exception to the general development of international law. It has developed internationally through binding treaties as well as through international practices that have risen to the level of custom-

---

10. RESTATEMENT (THIRD) OF FOREIGN RELATIONS LAW OF THE UNITED STATES §111, cmt. (i).
11. Medellín v. Texas, 552 U.S. 491 (2008).
12. Murray v. Charming Betsy, 6 U.S. 64 (1804).

ary international law. This section briefly presents the outer space treaties and relevant U.N. resolutions.

A. Major Outer Space Treaties

There are five major outer space treaties. Each treaty is reproduced in its entirety in Appendix 3. While not all space law practitioners will be confronted with issues arising under these treaties, attorneys in the space industry must be aware of their existence and recognize when they may be applicable to domestic space issues.

1. Treaty on Principles Governing the Activities of States in the Exploration and Use of Outer Space, including the Moon and Other Celestial Bodies (Outer Space Treaty)[13]

The Outer Space Treaty (Appendix 3-1), which became effective on October 10, 1967, is the foundational instrument of international space law. As of December 2011, 101 States are party to the Outer Space Treaty and 26 others have signed, but not ratified, the treaty.[14] The international community generally agrees that the Outer Space Treaty has risen to the level of customary international law and is therefore applicable to all States, even those that are not formal parties to the treaty. Hence, the international community gives great weight to the commitments under the treaty and expects States to adhere to them.[15]

One of the key provisions in the Outer Space Treaty is in Article I, which states, "The exploration and use of outer space, including the moon and other celestial bodies, shall be carried out for the benefit and in the interests of all countries . . . and shall be the *province of all mankind*" (emphasis added). According to the U.S. Senate Foreign Relations Committee, this phrase is essentially identical to "benefit of all

---

13. Treaty on Principles Governing the Activities of States in the Exploration and Use of Outer Space, including the Moon and Other Celestial Bodies, Jan. 27, 1967, 18 U.S.T. 2410, 610 U.N.T.S. 205.

14. For an up-to-date list of parties and signatories to the Outer Space Treaty, see U.N. Office for Outer Space Affairs, Treaty Signatures, http://www.oosa.unvienna.org/oosatdb/showTreatySignatures.do.

15. Ivan A. Vlasic, *The Growth of Space Law 1957–65: Achievements and Issues*, in YEARBOOK OF AIR AND SPACE LAW 1965, at 365, 374–80 (Rene H. Mankiewicz ed., 1967); Maj. Christopher M. Petras, *Space Force Alpha: Military Use of the International Space Station and the Concept of Peaceful Purposes*, 53 A.F. L. REV. 135, 153 (2002).

mankind."[16] The second paragraph of Article I stresses that exploration and use of space should be conducted "without discrimination."

Another important provision is in Article II, which provides that outer space "is not subject to national appropriation by claim of sovereignty, by means of use or occupation, or by any other means." This nonappropriation provision enshrines the status of outer space as *res communes*, or something that is owned by no one and subject to use by all. It is also the reason for the failure of a 1976 attempt by eight equatorial nations to declare that their sovereignty extended beyond their air space into outer space, and therefore into geostationary orbit. This declaration, known as the Bogotá Declaration, failed because of Article II.

Article III provides that international law, including the Charter of the United Nations, applies in outer space as it does on Earth. This article assures that legal questions regarding outer space not covered by any specific international space treaty will still be subject to general international law. In this way, outer space is by no means a legal void.

Article IV prevents States from placing weapons of mass destruction in orbit or on celestial bodies. It also states that the Moon and other celestial bodies will be used "exclusively for peaceful purposes." For example, a State cannot construct a military base on the Moon. The "peaceful purposes" clause does not apply to activities in Earth orbit.

Article V designates astronauts as "envoys of all mankind." In the interest of promoting international cooperation and collaboration in outer space, the international community sought to protect astronauts by requiring States to help those in danger. The general principle here was later expanded in the Rescue Agreement, discussed below.

Under Article VI, States are responsible for their national space activities, whether carried out by governmental or nongovernmental entities. Also, the "appropriate State" must authorize and supervise the space activities of its nongovernmental entities. The term "appropriate State" is undefined. However, because it was written in the singular, as contrasted with Article VII, it may imply that only one State is responsible for each nongovernmental entity operating in space.

Article VII holds launching States liable for damage caused by their "space objects." Here, the focus of "liability" is on space objects, whereas "responsibility" under Article VI focuses on national activi-

---

16. *Treaty on Outer Space: Hearings Before the Comm'n on Foreign Relations Senate Exec. D*, 90th Cong. 1st Session 56, 69-70 (1967).

ties. Under this article, a State qualifies as a "launching State" when it either launches or procures the launch of a space object or component thereof, or when the State launches an object either from its territory or facility. The Liability Convention, discussed below, was subsequently enacted to elaborate on this article.

A launching State retains its status for as long as the space object exists. Article VIII of the Outer Space Treaty states that "[o]wnership of objects launched into outer space, including objects landed or constructed on a celestial body, and of their component parts, is not affected by their presence in outer space or on a celestial body or by their return to the Earth." In short, once a State is designated as a launching State, it continues to be a launching State for that object indefinitely. The Registration Convention, also discussed below, elaborates on Article VIII.

The final substantive provision of the Outer Space Treaty, Article IX, is a general environmental provision stating that States shall operate with "due regard" for the interest of other States and avoid harmful contamination of Earth, outer space and celestial bodies. Article IX is discussed in more detail in chapter 11.

2. Agreement on the Rescue of Astronauts, the Return of Astronauts, and the Return of Objects Launched into Outer Space (Rescue Agreement)[17]

The Rescue Agreement (Appendix 3-2) entered into force on December 3, 1968, and was a result of the first U.N. convention elaborating on the Outer Space Treaty. Articles 2 through 4 address the rescue of spacecraft personnel, including obligations to notify the Secretary-General and launching authority of an emergency, to initiate the search and rescue operations, and to respond to distress signals.

The Rescue Agreement also addresses returning objects launched into outer space to their rightful owner. The preamble stresses that the purpose of the Rescue Agreement reflects a desire "to develop and give further concrete expression" to certain duties, including "the return of objects launched into outer space."[18] According to Article 5(3), the "launching authority" holds the right to request the return of their space objects. It states, "Upon request of the launching authority, objects launched into outer space or their component parts

---

17. Agreement on the Rescue of Astronauts, the Return of Astronauts and the Return of Objects Launched into Outer Space, 19 Dec. 1967, 19 U.S.T. 7570, 672 U.N.T.S. 119.
18. *Id.* at pmbl.

found beyond the territorial limits of the launching authority shall be returned to or held at the disposal of representatives of the launching authority, which shall, upon request, furnish identifying data prior to their return." Article 6 of the Rescue Agreement defines "launching authority" as "the State responsible for launching." This term is notably different from term "launching State" used in the Registration Convention and the Liability Convention. However, whether and to what extent this disparity would lead to legal inconsistencies in practice has so far not been resolved.

3. Convention on International Liability for Damage Caused by Space Objects (Liability Convention)[19]

The Liability Convention (Appendix 3-3), which entered into force on September 1, 1972, expands on Article VII of the Outer Space Treaty, including new definitions. First is a definition for damage: "loss of life, personal injury or other impairment of health; or loss of or damage to property of States or of Persons, natural or juridical, or property of international intergovernmental organizations." It also expands the term "launching" to include "attempted launching." Finally, the Liability Convention defines "launching State" so that up to four different States may be liable for any damage caused by a space object. A "launching State" may be

- a State that launches a space object;
- a State that procures the launching of a space object;
- a State from whose territory a space object is launched; or
- a State from whose facility a space object is launched.

While the term "liability" is left undefined, the Convention distinguishes between two situations. First, Article II imposes absolute liability for damage caused by a space object to the surface of Earth or to an aircraft in flight. The concept here is based on the theory that an innocent third-party victim should not be required to establish fault. Second, Article III imposes fault-based liability for damage caused by a space object anywhere other than the surface of Earth or to an aircraft in flight. However, liability does not extend to nationals of the launching State at fault. When a space object is launched by two or more States, the parties will be jointly and severally liable.

---

19. Convention on International Liability for Damage Caused by Space Objects, Mar. 29, 1972, 24 U.S.T. 2389, 961 U.N.T.S. 187.

The Convention recognizes three potential claimants in any liability action concerning a space object:

1. The State whose national suffered damage if in another State
2. The State where the damage occurred
3. The State whose permanent residents suffered damage

The Convention has no requirement for prior exhaustion of local remedies nor does it interfere with private claims. The Convention does not allow for private claims against a launching State, only claims brought by other States. It does not, however, prohibit private parties from asserting private claims under traditional national and international liability laws separate from the Convention. If an individual is first to bring a claim, the State must wait until it has been settled before asserting its own claim under the Convention. Because the Convention does not impose any limit on compensation, at least in theory, an individual may wish the State to pursue the claim if local courts enforce limits on damage awards.

4. Convention on Registration of Objects Launched into Outer Space (Registration Convention)[20]

The Registration Convention (Appendix 3-4) entered into force on September 15, 1976, and elaborated on Article VIII of the Outer Space Treaty. The Convention notably uses the same definition for launching State as the Liability Convention. It states in Article II(1), "When an object is launched into Earth orbit or beyond, the launching State shall register the space object" and "shall inform the Secretary-General of the United Nations." The Registration Convention is closely related to the Liability Convention in that its overall goal is to help identify objects in space so they can be attributed to a launching State for the purpose of assigning liability.

The Convention calls for the establishment of a National Register under Article II and an International Register under Article III. The International Register is administered by the U.N. Office of Outer Space Affairs, discussed below. Article IV lists the minimum information that States must provide to the International Register, which includes the name of launching State, an appropriate designator of the space object or its registration number, the date and location of launch, basic orbital

---

20. Convention on the Registration of Objects Launched into Outer Space, Jan. 14, 1975, 28 U.S.T. 695, 1023 U.N.T.S. 15.

parameters, and the general function of the space object. States are only required to furnish this information to the U.N. "as soon as practicable." The practice of most States has been to provide this information at least several months after the space object is launched.

5. Agreement Governing the Activities of States on the Moon and Other Celestial Bodies (Moon Agreement)[21]

The Moon Agreement (Appendix 3-5) was adopted by the U.N. General Assembly on December 18, 1979, but did not enter force until June 11, 1984. The Agreement in part aims to prevent the Moon and other celestial bodies from becoming areas of conflict. It also bans exploration and use of these bodies when the activities are not conducted for the "benefit of all mankind." The Moon Agreement states that natural lunar resources are the "common heritage of mankind," and establishes an international body to govern the exploitation of such resources when the technology to do so becomes available. None of the major space powers are party to the Moon Agreement, so it is generally considered dormant. The United States was especially concerned with the "common heritage" language because it could be interpreted to prevent the exploitation of natural resources for the benefit of a private entity or a single State.[22]

B. U.N. Outer Space Resolutions

In addition to the five outer space treaties, there are four U.N. General Assembly Resolutions of direct relevance to space operations, each of which is reproduced in Appendix 3:

- Principles Governing the Use by States of Artificial Earth Satellites for International Direct Television Broadcasting (Appendix 3-6)[23]
- Principles Relating to Remote Sensing of the Earth from Outer Space (Appendix 3-7)[24]

---

21. Agreement Governing the Activities of States on the Moon and Other Celestial Bodies, G.A. Res. 34/68, U.N. Doc. A/RES/34/68 (Dec. 5, 1979), *reprinted in* 18 I.L.M. 1434 (1979).

22. *See* Rosanna Sattler, *Transporting a Legal System for Property Rights: From the Earth to the Stars*, 6 CHIC. J. INT'L L. 23 (2005) (noting that the Moon Agreement "was not widely accepted, and no major space power has signed it because it further restricts ownership and prohibits and property rights until an international body is created").

23. U.N. Res. 37/92 (Dec. 10, 1982).

24. U.N. Res. 41/65 (Dec. 3, 1986).

- Principles Relevant to the Use of Nuclear Power Sources in Outer Space (Appendix 3-8)[25]
- Declaration on International Cooperation in the Exploration and Use of Outer Space for the Benefit and in the Interest of All States, Taking into Particular Account the Needs of Developing Countries (Appendix 3-9)[26]

Unlike the major space treaties, these resolutions are not binding. They are, nonetheless, generally followed by spacefaring nations and may have attained the status of customary international law, although this has not yet been tested judicially.

IV. INTERNATIONAL ORGANIZATIONS RELEVANT TO SPACE OPERATIONS

Many organizations have operations relevant to the space community. Some are solely concerned with space-related activities while others are only peripherally connected with space operations. This section provides a brief overview of the organizations that are generally most relevant to practicing space lawyers.

A. United Nations Committee on Peaceful Uses of Outer Space and Office of Outer Space Affairs

The United Nations established the Committee on the Peaceful Uses of Outer Space (COPUOS)[27] in 1958 as an ad hoc committee to assess the progress of international cooperation in space activities and to provide a venue for information gathering on space activities and programs. COPUOS was formalized as a permanent body in 1959. It is composed of 70 member states and has two permanent subcommittees: the Scientific and Technical Subcommittee and the Legal Subcommittee.

The United Nations Office of Outer Space Affairs (OOSA)[28] is the office responsible for implementing decisions of the U.N. General Assembly and COPUOS, as well for as assisting developing countries with furthering their space technology programs. OOSA is also responsible for maintaining the International Register of Objects Launched into Outer Space, as required by the Registration Convention.

---

25. U.N. Res 47/68 (Dec. 14, 1992).
26. U.N. Res. 51/122 (Dec. 13, 1996).
27. http://www.oosa.unvienna.org/oosa/en/COPUOS/copuos.html.
28. http://www.oosa.unvienna.org.

B. European Space Agency (ESA)

Another relevant international entity is the European Space Agency (ESA), a space agency whose status is equivalent to the National Aeronautics and Space Administration (NASA) in the United States or the Japan Aerospace Exploration Agency (JAXA) in Japan. According to the 1975 Convention establishing ESA, which did not enter into force until 1980, its purpose is "to provide for and to promote, for exclusively peaceful purposes, cooperation among European States in space research and technology and their space applications."[29] It is aimed at coordinating, elaborating, and implementing space activities of the Member States. Specifically, ESA strives to coordinate national space programs so that they are integrated into an overarching European space initiative. ESA is only involved in nonmilitary space activities.

Member States of ESA are not the same as the European Union (see table 3.1); Norway and Switzerland are members of ESA but are not members of the EU. Also, not all members of the EU are part of ESA. Nevertheless, the EU and ESA are legally connected through the 2004 Framework Agreement. The Framework Agreement aimed to develop a more formal cooperative relationship by identifying areas of cooperation, establishing joint initiatives and ad hoc agreements, and defining the terms of financial contributions.

C. International Telecommunications Union (ITU)

The International Telecommunications Union (ITU) is a specialized U.N. agency dedicated to communication and information technologies. It is responsible for allocating radio spectrum and the orbits of telecommunication satellites, as well as for developing standards to promote interconnectivity worldwide. The basic texts of the ITU are its Constitution, which presents general purposes and responsibilities, and the ITU Convention, which elaborates on the Constitution. These texts are amended by plenipotentiary conferences, which are held every four years and are the fundamental policymaking organ of the ITU. During world radiocommunication conferences, the ITU reviews and revises the radio regulations. The responsibilities of the ITU for space-related activities are discussed in detail in chapter 7.

---

29. Convention of Establishment of a European Space Agency, SP-1271(E), art. II (Purpose) (2003).

TABLE 3.1
ESA and EU Member States

ESA Member States		EU Member States	
Austria	Luxembourg	Austria	Latvia
Belgium	Netherlands	Belgium	Lithuania
Czech Republic	Norway	Bulgaria	Luxembourg
Denmark	Portugal	Cyprus	Malta
Finland	Romania	Czech Republic	Netherlands
France	Spain,	Denmark	Poland
Germany	Sweden	Estonia	Portugal
Greece	Switzerland	Finland	Romania
Ireland	United Kingdom	France	Slovakia
Italy		Germany	Slovenia
		Greece	Spain
		Hungary	Sweden
		Ireland	United Kingdom
		Italy	

D. Other International Organizations of Relevance

Numerous academic organizations also play an important role in international space activities. Through its 226 members, the International Astronautical Federation (IAF) brings together all space agencies, companies, and organizations to discuss space-related issues. The primary obligation of the IAF is to organize the annual International Astronautical Congress (IAC), which is by far the largest gathering of those in the space field. The IAC is a joint effort by the IAF, the International Institute of Space Law (IISL), and the International Academy of Astronautics (IAA). Established in 1960, IISL aims to foster the development of space law through colloquia, competitions, and the committees. It also publishes books and newsletters on space law. The IAA publications focus on the scientific aspects of space exploration. The Committee on Space Research (COSPAR) was formed by the International Council of Scientific Unions in 1958 to promote scientific research in outer space and to prepare scientific and technical standards related to space research. Finally, national space agencies have established several multinational forums for addressing specific issues in space operations, including the Space Frequency Coordination Group, the Inter-Agency Space Debris Coordination Committee, the Consultative Committee for Space Data Systems, and the Interagency Operations Advisory Group.

TABLE 3.2
Organizations with Connections to Space Law

Organization	Relevance to Space Law
International Institute for Unification of Private Law (UNIDROIT)	Draft Protocol to the Cape Town Convention on Matters Specific to Space Assets; Third Party Liability for Global Navigation Satellite Systems (GNSS) Services
International Civil Aviation Organization (ICAO)	Satellite technology for communication; interagency activities with ITU on spectrum and International Maritime Organization (IMO) on GNSS development
World Intellectual Property Organization (WIPO)	Addressing how to apply intellectual property laws to outer space
World Trade Organization (WTO)	Reference Paper and Basic Telecom Agreement (BTA); Lowering trade barriers in the satellite industry

In addition to organizations established to directly address space operations, there are many with indirect or tangential connections to space activities. Table 3.2 introduces some of these organizations and briefly states their relation to space law.

CHAPTER 4

The Development of U.S. Space Law

This chapter provides an overview of the development of space law in the United States. It is organized in decades and focuses on the major themes and developments for each of those decades. This chapter is not a comprehensive history of the United States' space policies and laws, but it will give the reader context in which to place current space developments and discourse.

I. THE 1950s: CREATION OF A CIVIL SPACE AGENCY

In 1958, President Dwight D. Eisenhower signed the National Aeronautics and Space Act (NAS Act).[1] The NAS Act was enacted in response to the Soviet Union's launch of *Sputnik* and created the U.S. civil space program, to be directed by "a civilian agency exercising control over aeronautical and space activities sponsored by the United States," to be known as the National Aeronautics and Space Administration (NASA).

Despite being a response to fears—military fears—created by *Sputnik*, the NAS Act contained language closely tracking language about peaceful uses of outer space that was emerging almost simultaneously from the U.N. General Assembly. For example, Congress declared that "activities in space should be devoted to peaceful purposes for the benefit of all mankind." The United States was attempting to straddle a delicate line: to promote peaceful use of space while not locking the United States into nonmilitarization should military action in space become necessary. Commentators have noted that the United States achieved this balance in two ways. First, Congress stated that space *should* be used for peaceful purposes, not that it *had to* be used for peaceful

---

1. National Aeronautics and Space Act of 1958, Pub. L. No. 85-568, 72 Stat. 426 (codified as amended at 51 U.S.C. §§ 20101 et seq.).

purposes, "Without knowing the full extent of the Soviet Union's space capabilities, Congress gave itself room to address the unknown as it unfolded by advocating rather than requiring 'peaceful purposes.'"[2] Second, Congress tackled the relationship between the military and civilian U.S. space programs. Congress delegated aeronautical and space activities to a civilian agency, NASA, but it carved out all activities "peculiar to or primarily associated with development of weapons systems, military operations, or the defense of the United States," to be under the direction and responsibility of the Department of Defense.

In 1959, NASA developed a long-range plan giving itself the following mission target dates from 1960 to 1970:

1960
- First launching of a Meteorological Satellite.
- First launching of a Passive Reflector Communications Satellite.
- First launching of a Scout vehicle.
- First launching of a Thor-Delta vehicle.
- First launching of an Altas-Agena-B vehicle (by the Department of Defense).
- First suborbital flight of an astronaut.

1961
- First launching of a lunar impact vehicle.

1961–1962
- Attain manned space flight Project Mercury.

1962
- First launching to the vicinity of Venus and/or Mars.

1963
- First launching of two-stage Saturn vehicle.

1963–1964
- First launching of unmanned vehicle for controlled landing on the moon.
- First launching of Orbiting Astronomical and Radio Astronomy Observatory.

---

2. Joanne Irene Gabrynowicz, *The Evolution of U.S. National Space Law and Three Long-Term Emerging Issues*, 4 HARV. L. & POL'Y REV. 405, 408 (2010).

1964
- First launching of unmanned lunar circumnavigation and return to earth vehicle.
- First reconnaissance of Mars and/or Venus by an unmanned vehicle.

1965–1967
- First launching in a program leading to manned circumlunar flight and to permanent near-earth space station.

Beyond 1970
- Manned flight to the moon. . . .[3]

Although the NASA Long Range Plan seemed quite aggressive at the time, especially considering how new the program was in 1960, the 1960s and 1970s did in fact prove to be among the most thrilling, innovative, and Earth-escaping decades for the U.S. space program. This is a testament both to the political will during those decades and to the vision, enthusiasm, and unrivaled proficiency of the people who made it possible.

II. THE 1960S: EXPLORING SPACE NOT BECAUSE IT IS EASY, BUT BECAUSE IT IS HARD

A. The Wiesner Report

President John F. Kennedy established a first-of-its-kind transition team to advise him on issues he would face as president, including those pertaining to space. The Ad Hoc Committee on Space was led by Massachusetts Institute of Technology Professor Jerome B. Wiesner, who had been a member of President Eisenhower's President's Science Advisory Committee, advised President Kennedy during the presidential campaign, and then became the president's science adviser. The team released the Report to the President-Elect of the Ad Hoc Committee on Space on January 10, 1961.[4] The report became known as the Wiesner Report.

---

3. NASA, Long Range Plan (Dec. 1959), http://www.hq.nasa.gov/office/pao/History/report59.html.
4. NASA, NASA History Office, Report to the President-Elect of the Ad Hoc Committee on Space (Jan. 10, 1961), http://www.hq.nasa.gov/office/pao/History/report61.html.

The Wiesner Report identified six major categories of space activities:

1. Ballistic missiles
2. Scientific observations from satellites
3. The exploration of the solar system with instruments carried in deep space probes
4. Military space systems
5. Man in orbit and in space
6. Nonmilitary applications of space technology

The Wiesner Report recognized that although most of the public's attention had been on the "overriding needs of our military security," the "sense of excitement and creativity" had moved away from the missile field to other listed categories.[5]

The Wiesner Report then identified five principal motivations for a vital and effective space program: (1) national prestige; (2) derivative contributions to national security; (3) scientific observation and experimenting; (4) nonmilitary applications such as satellite communications, broadcasting, satellite navigation, meteorology, and mapping; and (5) exciting possibilities for cooperation with all nations of the world.

The Wiesner Report went on to characterize the United States' space program as impressive, but in light of Soviet progress, "not ... impressive enough."[6] As a result, the Wiesner Report made several recommendations: (1) make the National Aeronautics and Space Council[7] an effective agency for managing the national space program; (2) establish a single responsibility within the military establishments for managing the military portion of the space program; (3) provide a vigorous, imaginative, and technically competent top management for NASA; (4) review the national space program and redefine the objectives in view of the experience gained during the past two years; and (5) establish the organizational machinery within the government to administer an industry-government civilian space program.

The Wiesner Report did not fall on deaf ears.

---

5. Id.
6. Id.
7. The National Aeronautics and Space Act of 1958 created the National Aeronautics and Space Council (NASC), which was an executive-level committee on space. It was chaired by the president of the United States and was allowed to employ a staff headed by a civilian executive secretary. The NASC was eventually abolished and then briefly revived in an alternate form, the National Space Council, from 1989 to 1993.

B. John F. Kennedy and the Apollo Program

On September 12, 1962, President Kennedy gave his famous address on the nation's space efforts to an audience at Rice University.[8] President Kennedy encapsulated the tremendous progress made by the U.S. program in these few words: "This is a breathtaking pace, and such a pace cannot help but create new ills as it dispels old, new ignorance, new problems, new dangers."[9] Unrelenting, however, President Kennedy set the United States on an increased pace to get the United States to the Moon. He announced not only that the United States would go to the Moon, but that it would do so by the end of the 1960s:

> We choose to go to the moon. We choose to go to the moon in this decade and do the other things, not because they are easy, but because they are hard, because that goal will serve to organize and measure the best of our energies and skills, because that challenge is one that we are willing to accept, one we are unwilling to postpone, and one which we intend to win, and the others, too.[10]

And so the seeds were planted for the Apollo Program.

C. Communications Satellite Act

The Communications Satellite Act of 1962 (Comsat Act) was part of an overall strategy to influence "emerging democracies" by providing them with services, such as telecommunications. The Comsat Act permitted the United States to participate in the development and operation of INTELSAT, an international telecommunications satellite organization. The effects of the Comsat Act have been far-ranging—the telecommunications satellite industry, largely spurred by INTELSAT, is the most lucrative space-based industry.

III. THE 1970s: THE SPACE SHUTTLE

On January 5, 1972, President Richard M. Nixon announced the development of the Space Transportation System, or Space Shuttle. He declared, "I have decided today that the United States should proceed

---

8. President John F. Kennedy, Address at Rice University on the Nation's Space Effort (Sept. 12, 1962), http://www.jfklibrary.org/Research/Ready-Reference/JFK-Speeches/Address-at-Rice-University-on-the-Nations-Space-Effort-September-12-1962.aspx.
9. Id.
10. Id.

at once with the development of an entirely new type of space transportation system designed to help transform the space frontier of the 1970s into familiar territory, easily accessible for human endeavor in the 1980s and '90s."[11] President Nixon announced that manned flights of the Space Shuttle would take place by 1978. The first flight of the Space Shuttle eventually took place in 1981. The Space Shuttle is described in detail in chapter 2.

IV. THE 1980s: DEVELOPING A FRAMEWORK FOR COMMERCIAL SPACE TRANSPORTATION AND REMOTE SENSING

In 1984, in an effort to spur the development of the commercial space industry, Congress passed the Commercial Space Launch Act (Launch Act) and the Land Remote-Sensing Commercialization Act (Remote Sensing Act).

A. Commercial Space Launch Act

Congress had two objectives in enacting the Launch Act: promoting the commercial space sector while, at the same time, developing licensing requirements to protect the public. Congress designated the Department of Transportation as the sole federal agency responsible for regulating the commercial space industry. The Launch Act and ensuing regulations created rules for licensing and regulation, liability insurance, and access to government launch facilities for commercial launch providers.

In 1988, Congress amended the Launch Act to partially protect the launch industry from third-party liability in the event of a catastrophic accident. The 1988 amendments authorize the U.S. government to indemnify FAA-licensed commercial space companies for third-party liability and, in return, require licensed launch providers to purchase launch insurance up to certain dollar amounts and obtain liability waivers from other participants in the licensed launch activity. The risk-sharing approach adopted by the Launch Act is further described in chapter 6.

---

11. President Richard M. Nixon, Statement on the Space Shuttle (Jan. 5, 1972), http://history.nasa.gov/stsnixon.htm.

B. Land Remote-Sensing Commercialization Act

Remote sensing refers to the use of satellites to image the surface of Earth and its atmosphere. As discussed in chapter 2, the first civil remote sensing satellite, *Landsat I*, was launched in 1972. *Landsat I* was followed by additional government-funded and government-launched civil remote sensing satellites. From 1972 to 1984, there was no regulatory regime for remote sensing. In 1984, Congress created such a regime in the Remote Sensing Act.[12]

The Remote Sensing Act created a three-phase process to privatize the government civil remote sensing satellites. Phase I consisted of privatizing the existing satellites by having the government contract a private company to operate them. Phase II would have entailed building and funding a system that would sell data to the U.S. government. Phase III would have seen the development of several private companies, operating remote sensing satellites, and selling the data therefrom. As discussed below, Phases II and III never came to fruition.

C. National Aeronautics and Space Act Amendment

The original version of the NAS Act contained no reference to commercial space activities. This is understandable considering that no such activities existed in 1958. As a result of this absence, commercial space operations were not a legally recognized sector of U.S. space activity. To remedy this, Congress, which had come to believe that commercial space was not only possible but should be encouraged, in 1984 amended the Declaration of Policy and Purpose of the NAS Act. The following declaration was added: "Congress declares that the general welfare of the United States requires that [NASA] seek and encourage to the maximum extent possible, the fullest commercial use of space."[13]

V. THE 1990s: WHERE DO WE GO FROM HERE?

As the 1990s began, the future direction of U.S. space policy was uncertain. The Cold War, which had served as the primary catalyst for much of the U.S. space program, had ended, and the commercialization of

---

12. Land Remote-Sensing Commercialization Act of 1984, Pub. L. No. 98-365, 98 Stat. 451 (1984) (codified at 15 U.S.C. § 4201 (repealed 1992)).

13. National Aeronautics and Space Appropriations Act of 1985, Pub. L. No. 98-361, 98 Stat. 426 (1984) (codified at 51 U.S.C. § 20102(c)).

space exploration expected in the 1980s had not yet materialized. The *Challenger* tragedy in 1986 called into question the ability of the Space Shuttle to fulfill its original mission of transforming outer space into a frontier "easily accessible for human endeavor." The mismanagement that led to the *Challenger* accident and the initial failure of the costly Hubble Space Telescope due to a flawed mirror also called into question NASA's ability to lead the U.S. civil space program. Such doubts led to numerous reports, some of which are discussed below.

A. The Augustine Report

In 1990, reacting to these concerns, Vice President Dan Quayle formed the Advisory Committee on the Future of the United States Space Program to evaluate the long-term future of NASA and the U.S. civil space program. The Committee was led by Norman Augustine, chief executive officer of Martin Marietta (now part of Lockheed Martin). The Committee's final report, known as the Augustine Report, recommended that the space program should include five activities: space science, Earth science, human spaceflight, space technology, and space transportation.[14] The Committee concluded that "the space science program warrants highest priority for funding" and should be the "fulcrum of the entire civil space effort" because space science "gives vision, imagination, and direction to the space program."[15]

The Augustine Report also described the Committee's conclusion that NASA, and the U.S. civil space program generally, was deserving neither of all the criticism leveled at it nor of excessive lauding. The report stated, "We conclude that the civil space program is neither as troubled as some would suggest nor nearly as strong as will be needed, given the magnitude of the challenges the program must undertake in the future."[16] The Augustine Report went on to describe nine specific concerns deserving attention: (1) lack of national consensus as to the goals of the civil space program; (2) NASA's overcommitment in terms of program obligations relative to resources available; (3) regular changes in project budgets; (4) institutional aging and NASA's insufficient responsiveness to valid criticisms and the need for change; (5) personnel policies that were incompatible with "long term mainte-

---

14. REPORT OF THE ADVISORY COMMITTEE ON THE FUTURE OF THE U.S. SPACE PROGRAM (Dec. 1990), http://www.hq.nasa.gov/office/pao/History/augustine/racfup1.htm.
15. *Id.*
16. *Id.*

nance of a leading-edge, aggressive, confident, and able workforce of technical specialists and technically trained managers"; (6) the natural tendency for projects to grow in scope, complexity and cost, which leads them to eventually collapse; (7) a starved NASA "technology base," which needed to be rebuilt and modernized; (8) the unforgiving nature of space projects with respect to human failings or neglect; and (9) excessive reliance on the Space Shuttle for access to space.[17]

The concerns described in the Augustine Report are still relevant today because none have been fully resolved and they all, in some form, run through the present-day discourse about NASA and the U.S. civil space program.

B. Land Remote Sensing Policy Act

In 1992, reacting to the failure of the U.S. government to successfully commercialize the Landsat system, Congress repealed the Remote Sensing Act and replaced it with the Land Remote Sensing Policy Act of 1992.[18] The Land Remote Sensing Policy Act recognizes the value of Landsat to "studying and understanding human impacts on the global environment in managing Earth's natural resources, in carrying out national security functions, and in planning and conducting many other activities of scientific, economic, and social importance."[19] The Act returned Landsat to the public sector and made unenhanced data available at cost to U.S. government-supported researches and agencies, and set a goal of eventually making this data available to a broader audience at the lowest possible cost. In 2008, Landsat data was finally made available to the general public for free. Landsat data is now mainly used by public sector activities such as forestry, land-management, and climate change monitoring.

The Land Remote Sensing Policy Act also authorized the Department of Commerce to license private remote sensing space systems, and in 1994 the Clinton administration issued a presidential directive establishing "U.S. industrial competitiveness in the field of remote sensing space capabilities" as a fundamental policy goal.[20] Together,

---

17. *Id.*
18. Land Remote Sensing Policy Act of 1992, Pub. L. No. 102-555, 106 Stat. 4163 (codified as amended at 15 U.S.C. §§ 5601–5672).
19. 15 U.S.C. § 5601(1).
20. The White House, Presidential Decision Directive 23: Foreign Access to Remote Sensing Space Capabilities, Fact Sheet (Mar. 10, 1994), http://www.fas.org/irp/offdocs/pdd23-2.htm.

these policies laid the foundation for U.S. companies to operate competitive, high-resolution imaging satellites. There are currently two federally licensed private systems providing commercial, high-resolution satellite data: GeoEye and DigitalGlobe. Nevertheless, even these two operators "rely heavily on the federal government as an anchor client, each having a $500 million government contract."[21] The licensing requirements for private Earth remote sensing space systems are described in chapter 8.

VI. THE EARLY 2000s: FOCUSING ON THE DETAILS OF COMMERCIAL SPACE REGULATION

A. Commercial Space Launch Amendments Act

In 2004, Congress amended the Launch Act with the Commercial Space Launch Amendments Act (CSLAA). The Launch Act, as amended by CSLAA, currently governs the launch and reentry of private spacecraft and shaped the regulations promulgated and enforced by the Federal Aviation Administration, the particulars of which are discussed in chapter 5. CSLAA's particular innovation was that it, for the first time, authorized private individuals to pay for, and commercial space entities to provide, space travel. Individuals who pay commercial providers for space travel, commonly referred to as "space tourists," are formally designated "space flight participants" (SFPs). The Code of Federal Regulations defines an SFP as "an individual, who is not crew, carried aboard a launch vehicle or reentry vehicle."[22] The CSLAA creates informed consent requirements for SFPs, insurance and indemnification requirements for operators, and procedures for licensing launches, reentries, and the operation of launch sites.

B. Regulating Commercial Remote Sensing Data

In 2000, the U.S. Commerce Department's National Oceanic and Atmospheric Administration issued the first regulations concerning the licensing of private remote sensing space systems, which were

---

21. Gabrynowicz, *supra* note 2, at 415 (citing Press Release, GeoEye, OrbImage Selected as NGA's Second NextView Provider (Sept. 20, 2004), http://geoeye.mediaroom.com/index.php?s=43&item=76; Press Release, DigitalGlobe, DigitalGlobe Awarded in Excess of $500 Million NextView Contract (Sept. 30, 2003), http://investor.digitalglobe.com/phoenix.zhtml?c=70788&p=irol-newsArticle&ID=1178620&highlight=).

22. 14 C.F.R. § 401.5.

subsequently revised in 2006.[23] The regulations address ground-based and space-based activities, and include licensing terms and conditions, requirements for annual operational auditing and record-keeping, rules for monitoring and compliance, and procedures for notification of foreign agreements.

In 2000, the federal government completed an interagency Memorandum of Understanding Concerning the Licensing of Private Remote Sensing Satellite Systems (Remote Sensing MOU).[24] The Remote Sensing MOU creates a "shutter control" decision-making process. Shutter control is a process whereby the federal government, through a series of determinations made at various levels, can prevent a licensed operator from acquiring or distributing remote sensing data. This happened shortly after September 11, 2001, when the federal government temporarily prevented a private operator from giving the public access to images of Afghanistan.

In 2003, the Bush administration issued a new U.S. Commercial Remote Sensing Policy, which superseded President Clinton's 1994 directive.[25] The "fundamental goal" of this policy was to "advance and protect U.S. national security and foreign policy interests by maintaining the nation's leadership in remote sensing space activities, and by sustaining and enhancing the U.S. remote sensing industry."[26] To accomplish this goal, U.S. government agencies were directed to "[r]ely to the maximum practical extent on U.S. commercial remote sensing space capabilities for filling imagery and geospatial needs for military, intelligence, foreign policy, homeland security, and civil users," and to "[p]rovide a timely and responsive regulatory environment for licensing the operations and exports of commercial remote sensing space systems."[27]

---

23. 15 U.S.C. § 5621.
24. Fact Sheet Regarding the Memorandum of Understanding Concerning the Licensing of Private Remote Sensing Satellite Systems, 15 C.F.R. § 960 app. 2 (2000).
25. The White House, U.S. Commercial Remote Sensing Policy, Fact Sheet (Apr. 25, 2003), http://www.whitehouse.gov/files/documents/ostp/press_release_files/fact_sheet_commercial_remote_sensing_policy_april_25_2003.pdf.
26. *Id.* at 2.
27. *Id.*

VII. THE LATE 2000s: CODIFYING SPACE LAW AS AN INDEPENDENT BODY OF LAW

In 2009, Congress created title 51 of the U.S. Code: National and Commercial Space Programs. Title 51 does not add new laws; rather, it recodifies existing laws relating to space operations into a single title of the U.S. Code. This recodification signals the existence of space law as an independent area of law in the United States. This is perhaps a natural development, a result of the increasing commercialization of space and the end of the Space Shuttle program. The consolidation and harmonization of U.S. space law also makes it easier to access and analyze for both U.S. and foreign practitioners. This alone makes it prime material for replication in other countries because it is now exportable as a comprehensive model of national space law.

U.S. space laws have developed over time as a reaction to national and international events and as an attempt to promote the growth of the U.S. space industry. U.S. space laws have been hailed for their comprehensiveness and their responsiveness to changing market demands. Even though at times reality has not quite risen to the expectations set by policymakers, the U.S. system of laws governing space—and more specifically commercial space—remains the most comprehensive in the world.

CHAPTER 5

Licensing Commercial Spaceflight

In the United States, the Federal Aviation Administration (FAA) regulates the launch and reentry of commercial launch vehicles and spaceports not operated by the federal government. The first FAA-licensed launch was a suborbital launch of a *Starfire* launch vehicle on March 29, 1989. Since then, the FAA has licensed more than 200 launches and the operation of eight commercial spaceports.[1]

This chapter provides an overview of the FAA licensing process for commercial space transportation. Section I discusses the FAA's jurisdiction over commercial spaceflight. Section II describes the six-step FAA licensing process for launch and reentry vehicles. Section III is a case study in waivers of FAA licensing requirements. Section IV discusses experimental launch permits. Section V addresses training, medical, and informed consent requirements for passengers and crew members on nongovernment spacecraft.

I. FAA JURISDICTION

The Commercial Space Launch Act of 1984, codified as amended in chapter 509 of title 51 of the U.S. Code, authorizes the FAA to issue licenses for nongovernmental space activities.[2] Such licenses include licenses to operate a launch site, launch vehicles into space, and reenter vehicles from space into Earth's atmosphere. The relevant regulations issued pursuant to the Launch Act are codified at 14 C.F.R. ch. III, parts 415, 420, 431, and 435 (hereinafter the Regulations).

---

1. Fed. Aviation Admin., Office of Comm. Space Trans., Launch Data and Information, http://www.faa.gov/about/office_org/headquarters_offices/ast/launch_license/.

2. Commercial Space Launch Act, 98 Stat. 3005, Pub. L. No. 98-575 (Oct. 30, 1984) (re-codified as amended in 51 U.S.C. §§ 50901 et seq.).

As expressed in the Launch Act, the Regulations have multiple purposes:

(1) to *promote economic growth and entrepreneurial activity* through use of the space environment for peaceful purposes; [and]
(2) to *encourage the United States private sector* to provide launch vehicles, reentry vehicles, and associated services by—
   (A) simplifying and expediting the issuance and transfer of commercial licenses;
   (B) facilitating and encouraging the use of Government-developed space technology; and
   (C) *promoting the continuous improvement of the safety of launch vehicles* designed to carry humans, including through the issuance of regulations, to the extent permitted by this chapter.[3]

These multiple purposes reflect competing perspectives about the proper role of government in regulating private spaceflight, particularly commercial human spaceflight, where the FAA's authority was clarified in 2004 with the passage of the Commercial Space Launch Amendments Act of 2004 (CSLAA). Some believe excessive regulation in the form of safety or design requirements would "only stifle innovation" and prevent the growth of the commercial human spaceflight industry. Proponents of this philosophy argue that people who engage in spaceflight "do not expect and should not expect to be protected by the government."[4] This position, however, is sometimes extended to argue that passengers, who are referred to as "spaceflight participants" (SFPs), should be denied government financial protection that is afforded to others involved in more traditional private space activities because "spaceflight participants wishing to ride on board a launch vehicle have chosen to undertake a risky venture of their own accord," and therefore they "do not merit the financial security provided by the promise of government indemnification."[5]

Others, however, are uncomfortable with the notion that the government does not do more to regulate the spaceflight industry. They

---

3. 52 U.S.C. § 50901(b) (emphasis added).
4. 150 Cong. Rec. H10048–49 (daily ed. Nov. 19, 2004).
5. H.R. Rep. No. 108-429, at 11 (2004).

call such "minimal" regulation a "tombstone mentality."[6] Although these perspectives continue to shape the development of commercial space laws in the United States, the first position has prevailed and is reflected in the changes to the Launch Act made by the CSLAA.

Under the Launch Act, a license from the U.S. government is required

(1) for a person to launch a launch vehicle or to operate a launch site or reentry site, or to reenter a reentry vehicle, in the United States;
(2) for a citizen of the United States to launch a launch vehicle or to operate a launch site or reentry site, or to reenter a reentry vehicle, outside the United States;
(3) for a citizen of the United States to launch a launch vehicle or to operate a launch site or reentry site, or to reenter a reentry vehicle, outside the United States and outside the territory of a foreign country unless there is an agreement between the United States Government and the government of the foreign country providing that the government of the foreign country has jurisdiction over the launch or operation or reentry.
(4) for a citizen of the United States to launch a launch vehicle or to operate a launch site or reentry site, or to reenter a reentry vehicle, in the territory of a foreign country if there is an agreement between the United States Government and the government of the foreign country providing that the United States Government has jurisdiction over the launch or operation or reentry.[7]

The U.S. Department of Transportation (DOT) is the lead agency for regulatory guidance pertaining to commercial space transportation activities. The Secretary of Transportation delegated commercial space licensing authority to the FAA. As a result, the FAA, through its Office of Commercial Space Transportation, licenses commercial launches,

---

6. T. Hughes & E. Rosenberg, *Space Travel Law (and Politics): The Evolution of the Commercial Space Launch Amendments Act of 2004*, 31 J. SPACE LAW 1 (2005) ("[T]he bill amounts to the codification of what has been come to be known in aviation safety parlance as the 'tombstone mentality': don't regulate until there are fatalities.... We should not legislate a tombstone mentality for safety oversight of this new space tourism industry.").

7. 51 U.S.C. § 50904(a). Note that a "citizen of the United States" is defined as "(A) an individual who is a citizen of the United States, (B) an entity organized or existing under the laws of the United States or a State; or (C) an entity organized or existing under the laws of a foreign country if the controlling interest (as defined by the Secretary of Transportation) is held by an individual or entity described in subclause (A) or (B) of this clause." 51 U.S.C. § 50902(1).

reentries, and the operation of launch and reentry sites pursuant to the Launch Act.

In order to carry out its duties, the FAA implements the Regulations found at 14 C.F.R. ch. III. The first commercial launch licensing regulations were issued in 1988 when no commercial launches had yet been conducted. Accordingly, the DOT developed what it considers to be "a flexible licensing process intended to be responsive to an emerging industry while ensuring public safety."[8] The DOT noted it would "continue to evaluate and, when necessary, reshape its program in response to growth, innovation, and diversity in this critically important industry."[9]

While the FAA has jurisdiction over the launch and reentry of space vehicles, it does not have jurisdiction over in-orbit activities. This is significant because, as discussed in chapter 6, activities subject to the FAA's jurisdiction enjoy the benefits of the Regulations' financial risk-sharing regime. The FAA's lack of jurisdiction over in-orbit activities means that the risk-sharing regime would not extend to cover an accident that occurred in orbit. In that case, the spaceflight operator could be liable for damages under traditional tort laws without the benefit of U.S. government indemnification, while the U.S. government could also be separately responsible for compensating other States under the Outer Space Treaty and the Liability Convention.

II. OBTAINING AN FAA LICENSE

The FAA has different procedures for issuing licenses for various types of launch and reentry activities and for launch and reentry sites. Most of the procedures are similar, so this section will use the procedures for licensing a reusable launch vehicle (RLV) as an example.[10] An RLV is a launch vehicle that is designed to return to Earth substantially intact and therefore may be launched more than one time, or that contains vehicle stages that may be recovered by a launch operator for future use in the operation of a substantially similar launch vehicle.

---

8. 65 Fed. Reg. 62,812, 62,813 (Oct. 19, 2000).
9. 65 Fed. Reg. 62,812, 62,813 (Oct. 19, 2000) (quoting 53 Fed. Reg. 11,006).
10. Operational safety requirements for expendable launch vehicles (ELVs) may be found in 14 C.F.R. part 417, for reentry vehicles that are not RLVs in part 435, for the operation of a launch site in part 420, and for reentry sites in part 433.

An RLV operator may apply for either a mission-specific license or an operator license. A mission-specific license "authorizes a licensee to launch and reenter, or otherwise land, one model or type of RLV from a launch site approved for the mission to a reentry site or other location approved for the mission."[11] An operator license "authorizes a licensee to launch and reenter, or otherwise land, any of a designated family of RLVs within authorized parameters, including launch sites and trajectories, transporting specified classes of payloads to any reentry site or other location designated in the license."[12] For either type of license, the FAA licensing process for launching and reentering an RLV has six steps: (1) preliminary consultation, (2) policy review, (3) safety review, (4) environmental review, (5) demonstration of financial responsibility, and (6) postlicensing requirements. The FAA also typically requires a separate payload review for the launch and reentry of any payload that will be carried by an RLV.

The FAA must complete its review 180 days after accepting a completed application. The agency must notify the applicant of any pending issue and action required to resolve the issue if the FAA has not made its licensing decision within 120 days after accepting the application.[13] The FAA may toll the review period if the applicant does not provide the FAA all required information.

A. Preliminary Consultation

The FAA requires prospective license applicants to consult with it before applying for a license "to discuss the application process and possible issues relevant to the FAA's licensing or permitting decision."[14] The preapplication consultation helps identify possible issues at the planning stages when the applicant can still make changes to the proposed activities. An applicant will therefore consult with the FAA to obtain input for drafting its license application. The preapplication consultation consists of informal conversations, either in person, in writing, or by telephone, where an FAA employee discusses the applicant's plans and flags possible problems that will need to be addressed. Once the applicant submits its application, the FAA performs policy, safety, payload, financial responsibility, and environmental reviews.

---

11. 14. C.F.R. § 431.3(a).
12. 14. C.F.R. § 431.3(b).
13. 51 U.S.C. § 50905.
14. 14 C.F.R. § 413.5.

Because the environmental review is the most time-intensive part of the process to complete, the applicant and the FAA often begin the environmental review process during the preapplication phase before the start of the 180-day license review period.

B. Policy Review

The FAA's policy review determines whether the proposed mission adversely affects U.S. national security, foreign policy interests, or international obligations.[15] To perform its review, the FAA requires the following information from the applicant: (1) the model, type, and configuration of the RLV; (2) identification of all vehicle systems, including structural, thermal, pneumatic, propulsion, electrical, aviations, and guidance systems as well as propellants; (3) information concerning any foreign ownership of the launching entity; and (4) and explanation of the launch and reentry flight profile, including the launch and reentry sites, the flight trajectory, and the sequence of planned events or maneuvers during the mission.[16]

Once it has received the required information, the FAA then consults with the Department of Defense, the Department of State, the National Aeronautics and Space Administration (NASA), and any other federal agencies it deems appropriate to determine the impact of the mission on the United States' various national and international policy interests.[17] Following these consultations, the FAA issues its decision. If the FAA issues a negative decision, the applicant may respond to the decision and argue its position that the mission does not negatively affect the United States' policy interests.[18]

C. Safety Review

Applicants must next satisfy the FAA's safety review. The goal of the safety review is to determine whether the launch and reentry can be conducted "without jeopardizing public health and safety and the safety of property."[19] Consistent with the regulatory philosophy behind the CSLAA described above, the FAA does not certify that the launch vehicle is safe for its crew, passengers, or payload. Rather, the FAA

---

15. 14 C.F.R. § 431.23.
16. 14 C.F.R. § 431.25.
17. 14 C.F.R. § 415.23.
18. 14 C.F.R. § 415.27.
19. 14 C.F.R. § 431.31(a).

safety review is only for the purpose of determining whether the launch and reentry will be conducted so that it is safe for the general public.

The applicant demonstrates it can meet the safety requirements by maintaining a safety organization, preparing a communication plan, and designating a safety official. The safety organization authorizes all mission decisions that may affect public safety.[20] The organization should facilitate communication between the applicant and the launch site operators, and between the applicant and the reentry site operators. The applicant must also describe the chain of command it plans to have in place to ensure compliance with the substantive terms and conditions stated in the mission license.

The communication plan identifies individuals responsible for monitoring safety-critical operations during the mission and how these people can quickly and effectively communicate time-sensitive and safety-critical information to each other.[21] The plan must describe the authority of vehicle safety operations personnel, by individual or position title, to issue hold/resume, go/no go, contingency abort, and emergency abort commands, and must include protocols for utilizing defined radio-communications terminology. Finally, the communication plan must include procedures that reflect how each of the relevant personnel receives a copy of the plan.

The safety official is authorized to examine all aspects of the applicant's operations as they relate to safety.[22] The Regulations dictate that the safety official will independently monitor safety parameters and report directly to the person responsible for the conduct of all licensed mission activities.

Having established and submitted proof of an internal structure that ensures compliance with the law, an applicant must next demonstrate to the FAA that its mission will meet the applicable risk standards.[23] The applicant does so by identifying and describing the structure of the RLV, including physical dimensions and weight; identifying and describing any hazardous materials, including radioactive materials, and their containers on the reusable launch vehicle; and providing data that verifies the risk elimination and mitigation measures resulting from the applicant's system safety analysis. The acceptable risk for a

---

20. 14 C.F.R. § 431.33(a).
21. 14 C.F.R. § 431.41.
22. 14 C.F.R. § 431.33.
23. 14 C.F.R. § 431.35.

proposed mission is measured in terms of the expected average number of casualties (E_c) per mission to the public and to an individual. The acceptable average number of casualties for the public at large is 0.00003 per mission (E_c 30 × 10⁻⁶) and to any one individual is less than 0.000001 per mission (E_c 1 × 10⁻⁶). Application of the safety requirements is discussed in the waiver case study below.

D. Environmental Review

In addition to policy and safety reviews, applicants for a license must pass the FAA's environmental review. Applicants provide the FAA information concerning the environmental impact of launching and reentering the particular vehicle at the particular launch and reentry site(s).[24] The FAA's environmental review determines whether the proposed mission complies with the National Environmental Policy Act (NEPA), the Council on Environmental Quality Regulations for implementing NEPA, and the FAA's Procedures for Considering Environmental Impacts.[25] Under NEPA, when the federal government takes major actions that affect the environment, it must conduct a review of those actions to determine whether the impact is significant and whether there are ways to mitigate those effects. In the context of commercial spaceflight, even though the private operator is engaging in the space activity, it is the government that is licensing that activity. Hence, when a company applies for a permit or a license, the federal agency in question—in this case the FAA—must evaluate the environmental impact of its permit or license decision under NEPA.[26]

The NEPA review process is centered around the Environmental Assessment (EA), unless the agency has previously issued a Categorical Exclusion finding that the action does not individually or cumulatively have a significant effect on the quality of the human environment. An EA determines the significance of the environmental effects of the proposed action and reviews alternative means of achieving the agency's objectives. The EA includes a discussion of (1) the need for the license, (2) alternative courses of action, (3) the environmental impacts of the proposed action, and (4) a list of agencies and persons consulted. The

---

24. 14 C.F.R. § 431.93.
25. 14 C.F.R. § 431.91.
26. COUNCIL ON ENVTL. QUALITY, A CITIZEN'S GUIDE TO THE NEPA (Dec. 2007), http://ceq.hss.doe.gov/nepa/Citizens_Guide_Dec07.pdf.

II. Obtaining an FAA License

EA focuses on the context and intensity of effects that may "significantly" affect the quality of the human environment.

At the conclusion of the EA process, the FAA decides whether to issue a Finding of No Significant Impact (FONSI), a document that presents the reasons why the agency has concluded that there are no significant environmental impacts projected to occur upon implementation of the action, or an Environmental Impact Statement (EIS), which described how the applicant's activity will significantly affect the quality of the human environment.[27] The environmental impact of spaceflight activities is discussed further in chapter 11.

E. Demonstration of Financial Responsibility

In addition to satisfying the licensing requirements described above, prior to conducting launch activities a licensee must demonstrate compliance with the "financial responsibility and allocation of risk requirements" set forth in the Regulations.[28] The licensee's financial responsibility requirements are included in the license and are based on the FAA's determination of the maximum probable loss (MPL) from the licensed launch activity. The process for determining the MPL is detailed in chapter 6. The licensee demonstrates compliance with the financial responsibility requirements by submitting to the FAA the three-party reciprocal waiver of claims agreement required under part 440.17(c) of the Regulations and evidence of insurance coverage or other form of financial responsibility required by the license.[29]

F. Postlicensing Requirements

The Regulations also impose postlicensing requirements on the licensee. For example, the licensee must ensure that any representations made to the FAA continue to be accurate. If there are any changes, for example, to the payload, flight trajectory, flight-critical systems, or the launch or the reentry site, the licensee is required to request a modification of its license from the FAA.[30] The licensee must also organize issuance of Notices to Airmen and Mariners with the local offices of the FAA and

---

27. For examples of FONSIs, EAs, and EISs related to commercial spaceflight, visit FAA, AST NEPA Documents, http://www.faa.gov/about/office_org/headquarters_offices/ast/environmental/nepa_docs/.
28. 14 C.F.R. § 440.5(a).
29. 14 C.F.R. § 440.15.
30. 14 C.F.R. § 431.93.

the U.S. Coast Guard.[31] The licensee must maintain all records pertaining to the mission for three years after its completion, and provide specific information to the FAA about the mission at specific intervals before the start of the licensed activity.[32]

G. Payload Reviews

In addition to the RLV mission or operator license, the FAA may also require payload and payload reentry reviews.[33] The FAA reviews a payload proposed for launch to determine whether a license applicant or payload owner or operator has obtained all required licenses, authorizations, and permits, and to determine whether its launch would jeopardize public health and safety, safety of property, U.S. national security or foreign policy interests, or international obligations of the United States.[34] A payload reentry review is conducted to examine the policy and safety issues related to the proposed reentry of a payload, other than a U.S. government payload or a payload whose reentry is subject to regulation by another federal agency. A payload reentry review may be conducted as part of an RLV mission license application review or may be requested by a payload owner or operator in advance of or separate from an RLV mission license application.[35]

III. WAIVERS: A CASE STUDY

The FAA may waive any of its licensing requirements. To determine whether it can waive safety requirements, the FAA will analyze whether the waiver (1) would jeopardize public health and safety or safety of property; (2) would jeopardize national security and foreign policy interests of the United States; and (3) would be in the public interest.[36]

Demonstrating how difficult it can be to write detailed regulations for an industry that is still in its infancy, the FAA had to waive two of its safety requirements the first time it licensed a private spacecraft to reenter Earth's atmosphere. This waiver provides a useful case study of the FAA's decision-making process.

---

31. 14 C.F.R. § 431.93.
32. Id.
33. 14 C.F.R. § 431.7.
34. 14 C.F.R. § 415.51.
35. 14 C.F.R. § 431.51.
36. 14 C.F.R. § 404.5(b); 51 U.S.C § 50905(b)(3).

On December 8, 2010, Space Exploration Technologies Corp. (SpaceX) conducted its first demonstration flight under NASA's Commercial Orbital Transportation Service (COTS), which consisted of launching and then safely retrieving the *Dragon* spacecraft from space. In order to conduct its COTS demonstration flight, SpaceX applied for, and received, a license to "reenter" *Dragon* as a "reentry vehicle." When the FAA issued a reentry license to SpaceX, it was the first time the FAA issued such a license to a private company. In the process of issuing the license the FAA also had to waive two safety requirements.

The Regulations define the term "reentry" and "reenter" as the purposeful return or attempt to return a reentry vehicle and its payload, if any, from orbit or from outer space to Earth.[37] The term includes activities conducted in orbit to determine reentry readiness and that are critical to ensuring public health and safety and the safety of property during reentry. The term also includes activities conducted on the ground after the vehicle lands to ensure the reentry vehicle does not pose a threat to public health and safety. The Regulations also define a "reentry vehicle" as "a vehicle designed to return from Earth orbit or outer space to Earth substantially intact."[38] If the reentry vehicle can be used more than once it is RLV. When a reentry vehicle is not an RLV, it is a "reentry vehicle other than a reusable launch vehicle."[39]

SpaceX petitioned the FAA for two substantive waivers in connection with its COTS mission. First, SpaceX applied for a waiver of the acceptable risk requirements.[40] Second, SpaceX petitioned the FAA to allow *Dragon* to "autonomously reenter" Earth's atmosphere by guiding itself back to Earth, rather than reenter *only* if guided by ground-based operators.[41]

The combined risk for both the launch of the Falcon 9 launch vehicle and reentry of *Dragon* was 0.000047, above the regulatory acceptable risk standard. However, the FAA analyzed the reasons for the increased risks, SpaceX's ability to mitigate the risks, and the fact that the launch and reentry, if reviewed separately, were each within the acceptable risk standards. Consequently, the FAA determined that

---

37. 14 C.F.R. § 401.5.
38. *Id.*
39. 14 C.F.R. § 435.
40. 75 Fed. Reg. 75,619–21 (Dec. 6, 2010).
41. 75 Fed. Reg. 75,621–24 (Dec. 6, 2010).

granting a waiver would be "consistent with the safety rationale underlying section 431.35."[42]

The Regulations normally do not allow autonomous reentries.[43] Reentries must be initiated by human command. SpaceX petitioned the FAA to waive this requirement. Under normal circumstances, once *Dragon* entered orbit, SpaceX planned to check *Dragon*'s health from Earth. The ground operators reviewing the health check would then decide whether to reenter *Dragon* based on the results of this test. However, SpaceX also made it possible for *Dragon* to perform a self-check and determine its own health. *Dragon* would then be able to determine on its own whether it was healthy enough to proceed with reentry. *Dragon*'s ability to autonomously reenter Earth's atmosphere would be critical if, for some reason, SpaceX's ground-based operators lost contact with *Dragon* during the mission.

In order to determine whether to waive the guided reentry requirement, the FAA considered whether *Dragon*'s safety features would allow for a safe autonomous reentry in the event of a communications failure. *Dragon*'s safety features included the following:

- The vehicle automatically reduces itself to its lowest energy level in the case of an off-nominal burn.
- The vehicle has the ability to autonomously guide itself to the same predetermined landing site located more than 780 kilometers from the coastline.
- The vehicle has the ability to monitor its safety-critical systems in real-time.
- The vehicle has over 100 percent margin on both power and propellant budgets.
- The vehicle has a space-grade inertial measurement unit (IMU).
- The vehicle has a space-grade flight computer.
- The vehicle has redundant drogue parachutes and dual redundant main parachutes.

The FAA ultimately determined that these safety features were sufficient for the FAA to waive the guided reentry requirement.[44]

---

42. *Id.*
43. 14 C.F.R. § 431.43(e).
44. 75 Fed. Reg. 75,621–24 (Dec. 6, 2010).

IV. EXPERIMENTAL PERMITS

Obtaining a waiver of licensing requirements is time-consuming and cumbersome. Therefore, Congress granted the FAA the authority to issue experimental permits for certain types of reusable suborbital rocket flights as a way for operators to perform multiple test flights and conduct further research and development without applying for multiple licenses.[45] Experimental permits are only available for suborbital rockets operated for one or more of the following purposes: research and development to test new design concepts, equipment, or operating techniques; demonstrating compliance with licensing requirements; and crew training.[46] A rocket operated under an experimental permit may carry SFPs, but "it may not be operated for compensation or hire."[47] A rocket operated under an experimental permit is also not eligible for U.S. government indemnification.

V. TRAINING, MEDICAL, AND INFORMED CONSENT REQUIREMENTS

The Launch Act and its implementing Regulations impose different training, medical, and informed consent requirements for SFPs and crew on commercial spacecraft. Under the Launch Act, "crew" is defined as any employee of a licensee or of a contractor or subcontractor of a licensee "who performs activities in the course of that employment directly relating to the launch, reentry, or other operation of or in a launch vehicle or reentry vehicle that carries human beings."[48] An SFP is "an individual, who is not crew, carried within a launch vehicle or reentry vehicle."[49]

A. SFPs

The Regulations do not impose any medical standards for SFPs. This is consistent with the FAA's overall safety regime that limits itself to protecting the safety of the public. In 2004, the FAA was given authority to create training and medical standards for SFPs three years after passage of CSLAA. Part 460.51 of the Regulations requires that the each

---

45. Hughes & Rosenberg, *supra* note 6.
46. 51 U.S.C. § 50906.
47. Hughes & Rosenberg, *supra* note 6.
48. 51 U.S.C. § 50902(2).
49. 51 U.S.C. § 50902(17).

SFP must receive training "on how to respond to emergency situations, including smoke, fire, loss of cabin pressure, and emergency exit." The Launch Act prohibits the FAA from proposing regulations governing the design or operation of a launch vehicle to protect the health and safety of crew and SFPs until October 1, 2015, or until a design feature or operating practice has resulted in a serious or fatal injury, or contributed to an event that posed a high risk to crew or SFPs during a licensed commercial human spaceflight.[50]

The Launch Act, however, does require that all SFPs provide written informed consent to the technical risks of human spaceflight. The Launch Act mandates that an operator may launch or reenter an SFP only if the operator (1) "informed the spaceflight participant in writing about the risks of the launch and reentry, including the safety record of the launch or reentry vehicle type"; (2) informed the SFP "that the United States Government has not certified the launch vehicle as safe for carrying crew or SFPs"; and (3) "the spaceflight participant has provided written informed consent to participate in the launch and reentry."[51] The Regulations flesh out these requirements.

Before agreeing to fly an SFP, a commercial human spaceflight (CHSF) operator must discuss the following six topics with the SFP: (1) hazards associated with suborbital flights generally, (2) lack of safety certification by the U.S. government for carrying crew or SFPs, (3) the safety record of launch and reentry vehicles generally, (4) the safety record of the CHSF operators' particular vehicle, (5) the availability of additional information if the SFP desires it, and (6) an opportunity for the SFP to ask additional questions.[52]

As part of the process, the SFP must receive a written disclosure of the known hazards for each mission "that could result in serious injury, death, disability, or total or partial loss of physical and mental function." Additionally, an SFP must be informed in writing that there are unknown hazards and participation in spaceflight may result in death, serious injury, or total or partial loss of physical or mental function.

When discussing the safety record of all launch or reentry vehicles, an SFP must receive the following information: (1) the total number of people who have been on a suborbital or orbital spaceflight and the total number of people who have died or been seriously injured

---

50. 51 U.S.C. § 50905(c)(2), (3), as amended by H.R. 658, 112th Cong. § 827 (2012).
51. 51 U.S.C. § 50905.
52. 14 C.F.R. § 460.45.

on these flights, and (2) the total number of launches or reentries conducted with people on board and the number of catastrophic failures of those launches or reentries.

When describing the safety record of its vehicle to each SFP, the operator's safety record will include (1) the number of vehicle flights, (2) the number of accidents and human spaceflight incidents, and (3) whether any corrective actions were taken to resolve these accidents and human spaceflight incidents.

Lastly, a CHSF operator must inform the SFP that the SFP can ask for additional information regarding accidents and human spaceflight incidents. In the same context, the SFP must be given an opportunity to ask additional questions. The final written consent must identify the space launch vehicle it covers, state that the SFP understands the risk and that their presence on board the vehicle is voluntary, and be signed and dated by the SFP.

B. Crew

The Regulations include certain training and medical requirements for CHSF crew.[53] Each crew member must complete training on how to carry out his or her role on board or on the ground so that the vehicle will not harm the public. This training must be under both nominal and nonnominal conditions and include abort scenarios and emergency operations. The pilot of a CHSF spacecraft must have at least an FAA pilot certificate with an instrument rating; possess aeronautical knowledge, experience, and skills necessary to pilot and control the launch or reentry vehicle; and receive vehicle and mission-specific training for each phase of flight.

In 2006, the FAA determined that requiring second-class medical certification for nonsafety critical crew members was not necessary. But the FAA also reasoned that requiring second-class medical certification only on the basis of title would allow certain safety-critical crew members to operate without adequate medical certification. Therefore, the FAA requires crew members who have a safety-critical role to have a second-class airman medical certificate issued at least 12 months before the month of launch or reentry and demonstrate the ability to "withstand the stresses of spaceflight, which may include high acceleration or deceleration, microgravity, and vibration, as well as be in

---

53. 14 C.F.R. § 460.5.

sufficient condition to safely carry out his or her duties so that the vehicle will not harm the public."[54]

Unlike the detailed informed consent requirements for SFPs, CHSF operators must only inform their crew that the U.S. government has not certified the launch and reentry vehicle as safe for carrying flight crew or spaceflight participants.[55] In addition, the Regulations' notification requirement requires only that an operator inform the crew that risks exist, not that it identify all potential operational and design hazards.[56]

The Regulations mandate that each member of a flight crew and any remote operator must execute a reciprocal waiver of claims with the FAA.[57] There are no such mandatory cross-waivers for the benefit of licensed CHSF operators.[58] The absence of mandatory cross-waivers means that crew and CHSF operators are entitled to, and should, address issues of liability contractually. For an example of such cross-waivers, and possible contractual language, see Appendix D of the Regulations.[59]

GLOSSARY

Commercial Space Launch Act of 1984 (the "Launch Act"): The statute governing the licensing of the launch and reentry of nongovernmental spacecraft and the operation of non-governmental spaceports.

Commercial Space Launch Amendments Act of 2004 (CSLAA): A major revision to the Launch Act that authorized the licensing of commercial human spaceflight.

Crew: Any employee or independent contractor of a licensee, transferee, or permittee, or of a contractor or subcontractor of a licensee, transferee, or permittee, who performs activities in the course of that employment or contract directly relating to the launch, reentry, or

---

54. 14 C.F.R. § 460.5(b).
55. 14 C.F.R. § 460.9.
56. Human Spaceflight Requirements for Crew and Spaceflight Participants, 71 Fed. Reg. 241 (2006).
57. 14 C.F.R. § 460.19.
58. M. Mineiro, *Assessing the Risks: Tort Liability and Risk Management in the Event of a Commercial Human Spaceflight Accident*, 74 J. AIR LAW & COMM. 392 (2008).
59. App'x D, 14 C.F.R. §§ 460.1 *et seq.*

other operation of or in a launch vehicle or reentry vehicle that carries human beings. A crew consists of flight crew and any remote operator.

Expendable launch vehicle: A launch vehicle whose propulsive stages are flown only once.

Experimental permit or permit: An authorization by the FAA to a person to launch or reenter a reusable suborbital rocket.

Flight crew: Crew that is on board a vehicle during a launch or reentry.

Flight safety system: A system designed to limit or restrict the hazards to public health and safety and the safety of property presented by a launch vehicle or reentry vehicle while in flight by initiating and accomplishing a controlled ending to vehicle flight. A flight safety system may be destructive resulting in intentional break up of a vehicle or nondestructive, such as engine thrust termination enabling vehicle landing or safe abort capability.

Launch site safety assessment: An FAA assessment of a Federal launch range to determine if the range meets FAA safety requirements. A difference between range practice and FAA requirements is documented in the LSSA.

Launch site: The location on Earth from which a launch takes place (as defined in a license the Secretary issues or transfers under this chapter) and necessary facilities at that location.

Launch vehicle: A vehicle built to operate in, or place a payload in, outer space or a suborbital rocket.

Launch: To place or try to place a launch vehicle or reentry vehicle and any payload from Earth in a suborbital trajectory, in Earth orbit in outer space, or otherwise in outer space, and includes preparing a launch vehicle for flight at a launch site in the United States. Launch includes the flight of a launch vehicle and includes pre- and post-flight ground operations as follows:

(1) Beginning of launch. (i) Under a license, launch begins with the arrival of a launch vehicle or payload at a U.S. launch site.

(ii) Under a permit, launch begins when any pre-flight ground operation at a U.S. launch site meets all of the following criteria:

(A) Is closely proximate in time to flight,

(B) Entails critical steps preparatory to initiating flight,

(C) Is unique to space launch, and

(D) Is inherently so hazardous as to warrant the FAA's regulatory oversight.

(2)End of launch. (i) For launch of an orbital expendable launch vehicle (ELV), launch ends after the licensee's last exercise of control over its launch vehicle.

(ii) For launch of an orbital reusable launch vehicle (RLV) with a payload, launch ends after deployment of the payload. For any other orbital RLV, launch ends upon completion of the first sustained, steady-state orbit of an RLV at its intended location.

(iii) For a suborbital ELV or RLV launch, launch ends after reaching apogee if the flight includes a reentry, or otherwise after vehicle landing or impact on Earth, and after activities necessary to return the vehicle to a safe condition on the ground.

Office of Commercial Space Transportation (AST): The FAA line of business responsible for regulating commercial space activities under the Launch Act.

Payload: An object that a person undertakes to place in outer space by means of a launch vehicle, including components of the vehicle specifically designed or adapted for that object.

Public safety: For a particular licensed launch, the safety of people and property that are not involved in supporting the launch and includes those people and property that may be located within the boundary of a launch site, such as visitors, individuals providing goods or services not related to launch processing or flight, and any other launch operator and its personnel.

Reenter or reentry: To return or attempt to return, purposefully, a reentry vehicle and its payload, if any, from Earth orbit or from outer space to Earth. The term "reenter; reentry" includes activities conducted in Earth orbit or outer space to determine reentry readiness and that are critical to ensuring public health and safety and the safety of property during reentry flight. The term "reenter; reentry" also includes activities conducted on the ground after vehicle landing on Earth to ensure the reentry vehicle does not pose a threat to public health and safety or the safety of property.

Reentry site: The location on Earth where a reentry vehicle is intended to return. It includes the area within three standard deviations of the intended landing point (the predicted three-sigma footprint).

Reentry vehicle: A vehicle designed to return from Earth orbit or outer space to Earth substantially intact. A reusable launch vehicle that is designed to return from Earth orbit or outer space to Earth substantially intact is a reentry vehicle.

Reusable launch vehicle (RLV): A launch vehicle that is designed to return to Earth substantially intact and therefore may be launched more than one time or that contains vehicle stages that may be recovered by a launch operator for future use in the operation of a substantially similar launch vehicle.

Risk: A measure that accounts for both the probability of occurrence of a hazardous event and the consequence of that event to persons or property.

Space flight participant (SFP): An individual, who is not crew, carried aboard a launch vehicle or reentry vehicle.

Suborbital rocket: A vehicle, rocket-propelled in whole or in part, intended for flight on a suborbital trajectory, and the thrust of which is greater than its lift for the majority of the rocket-powered portion of its ascent.

Suborbital trajectory: The intentional flight path of a launch vehicle, reentry vehicle, or any portion thereof, whose vacuum instantaneous impact point does not leave the surface of Earth.

CHAPTER 6

Liability and Insurance Issues for Private Spaceflight

Risk is inherent in everything we do. Even decisions to refrain from doing something can entail risk. No matter what precautions are taken, there is no way to eliminate risk completely in any activity. The goal is to mitigate and reduce risk as much as possible while balancing those efforts with cost and practicality. The challenge is how to identify the level of risk we are willing to accept. Space activities are no different.

This chapter provides an overview of liability issues raised by private spaceflight. Section I introduces the topic by providing a high-level perspective of potential sources of risk in spaceflight. Section II presents the liability regimes at the international, national, and state levels. Section III reviews insurance practices in the private spaceflight industry. Section IV discusses contractual risk mitigation practices.

I. IDENTIFYING SOURCES OF RISK

Spaceflight is an inherently risky endeavor. Harm can occur at every stage of flight. A fire during a preflight test of Apollo 1 resulted in the death of three astronauts in 1967. The launch phase is generally the most risky because of the technological complexities involved in launching a vehicle into space and the use of tremendous amounts of highly combustible fuels. The potential dangers during launch were vividly illustrated by the tragic *Challenger* accident in 1986, attributed to multiple causes, including faulty design, inclement weather, and poor communication within the National Aeronautics and Space Administration (NASA).[1]

---

1. Presidential Comm'n, Report on the Space Shuttle *Challenger* Accident (1986), http://history.nasa.gov/rogersrep/genindex.htm.

While manned spaceflight has obvious risks to the crew onboard, launches also impose risks on others. Employees working at the launch site and spectators observing the launch are within close proximity to the rocket and may be harmed in an accident. Communities located below the flight trajectory face risk of harm from the launch vehicle, destruction of the payload, or leaked fuel from a crash.[2]

A successful launch does not mark the end to potential risk. Damage to the spacecraft in orbit can be sustained in many different ways, such as malfunctioning equipment and impacts with space debris. If damaged, a spacecraft could pose risks to other space objects. The failed reentry of spacecraft could result in harm to people onboard, such as with the *Columbia* tragedy in 2003, or to people and property on the ground.

Considering all of these potential risks, it comes as no surprise that liability regimes have developed in the shadows of endeavors to pursue new goals in spaceflight and space exploration. Because the forms of risk are so varied and the uncertainty associated with that risk is so great, developing a comprehensive liability regime and a successful insurance infrastructure to foster the private spaceflight industry is a significant challenge.

II. LIABILITY REGIMES WITHIN THE UNITED STATES

Liability in spaceflight could arise internationally or domestically. Internationally, through the Treaty on Principles Governing the Activities of States in the Exploration and Use of Outer Space, including the Moon and Other Celestial Bodies (Outer Space Treaty) and the Convention on International Liability for Damage Caused by Space Objects (Liability Convention), a launching State can be liable for damage caused by its space objects. As discussed in chapter 3, the Liability Convention (Appendix 3-3) defines "launching State" very broadly to include potentially multiple States involved in a launch. A launching State is any of the following:

- a State that launches a space object
- a State that procures the launching of a space object

---

2. *See, e.g.,* Appalachian Ins. Co. v. McDonnell Douglas Corp., 214 Cal. App. 3d 1 (4th Dist. 1989) (complete loss of the satellite payload); Martin Marietta Corp. v. INTELSAT, 991 F.2d 94 (4th Cir. 1992) (failure of launch vehicle to place a satellite in the proper orbit).

- a State from whose territory a space object is launched
- a State from whose facility a space object is launched

Therefore, the United States can be liable for damage caused by a space object if it meets any of the above criteria. If damage is to the surface of Earth or to an aircraft in flight, the United States is absolutely liable. If damage is not on Earth's surface or to an aircraft in flight, the Liability Convention imposes negligence, or fault, liability.

Although the United States, not individual providers, is liable for damage to non-U.S. parties under the Liability Convention, nothing prevents it from, in turn, imposing liability on commercial launch providers in the event that one of their space objects causes damage to a foreign party. In addition, the remedies under the Liability Convention are nonexclusive, so international plaintiffs may still bring suit against the U.S. launch provider directly under foreign and domestic tort laws.

A. Liability at the National Level

For commercial launches, liability at the national level is addressed in the Commercial Space Launch Act.[3] The Launch Act provides a three-tier risk-sharing regime for launch and reentry licensees. The adoption of this risk-sharing mechanism was largely to help the U.S. private launch industry compete in the international market. Specifically, U.S. companies argued that they had difficulty competing with European launch provider Arianespace, which benefited greatly from government indemnification.[4]

As shown in table 6.1, the Launch Act first requires that operators either obtain third-party liability insurance or show financial responsibility sufficient to compensate third parties for the maximum probable loss (MPL) when issued a launch or reentry license. The MPL is determined on a case-by-case basis in connection with each license grant, but is capped at $500 million in 1988 dollars, as adjusted for inflation.[5] The United States is thus protected from liability for claims up to that amount. If the licensee is unable to obtain insurance for this amount,

---

3. Commercial Space Launch Act, 98 Stat. 3005, Pub. L. No. 98-575 (Oct. 30, 1984) (codified as amended in 51 U.S.C. §§ 50901 et seq.).

4. Timothy Hughes & Esta Rosenberg, *Space Travel Law (and Politics): The Evolution of the Commercial Space Launch Amendments Act of 2004*, 31 J. SPACE L. 1 (2005); Michael S. Straubel, *The Commercial Space Launch Act: The Regulation of Private Space Transportation*, 52 J. AIR L. & COMM. 941 (1986–1987).

5. When adjusted for inflation as of 2012, the MPL cap is over $950 million.

TABLE 6.1
Three-Tier Liability Regime under the Launch Act (in 1988 dollars)

Amount of Liability	Responsible Party
Up to MPL (capped at $500 million)	Licensee
Between MPL and $2 billion	U.S. government
Greater than $2 billion	Licensee

the Launch Act provides that it may satisfy the MPL requirement by obtaining the maximum insurance available on the world market at a reasonable cost.[6]

Second, Congress may indemnify the licensee by appropriating up to $1.5 billion, also in 1988 dollars, for third-party liability greater than the MPL. In this way, the Launch Act has protected the licensee from unlimited liability. Finally, any liability greater than the first two tiers is the responsibility of the licensee.[7]

In addition to the risk-sharing mechanism, the Launch Act requires that two cross-waivers of liability be entered into in connection with each licensed launch or reentry. First, each licensee must enter into a reciprocal waiver of claims with the various commercial entities involved in the launch or reentry activity, including the licensee's contractors, subcontractors, and customers, and their respective contractors and subcontractors. Under this cross-waiver, each party must agree "to be responsible for property damage or loss it sustains or for personal injury to, death of, or property damage or loss sustained by its own employees resulting from an activity carried out under the applicable license."[8] Second, the licensee, contractors, subcontractors, crew, space flight participants (SFPs), and customers of the licensee must enter into a similar cross-waiver of claims with the U.S. government and its contractors and subcontractors, to the extent the amount of the claims exceeds the amount of insurance the licensee is required to obtain pursuant to its license.

The cross-waiver requirement is meant to stimulate private launch and reentry activities by minimizing the scope of interparty litigation and liability following an accident. Nevertheless, while the Launch Act requires crew and SFPs to waive any claims against the U.S. govern-

---

6. 51 U.S.C. § 50914.

7. 51 U.S.C. § 50915. As of this writing, this provision is scheduled to expire on December 31, 2012. The U.S. government indemnification limit is approximately $2.7 billion when adjusted for inflation as of 2012.

8. 51 U.S.C. § 50914(b)(1).

ment, nothing in the Launch Act currently requires crew and SFPs to waive claims against the private parties involved in the licensed or permitted activity, including the launch provider, its contractors and subcontractors, or each other.[9] These entities are therefore not protected by federal law against claims by crew and SFPs, or their heirs, in the event of an accident. Liability that is not covered by the Launch Act protections must instead be addressed by State law liability shields, insurance, and contractual liability allocation.

B. State Space Tourism Liability Laws
1. Purpose of State Legislation
Only a handful of U.S. states have enacted laws that address liability issues in commercial human spaceflight activities, but others are moving in that direction. Some states are more proactive in developing legislation than others. Heightened interest in attracting spaceflight providers is often due to a historic connection with the space industry, a desire to attract high-tech, well-paying space industry jobs, and geographic attributes that make it easier to find appropriate launch sites, such as lower population density and flat landscapes. Also, if a launch provider wishes to use expendable launch vehicles (ELVs), it will look for launch sites in a coastal state so the components can fall into the ocean. A state will also have a greater interest in enacting liability shield legislation if it has already established a space authority or built a spaceport.

The drive behind this state level legislation reflects states' interest in attracting a new and highly evolving field. In this way, the purpose here mirrors the purpose of the 2004 amendments to the Launch Act: to promote economic growth and encourage a new industry. States are recognizing that the industry has the potential for developing into a profitable tourist industry that will also bring in a highly educated workforce. State laws that limit operator liability make the state more appealing to launch providers as they search for locations to set up shop.

2. Specific State Laws
This section discusses specific state liability laws, highlighting the laws enacted by Florida, New Mexico, Texas, and Virginia.

---

9. *See, e.g.*, Hughes & Rosenberg, *supra* note 4, at 59–64.

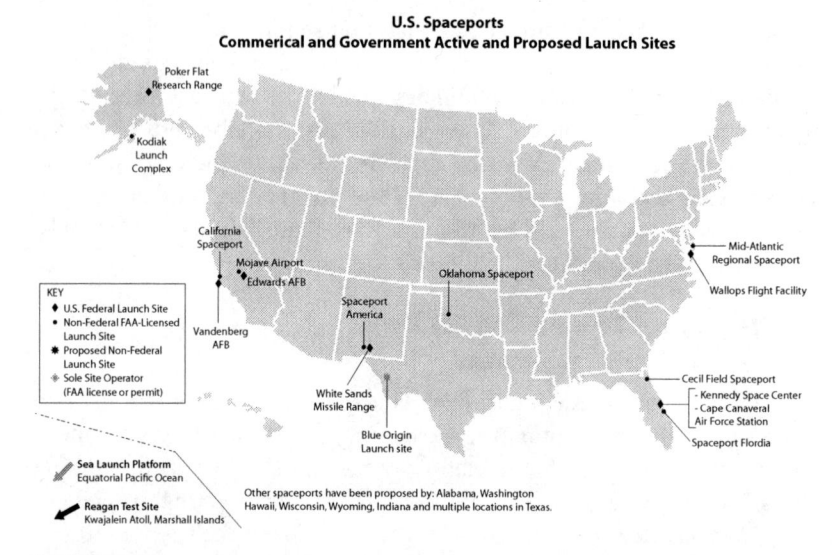

FIGURE 6.1
Spaceports in the United States as of July 2010

Credit: FAA/AST

a. Liability Issues Regarding Spaceports

A spaceport is a launch and reentry site for space vehicles, including facilities and supporting infrastructure related to launch, landing, or payload processing (see figure 6.1). Spaceports operated by the U.S. government do not require a Federal Aviation Administration (FAA) license to operate. Some of these spaceports, such as Cape Canaveral, are accessible to commercial launch providers using FAA-licensed vehicles. Nonfederal spaceports, however, do require an FAA license to operate.

Spaceports offer various and differing services. For example, Mid-Atlantic Regional Spaceport (MARS) in Virginia provides both foreign and domestic launch services. It has two medium lift launch pads to launch into low Earth orbit. In contrast, Spaceport America in New Mexico has a runway for horizontal launches like airplanes and has generated significant interest in the space tourism industry, with Virgin Galactic as the spaceport's anchor tenant.

The Launch Act does not require operators of spaceports either to obtain insurance or demonstrate financial responsibility, as it requires for licensees of launch and reentry vehicles. Originally, this was because any significant harm or damage was most likely to occur on the vehicle or as a result of a failed launch. Spaceports are turning into tourist destinations and places the general public will be visiting regularly, which may create more opportunity for injuries to third parties.

b. Uncertainty over Informed Consent

The Launch Act imposes on licensees a duty to disclose to SFPs the risks associated with spaceflight. SFPs are required to give their informed consent before participating in spaceflight activities. Uncertainty over how much is enough to satisfy the informed consent requirement reflects the question of whether the emerging human spaceflight industry should be considered legally equivalent to, on the one hand, high-risk adventure sports or, on the other hand, the highly regulated airline industry. As discussed in chapter 5, some believe that the industry should be highly regulated to ensure safety. States adopting this philosophy may go so far as to argue that SFPs should not be allowed to waive liability of spaceflight entities and may therefore find any signed statements indicating informed consent invalid. On the other hand, states interested in promoting the spaceflight industry within their borders will determine that spaceflight is akin to adventure sports and will impose minimal informed consent requirements. The following section discusses state immunity legislation aimed at promoting the latter policy.

c. Immunity Legislation at the State Level

As previously mentioned, the Launch Act does not extend the cross-waiver requirement to crew or SFPs. And while the Launch Act requires the operator to provide enough safety information to obtain the informed consent of the SFP prior to flight, the Launch Act does not preempt causes of action under state liability laws. Therefore, states have taken the initiative to shield commercial operators from liability for personal injury and wrongful death claims.

Virginia's Space Flight Liability and Immunity Act of 2007[10] granted conditional liability immunity to companies providing human spaceflight services in the event of an injury resulting from the risks inherent in spaceflight. This was the first state to do so in the wake of the passage of the 2004 Commercial Space Launch Amendments Act, which authorized the FAA to license commercial suborbital space tourism.

---

10. VA. CODE ANN. §§ 8.01-227.8–.10 (2007).

Florida,[11] New Mexico,[12] and Texas[13] have followed Virginia's lead and adopted comparable legislation.[14]

While similar, the statutes are not identical. One difference is scope of the term "spaceflight entity." The Virginia, Florida, and Texas laws apply to a wider range of spaceflight entities than the New Mexico law does. Virginia and Florida extend the definition of "spaceflight entity" to include manufacturers or suppliers merely reviewed by the FAA during the licensing process. Texas includes those manufacturers or suppliers used by the entity and reviewed by the FAA during the licensing process. It also includes any employees, officers, directors, owners, stockholders, members, and partners of entities, manufacturers, and suppliers. New Mexico, however, extends immunity only to public or private entities with an FAA launch license.

The purpose of these laws is to create some security for companies that they will not be sued by SFPs or their heirs for spaceflight activities undertaken at their own risk. Nevertheless, the immunity offered by the Florida, New Mexico, Virginia, and Texas laws is not absolute. No immunity is offered where parties have acted with gross negligence or intentionally caused the injury or damage. Florida and New Mexico extend this exception to situations where the spaceflight operator had actual knowledge or reasonably should have known of the danger present.

Honoring the requirements of the Launch Act, the state laws require operators to obtain signed warning statements from SFPs. The state laws specify the language necessary for an acceptable statement. While the specific language differs slightly among states, in general it asserts that the SFP has received a description of the risks associated with spaceflight activities and the risk of the SFP waiving liability for any risks inherent in spaceflight activities.

To illustrate the differences in language, the text of the warning statements as it appears in each state statute is reprinted in the Sample Warning Statements box.

---

11. FLA. STAT. § 331.501 (2009).
12. S.B. 9, 49th Leg., Reg. Sess. (N.M. 2010).
13. S.B. 115, 82d Leg. (Tex. 2011).
14. *See* Meredith Blasingame, *Nurturing the United States Commercial Space Industry in an International World: Conflicting State, Federal, and International Law*, 80 MISS. L.J. 741 (2010).

Sample Warning Statements

Virginia

WARNING AND ACKNOWLEDGEMENT: I understand and acknowledge that, under Virginia law, there is no civil liability for bodily injury, including death, emotional injury, or property damage sustained by a participant in space flight activities provided by a space flight entity if such injury or damage results from the risks of the space flight activity. I have given my informed consent to participate in space flight activities after receiving a description of the risks of space flight activities as required by federal law pursuant to 49 U.S.C. § 70105 and 14 C.F.R. § 460.45. The consent that I have given acknowledges that the risks of space flight activities include, but are not limited to, risks of bodily injury, including death, emotional injury, and property damage. I understand and acknowledge that I am participating in space flight activities at my own risk. I have been given the opportunity to consult with an attorney before signing this statement.

Florida

WARNING: Under Florida law, there is no liability for an injury to or death of a participant in a spaceflight activity provided by a spaceflight entity if such injury or death results from the inherent risks of the spaceflight activity. Injuries caused by the inherent risks of spaceflight activities may include, among others, injury to land, equipment, persons, and animals, as well as the potential for you to act in a negligent manner that may contribute to your injury or death. You are assuming the risk of participating in this spaceflight activity.

New Mexico

WARNING OF RISKS AND RELEASE OF LIABILITY:

1. I understand that the commercial human space flight industry is an emerging industry and that private industry has begun to develop vehicles capable of carrying human beings into space.
2. I understand that commercial human space flight activities involve inherent risks that cannot be eliminated or controlled through the exercise of reasonable care.

3. I therefore understand, acknowledge and agree that I am waiving all claims for any loss, damage or injury, including bodily injury, emotional injury, death or property damage, that I sustain in space flight activities provided by a space flight entity if the loss, damage or injury results from the risks of the space flight activity.
4. I understand, acknowledge and agree that this waiver shall also be binding on my representatives, including my heirs, administrators, executors, assignees, next of kin and estate, or any other person who attempts to bring a claim on my behalf.
5. I have been informed of the risks of space flight activities as required by federal law pursuant to 49 U.S.C. Section 70105 and 14 C.F.R. Section 460.45, and I consent to participate in space flight activities after receiving a description of risks.
6. I acknowledge that the risks of space flight activities include, but are not limited to, risks of bodily injury, including death, emotional injury and property damage. I understand, acknowledge and agree that I am participating in space flight activities at my own risk.
7. I have been given adequate opportunity to consult with an attorney of my own choosing before signing this warning of risks and release of liability.

Texas
AGREEMENT AND WARNING: I understand and acknowledge that a space flight entity is not liable for any injury to or death of a space flight participant resulting from space flight activities. I understand that I have accepted all risk of injury, death, property damage, and other loss that may result from space flight activities.

3. Federal Preemption Debate

In response to states' initiatives to enact launch legislation, there is a debate over whether the Launch Act preempts states from enacting laws governing launches and, if it does not, whether it should. The Launch Act expressly allows states to enact laws that are in addition to, or more stringent than, the Launch Act requirements as long as they are consistent with the Launch Act. Those arguing in favor of federal preemption include many in the private space launch industry because it would remove some of the uncertainty, create uniformity, and help further the overall purposes of the Launch Act by promoting the private space launch industry. Federal preemption would be particularly helpful where accidents are likely to cross state boarders, as in spaceflight operations.[15]

III. INSURANCE PRACTICES

The first two sections of this chapter discussed the types of risk involved in spaceflight and the liability regimes established to respond to those risks. This section briefly discusses the insurance practices of the spaceflight industry, including what stages of spaceflight operators can insure and how they can go about obtaining that insurance.[16]

A. Stages of Spaceflight to Insure

There are four major stages of spaceflight to insure in the commercial spaceflight industry:

- Manufacturing and prelaunch
- Launch
- In-orbit life
- Recovery of failed ELV launch

The more common stages insured are the launch and in-orbit life phases. Launch insurance serves to indemnify the owner of a payload for a failed launch or for failure of a launch vehicle to place the payload into its proper orbit. While premium amounts fluctuate, they can be as low at 10 percent and as high as 30 percent of the total mission cost.[17] In-orbit insurance can include coverage for the first year or for the

---

15. *See, e.g.*, *Office of Commercial Space Transportation's Fiscal Year 2012 Budget Request: Hearing Before the H. Subcomm. on Space and Aeronautics* (2011) (statement of Henry R. Hertzfeld, Space Policy Institute), http://science.house.gov/sites/republicans.science.house.gov/files/documents/hearings/050511_Hertzfeld.pdf; PAUL ECKERT, JAMES R. MCMURRY & ROSANNA SATTLER, LIABILITY LIMITATION IN COMMERCIAL HUMAN SPACEFLIGHT: BENEFITS FOR ENTREPRENEURSHIP, PARTNERSHIP, AND POLICY, (Boeing Co. 2009) http://216.70.69.189/media/pnc/6/media.216.pdf; *see generally* Blasingame, *supra* note 14, at 742 (arguing that a complex web of law at the state, federal, and international level ensures that the policy objectives of all are represented).

16. *See, e.g.*, Pamela Meredith, *Space Insurance Law—with a Special Focus on Satellite Launch and In-Orbit Policies*, 21 AIR & SPACE LAW. 13 (2008).

17. FED. AVIATION ADMIN., SECOND QUARTER 2006 QUARTERLY LAUNCH REPORT (2006), http://www.faa.gov/about/office_org/headquarters_offices/ast/media/2Q2006_QLR.pdf.

entire life of a commercial satellite. Policies covering the manufacturing and prelaunch stages would cover damage during construction or transporting the vehicle or payload to the launch site. Policies covering the recovery of a failed ELV launch would pay for the actions necessary to return the vehicle to its owner.

B. Obtaining Insurance

In addition to the stages of spaceflight a company can insure, there are different types of insurance available as well. Because of the extraordinary expense and risk of every stage of spaceflight, insurance is essential to protect against the loss of launch vehicles and payloads as well as against third-party claims. Therefore, companies can obtain two different types of insurance: insurance for space objects and liability insurance.

The first step is to decide on an insurance broker. Insurance brokers are often former aerospace engineers who have developed an intricate understanding of space operations and are able to understand the technical aspects of the insured objects.

Once a broker is chosen, the underwriting process requires that the owner of the space object provide technical and financial information for review. The nature of the project, the intended destination of the object, and the location of launch are just a few of the many factors reviewed. Then the different insurance providers respond with proposals that include the terms and conditions, premiums, and capacities based on the information received.[18]

Insurers and reinsurers in the space business tend to be very large multinational corporations that offer a specialty line of space insurance. Specialty lines are needed in this field in part because of the high costs involved and also because of the expertise necessary to understand underwriting space launch vehicles, satellites, and other payloads. The risk taken on by underwriters is often spread among multiple reinsurers and financiers.

C. Global Insurance Market

The space insurance market for communication satellites has been in existence for over 30 years, but it took some time to become profitable and attract underwriters willing to accept the risk. On the other hand,

---

18. *Id.*

the insurance market for suborbital tourism is virtually nonexistent because the industry is still in its infancy.

The insurance market is continually shifting between hard and soft markets. A hard market has high premiums and low capacity. A soft market has low premiums and high capacity. The space insurance industry will tend toward a hard market after a series of unsuccessful launches, due to the need to increase premiums after making large payouts on the policies. After a period of successful launches, the insurance market will recover and move back toward a soft market. The FAA reported in 2002 a movement to a hard market that did not soften until 2005, which was largely a result of the flawed Boeing 702 satellite series that amounted to claims of about $1.5 billion.[19]

IV. CONTRACTUAL RISK ALLOCATION

In addition to federal and state liability laws and insurance policies, risk can be allocated among the various nongovernmental participants in spaceflight activities through contractual liability provisions. SFPs may be asked to waive all claims, on behalf of themselves and their heirs, against a spaceflight operator and its contractors as a condition of participating in a space tourism flight, similar to the waivers that participants in most high-risk activities are asked to sign. Spaceflight operators and their suppliers and customers allocate responsibility for liability resulting from an accident through indemnification, warranty, liability waiver, and insurance clauses in their service contracts. For example, a supplier might assume responsibility if an accident is caused by a manufacturing defect in its product, whereas the spaceflight operator might assume responsibility if an accident is caused by an error integrating the product into the spacecraft following delivery. A launch services provider may agree to launch a replacement satellite within a certain period of time if the original satellite is destroyed or placed in a useless orbit due to launch failure (re-flight costs would ordinarily be covered by insurance). To demonstrate how these provisions are implemented in practice, Appendix 6-1 contains an excerpt from a 2010 launch services contract between Iridium Satellite LLC and Space Exploration Technologies Corp. (SpaceX).

---

19. *Id.*

Spaceflight is inherently risky and the sources of risk are complex and often very challenging to overcome. The United States has helped the private industry by curbing this risk in the Launch Act, which allows for partial indemnification for third-party liability. Additionally, states have begun to compete to attract the spaceflight industry and have enacted legislation to supplement the policy goals of the Launch Act. Companies further reduce the risk of spaceflight through the international insurance market to protect their space objects during different stages of space operations and to protect themselves from third-party liability claims. Finally, private participants in spaceflight operators can further reduce financial risk through contractual liability clauses.

One of the major challenges for commercial spaceflight companies is moving forward in spite of the many legal uncertainties concerning liability, particularly with the advent of commercial human spaceflight. Without the international insurance market, the commercial spaceflight industry would have little chance of surviving.

CHAPTER 7

Licensing Private Telecommunication Satellites

The operation of private telecommunications satellites is, by far, the most successful, widespread, and profitable commercial use of outer space to date. As of June 2011, 37 percent of all operational satellites in Earth orbit were commercial telecommunications satellites.[1] The telecommunications satellite industry is a nearly $180 billion industry, which represents 61 percent of total space industry revenues.[2] The largest segment of the telecommunications sector is satellite television broadcasting, which in 2010 represented almost 80 percent of all satellite service revenues.[3]

Telecommunications satellites are especially useful when they remain stationary with respect to a geographic point on Earth. As described in chapter 1, as few as three satellites arranged equidistantly around Earth in geostationary orbit (GSO) would provide worldwide radio coverage. Accordingly, GSO is valuable orbital real estate for telecommunication purposes, and as a result it is already overcrowded. This overcrowding increases the risk of accidental satellite collisions and signal interference if satellites in GSO operate too closely together. The positioning and frequency use of telecommunication satellites are therefore highly regulated at the international and national levels.

---

1. Futron Corp. & Satellite Indus. Ass'n, State of the Satellite Industry Report (Aug. 2011), http://www.sia.org/PDF/2011_State_of_Satellite_Industry_Report_(Aug. 2011).pdf.
2. *Id.*
3. *Id.*

I. INTERNATIONAL COORDINATION OF SATELLITE SERVICES

The most important international entity in telecommunications is the International Telecommunication Union (ITU). The ITU is responsible for harmonizing, coordinating, and standardizing telecommunication activities across the globe.[4] Specifically regarding satellite-related communications, the ITU Constitution provides that it shall

> (a) effect allocation of bands of the radio-frequency spectrum, the allotment of radio frequencies and the registration of radio-frequency assignments and, for space services, of any associated orbital position in the geostationary-satellite orbit or of any associated characteristics of satellites in other orbits, in order to avoid harmful interference between radio stations of different countries;
> (b) coordinate efforts to eliminate harmful interference between radio stations of different countries and to improve the use made of the radio-frequency spectrum for radiocommunication services and of the geostationary-satellite and other satellite orbits.[5]

Having an organization to coordinate these activities at the international level is essential. Not only are the most useful orbits already crowded, but spectrum is also a finite resource. The electromagnetic spectrum refers to the range of radiation extending across different frequencies. Based on the characteristics of wavelengths, ranges of frequencies are used for various activities. Bandwidths specifically relevant for satellite communications are described in table 7.1. Generally speaking, satellite communications largely use ultra-high frequency (UHF), super-high frequency (SHF), and extremely high frequency (EHF) bandwidths.[6]

Spectrum is a limited resource in high demand. In consideration of this, the ITU Constitution specifies that the radio frequency spectrum should be "used to the minimum essential to provide in a satisfactory manner the necessary services."[7] It is also subject to an equitable access provision to ensure that developing countries will not be left without

---

4. ITU Constitution, http://www.itu.int/net/about/basic-texts/index.aspx.
5. *Id.* at ch. 1, art. 1, § 2(a) & (b).
6. *See generally* GERALD MARAL, MICHEL BOUSQUET & ZHILI SUN, SATELLITE COMMUNICATIONS SYSTEMS: SYSTEMS, TECHNIQUES, AND TECHNOLOGY (Wiley 2011).
7. ITU Constitution, art. 44(1).

TABLE 7.1
Electromagnetic Bands and Their Utilization

Band	Frequencies	Utilization
UHF	300 MHz to 3 GHz	TV broadcast, mobile satellite, land mobile, radio astronomy, air traffic control radar, global positioning systems, Mobile User Objective System (MUOS), UHF follow-on
L band	1 to 2 GHz	Aeronautical radio navigation, radio astronomy, Earth exploration satellites
S band	2 to 4 GHz	Space research, fixed satellite communication
SHF	3 to 30 GHz	Satellite TV, Defense Satellite Communications System, Wideband Global SATCOM
C band	4 to 8 GHz	Fixed satellite communication, meteorological satellite communication
X band	8 to 12 GHz	Fixed satellite broadcast, space research
Kurtz-under (Ku) band	12 to 18 GHz	Mobile and fixed satellite communication, satellite broadcast
K band	18 to 27 GHz	Mobile and fixed satellite communication
Kurtz-above (Ka) band	27 to 40 GHz	Intersatellite communication, mobile satellite communication
EHF	30 to 300 GHz	Remote sensing, military strategic and tactical relay (Milstar), Advanced Extremely High Frequency System, Transformational Satellite Communications System
Millimeter	40 to 300 GHz	Space research, intersatellite communications

Source: AIR COMMAND & STAFF COLL., AU-18 SPACE PRIMER (Air Univ. Press, 2009), available at http://space.maxwell.af.mil/au-18-2009/index.htm

any available spectrum come time when they obtain the necessary technology to utilize that spectrum.[8]

The ITU Constitution also specifically addresses national security concerns of States. While the ITU is involved only with commercial services, it maintains that States have a right to cut off private telecommunications when necessary to protect their national security.[9] Also, States "retain their entire freedom with regard to military

---

8. *Id.* art. 44(2).
9. *Id.* art. 34.

radio installations" and such installations must avoid harmful interference only "so far as possible."[10]

International cooperation through the ITU works because it is in the best interest of States and private entities. It helps protect their investments by minimizing harmful interference and establishes organization in a cluttered space.

The ITU process for coordinating the frequency usage is generally broken down into three stages: allocation, allotment, and assignment. The first two stages take place at the international level and involve States only, not intergovernmental organizations or private entities. The last stage, assignment, occurs at the domestic level.

II. ITU ALLOCATION AND ALLOTMENT

Plenipotentiary conferences are large conferences held every four years during which Member States determine how the ITU will carry out its policy objectives.[11] Since the 1994 Kyoto conference, nongovernmental entities have been granted access and at least some level of participation, although they do not have the official input of Member States. At the World Radiocommunication Conferences (WRCs), States discuss on a global level the allocation of spectrum. WRCs may also revise the Radio Regulations, which is the international treaty governing the use of the radio frequency spectrum and the GSO and non-GSO (NGSO) satellite orbits, along with any related assignment or allotment plans.[12] The World Conference on International Telecommunications, to be held in Dubai in December 2012, will revisit the International Telecommunications Regulations.[13]

Allocation is the process by which the ITU dedicates frequency bands to specific telecommunications services. The international community collectively determines the most efficient use of bandwidths

---

10. *Id.* art. 48.

11. ITU, About the Plenipotentiary Conference, http://www.itu.int/plenipotentiary/2010/about.html.

12. ITU, World Radio Telecommunication Conferences (WRC), http://www.itu.int/ITU-R/index.asp?category=conferences&rlink=wrc&lang=en.

13. ITU, World Conference on International Telecommunications (WCIT-12), http://www.itu.int/en/wcit-12/Pages/default.aspx.

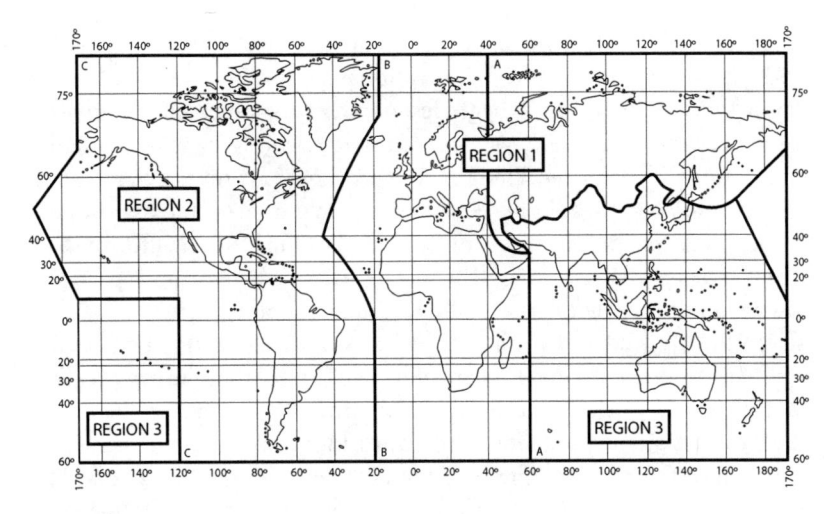

FIGURE 7.1
Regions as Defined in the Table of Frequency Allocations
Credit: ITU

so that they can be properly allocated. The Radio Regulations separate satellite services into the following categories:

- Fixed Satellite Services (FSS)
- Mobile Satellite Services (MSS)
- Broadcasting Satellite Services (BSS)
- Earth Exploration Services (EES)
- Space Research Services (SRS)
- Space Operation Services (SOS)
- Radiodetermination Satellite Services (RSS)
- Inter-Satellite Services (ISS)
- Amateur Satellite Services (ASS)

The Radio Regulations also divide the globe into three main regulatory regions, as shown in figure 7.1. Each region handles frequency allocation individually.

Satellites that are too close together can interfere with each other's radio transmissions. The ITU Constitution addresses this concern by mandating that telecommunication satellites be "operated in such a manner as not to cause harmful interference to the radio services or communications of other Member States or of recognized operating agencies."[14] "Harmful interference" is defined as "interference which endangers the functioning of a radionavigation service or of other safety services or seriously degrades, obstructs or repeatedly interrupts a radiocommunication service operating in accordance with the Radio Regulations."[15] This requirement is necessary to prevent overlaps in spectrum use that would hinder or completely prevent the use of spectrum. Satellites that are transmitting on the same bandwidth must be placed farther apart. To make the most efficient use of orbital space, the ITU will commonly alternate satellites. For example, it may allocate a satellite using Ku bandwidth in between two others using L bandwidth.

Once frequencies have been allocated for certain uses, Member States apply to the ITU for licenses. Allotment is the process by which the ITU assigns orbital slots and bandwidths to certain States. Only States have the right to approach the ITU and request allotments. The ITU Master International Frequency Register catalogs allotment of bandwidths as well as frequency assignments as they are reported to the ITU pursuant to requirements under the Radio Regulations.[16]

III. FCC LICENSING OF PRIVATE TELECOMMUNICATIONS SATELLITES

Once the ITU has allotted certain frequencies and slots to specific States based on their requests, the States determine how they wish to assign those licenses at the domestic level. In the United States, assignment is the responsibility of the Federal Communications Commission (FCC) for all users of spectrum other than the federal government,[17] and of the National Telecommunications and Information Administration (NTIA) within the Department of Commerce for the federal government's use of spectrum.[18]

A private operator must obtain FCC approval before using radio frequency spectrum. The FCC promulgates its satellite licensing regulations pursuant to the Communications Satellite Act of 1962 and the

---

14. ITU Constitution, art. 45.
15. *Id.* at Annex.
16. Radio Regulations ch. III, art. 4.3; *see also id.* art. 8: Status of Frequency Assignments Recorded in the Master International Frequency Register.
17. The complete FCC Table of Frequency Allocations for the United States, 47 C.F.R. § 2.106 (2012), is available at http://transition.fcc.gov/oet/spectrum/table/fcctable.pdf.
18. NTIA, Spectrum Management, http://www.ntia.doc.gov/category/spectrum-management.

Communications Act of 1934.[19] Reflecting the same concerns as at the international level, the FCC closely manages the use of spectrum within the United States with the goal of using this limited and highly desirable resource to its full potential. Note that the FCC is a unique agency because it reports directly to Congress rather than to the Executive Branch.

A. Underlying Spectrum Management Principles

The FCC has explored various methods available to issue licenses for radio spectrum in the United States. It has conducted "beauty contests" where it selects the best licensee for the spectrum. It has also conducted lotteries to determine which applicant receives a license, where the winning licensee had the right either to use or resell the license. The FCC has also implemented auctions, where the entity willing to pay the most gets the license.

In response to the perceived failure of previous licensing methods, the FCC adopted a new approach to satellite licensing in 2003. The FCC recognized that the maturation of the satellite industry called for a more efficient process that was less costly.[20] This new method redefined spectrum licenses in terms of property rights and made these rights freely transferable by licensees.

A fundamental principle of spectrum management is the Coase theorem of economic efficiency, which states that in the absence of transaction costs, whoever values a good most will eventually end up with it regardless of to whom the good is initially given.[21] Applying this to radio spectrum, which person or entity the FCC assigns an initial license for a designated bandwidth does not matter because, ultimately, the person who values that spectrum most will ultimately obtain it.

This overarching principle, however, does not speak to the transaction costs involved with transferring licenses. Therefore, the FCC decided to reduce transaction costs by defining spectrum licenses in terms of property rights. Property rights offer certain advantages because they have defined boundaries and are thus easier to transfer. In essence, Coase argued that markets should be used to sort out how

---

19. Communications Satellite Act art. 201(c)(11); Communications Act tits. I–III (1934).
20. Amendment of the Commission's Space Station Licensing Rules and Policies, Report and Order, FCC 03-102, 18 FCC Rcd. 10,760 (2003), *available at* http://hraunfoss.fcc.gov/edocs_public/attachmatch/FCC-03-102A1.pdf.
21. Ronald H. Coase, *The Problem of Social Cost*, 3 J. LAW & ECON. 1 (1960).

to regulate scarce resources and to do so by creating property rights in the spectrum so that an owner could buy, sell, or lease its right to use a portion of spectrum.

The FCC regulations also reflect two additional principles relating to satellite licensing known as the Efficient Use and Open Skies policies. The Efficient Use policy is intended to reduce interference while maximizing the number of licenses possible to allow for a large variety of services. To implement this policy, the FCC adopted certain technical rules establishing minimum requirements to prevent interference.[22] The FCC has determined that, in accordance with the Coase theorem, the market offers the best possible means to ensure efficient use of the spectrum and therefore the regulations provide licensees flexibility in their license criteria.

The Open Skies policy reflects the same goals highlighted by the Efficient Use policy. The overall purpose is to develop rules and regulations with few constraints to allow licensees maximum flexibility in operating their systems. The FCC actively promotes competition and attempts to expand the variety of services offered in the satellite industry through this policy.[23]

B. Licensing the Space Station Segment of Satellite Systems

The FCC has different methods for issuing satellite licenses for the space segments of a satellite system depending on whether it is a GSO-like satellite system or an NGSO-like satellite system.

1. GSO-Like Satellite Systems

The FCC Regulations for Satellite Communications (the FCC Regulations) define "GSO-like satellite system" as "a GSO satellite designed to communicate with Earth stations with directional antennas."[24] Examples of these systems "are those which use Earth stations with antennas with directivity toward the satellites, such as [Fixed Satellite Services (FSS)], and [Mobile Satellite Services (MSS)] feeder links which use GSO satellites."[25]

---

22. *See* 47 C.F.R. § 25.217 (default service rules).
23. *See* FCC, Regulating Satellite Networks: Principles and Policies, http://transition.fcc.gov/connectglobe/sec8.html.
24. 47 C.F.R. §25.158.
25. *Id.*

III. FCC Licensing of Private Telecommunications Satellites

For GSO-like satellite systems, the FCC uses a first-come, first-served method that considers applications in the order they are filed.[26] Once an application is determined acceptable for filing, it is placed on public notice to give interested parties an opportunity to file pleadings within 30 days. The application will be granted if the application meets general FCC standards and the proposed satellite will not cause harmful interference to any previously licensed operations. In a departure from the Coase theorem, once granted a license, the licensee cannot "transfer, assign, or otherwise permit any other entity to assume its place in any queue."[27]

The first qualified application is granted the license. However, applications may be split if they are submitted at the same time. If the FCC determines that the two applications are "mutually exclusive" and they both meet the above criteria, then each application will be granted and "the available bandwidth at the orbital location or locations in question will be divided equally among those licensees."[28] The FCC will find applications to be mutually exclusive if their conflicts are such that the grant of one application would effectively preclude by reason of harmful interference or other practical reason the grant of one or more other applications.[29]

2. NGSO-Like Satellite System Licensing

NGSO-like satellite systems include all NGSO satellite systems and "GSO MSS satellite systems, in which the satellites are designed to communicate with Earth stations with omni-directional antennas."[30] These refer to satellites in an orbit below GSO that do not stay in a fixed position relative to the Earth station, so the Earth station must follow the satellite in any direction. Satellite telephone systems are an example of NGSO-like satellite systems.

For NGSO-like satellite systems, the FCC uses a modified processing round method that distinguishes between "lead applications" and "competing applications."[31] Lead applications "initiate a processing round, establish a cut-off date for competing NGSO-like satellite

---

26. *Id.*
27. 47 C.F.R. § 25.158(b)–(c).
28. 47 C.F.R. §§ 25.155, 25.158(d).
29. 47 C.F.R. § 25.155.
30. 47 C.F.R. § 25.157(a).
31. 47 C.F.R. § 25.157

system applications, and provide interested parties an opportunity to file pleadings in response to the application."[32] After the cut-off date, the FCC will no longer accept additional competing applications. Those competing applications accepted for filing are put on public notice so that interested parties may file response pleadings. The FCC will then grant all applications in the processing round that meet FCC Regulations standards.[33]

If there is insufficient spectrum available for all qualified applications, the FCC will divide the spectrum equally. The applicant who values the spectrum most will then presumably purchase sufficient spectrum from other applicants who value it less. If the processing round has fewer than three applications, the FCC will assign one third of the spectrum and reserve the rest for other licenses in additional processing rounds.[34] The FCC presumes that three satellite licenses in a given frequency band are enough "to make reasonably efficient use" of the spectrum.[35]

Here, unlike with GSO-like systems, licenses are transferable. Therefore, the FCC relies on the secondary market to give providers sufficient spectrum. It rejected the first-come first-served method as that method would preclude other operators from providing service in the same frequency band.[36]

3. Orbit Act Auctions Exemption and Direct Broadcasting Satellites

In 2000, the Open-market Reorganization for the Betterment of International Telecommunications Act (ORBIT Act) removed the FCC's authority to use competitive bidding as a method for assigning spectrum licenses. The ORBIT Act states:

> Notwithstanding any other provision of law, the [FCC] shall not have the authority to assign by competitive bidding orbital locations or spectrum used for the provision of international or global satellite communications services. The President shall oppose in the [ITU] and in other bilateral and multilateral fora any assignment by competitive bidding of orbital locations or spectrum used for the provision of such services.[37]

---

32. 47 C.F.R. § 25.157(c).
33. The standards are detailed at 47 C.F.R. § 25.156(a).
34. 47 C.F.R. § 25.157(d)–(e).
35. 47 C.F.R. § 25.157(g).
36. Amendment of the Commission's Space Station Licensing Rules and Policies, Report and Order, *supra* note 22.
37. ORBIT Act, Pub. L. No. 106-180, § 647, 114 Stat. 48 (2000).

III. FCC Licensing of Private Telecommunications Satellites 127

This section of the ORBIT Act prevents the FCC from holding auctions to assign licenses. However, the FCC used this language to justify an exemption for auctions specifically for direct broadcast satellite (DBS) service in the DBS Auction Order.[38] DBS service is satellite television broadcast specifically for home use. In ITU Region 2, which covers the United States, the DBS spectrum is the 12.2–12.7 GHz bandwidth. The FCC reasoned that DBS was not subject to the ORBIT Act restrictions because DBS was not "for the purpose of international or global satellite communications services."[39]

Northpoint Technology Ltd. challenged the FCC's decision in the DBS Auction Order in the D.C. Circuit Court of Appeals. While the court agreed with the FCC's interpretation that a "strictly domestic satellite communications service . . . has nothing to do with multiple spectrum auctions in foreign jurisdictions," it rejected the FCC's exemption for DBS because, in part, the FCC had previously emphasized the potential for international DBS satellite service.[40] In the earlier order, known as the DISCO I Order, the FCC had stated that it would allow all U.S. licensed FSS systems, MSS systems, and DBS systems to offer both domestic and international service.[41]

The FCC responded with the DISCO II Order, which implemented U.S. commitments to the World Trade Organization (WTO) Agreement on Basic Telecommunications Services. The United States made binding obligations to open the U.S. satellite market to foreign competition.[42] Significantly, the United States only made market access commitments for FSS and MSS systems. It did not make commitments for

---

38. Auction of Direct Broadcast Satellite Licenses, Order, 19 FCC Rcd. 820 (2004).

39. *See* Auction of Direct Broadcast Satellite Service Licenses Scheduled for August 6, 2003; Comment Sought on Reserve Prices or Minimum Opening Bids and Other Auction Procedures, 68 Fed. Reg. 12,906 (Mar. 18, 2003).

40. Northpoint Tech., Ltd. & Compass Sys., Inc. v. FCC, 412 F.3d 145 (D.C. Cir. 2005).

41. Amendment to the Commission's Regulatory Policies Governing Domestic Fixed Satellites and Separate International Satellite Systems, Report and Order, IB Docket No. 95-41, 11 FCC Rcd. 2429 (1996); *see also* In the Matter of Amendment to the Commission's Regulatory Policies Governing Domestic Fixed Satellites and Separate International Satellite Systems, IB Docket 95-41 (FCC 96-14) (released Jan. 22, 1996).

42. WTO, Telecommunications Agreement, http://www.wto.org/english/tratop_e/serv_e/telecom_e/telecom_e.htm. Note also U.N. General Assembly Resolution 37/92, which recommends that activities in the field of international direct broadcasting of satellite television "should promote the free dissemination and mutual exchange of information and knowledge in cultural and scientific fields, assist in educational, social and economic development, particularly in the developing countries, enhance the qualities of life of all peoples and provide recreation with due respect to the political and cultural integrity of States."

DBS, Direct to Home, or Digital Audio Radio Service. The DISCO II Order also adopted a presumption favoring WTO Members by presuming that satellite systems licensed by WTO Members providing covered services met the competition part of the public interest analysis.[43] For non-WTO Members, the FCC continues to apply what is known as the ECO-Sat Test, requiring that parties must demonstrate that the U.S. satellites have competitive opportunities in their country.[44]

The DISCO I Order only addressed U.S.-licensed systems. The FCC next focused on systems authorized outside of the United States that were providing service within the United States. In its DISCO II Order, the FCC adopted procedures for permitting these non-U.S. operators to operate within the United States.[45] The approached used by the FCC takes into consideration spectrum availability, technical requirements, national security, law enforcement, and trade concerns.[46]

The FCC established two procedures for hearing requests of a satellite provider to allow a non-U.S. satellite to provide service within the United States.[47] The first procedure applies when a non-U.S. satellite provider wishes to participate in an NGSO processing round. It may either file an Earth station application to communicate with the non-U.S. satellite or it may file a "letter of intent" to use the non-U.S. satellite to provide service via future Earth stations yet to be licensed.[48]

The second procedure applies when a non-U.S. satellite operator desires immediate access to the U.S. market through the coordination efforts in the ITU Radio Regulations for Region 2.[49] Here, if a potential Earth station operator desires to communicate with a non-U.S. satellite, it must apply for an initial Earth station license. If an existing Earth station operator desires to communicate with a non-U.S. satellite, it must file a modification application.[50]

---

43. Amendment of the Commission's Regulatory Policies to Allow Non-U.S.-Licensed Space Stations to Provide Domestic and International Satellite Service in the United States, IB Docket No. 96-111, Report and Order, FCC 97-399, 12 FCC Rcd. 24,094, 24,112–13 (para. 40) (1997) [hereinafter DISCO II Order]; 47 C.F.R. § 25.137(a).
44. DISCO II, 12 FCC Rcd. at 24,127 (para. 72).
45. Id. at 24,174 (para. 188).
46. In the Matter of Removal of Approved Non-U.S.-Licensed Space Stations from the Section 214 Exclusion List, IB Docket No. 07-23, Order (2011).
47. See DISCO II Order, 12 FCC Rcd. at 24,174 (para. 188).
48. Id. at 24,173–74 (paras. 184–185, 188).
49. Id. at 24,174 (para. 186).
50. See 47 C.F.R. § 25.137; DISCO II Order, 12 FCC Rcd. at 24,175–76 (paras. 191–192).

III. FCC Licensing of Private Telecommunications Satellites

To promote competition in the U.S. market, the FCC subsequently streamlined the process for non-U.S. fixed satellite service providers to obtain permission to serve the U.S. satellite market in the First Reconsideration Order of DISCO II. This Order adopted two new procedures. First, operators of in-orbit non-U.S. satellites are now permitted to request authority to provide space segment capacity service to licensed Earth stations in the United States, an authority previously warranted only by an Earth station operator under DISCO I. Second, it reduced regulatory constraints on the procedure by which Earth station licensees can obtain approval to access non-U.S. satellites.[51]

C. Licensing Earth Stations

The FCC is also responsible for licensing Earth stations. The FCC Regulations distinguish between different classifications of Earth stations:

- fixed Earth stations that transmit and receive
- temporary fixed Earth stations that transmit and receive
- fixed Earth stations that only receive
- fixed Earth stations using a very-small-aperture terminal (VSAT) network at 12–14 GHz
- developmental Earth stations that are fixed or temporary fixed
- mobile Earth stations for hand-held and vehicle-mounted units

The process for obtaining a license to operate an Earth station is contained in 47 C.F.R. §§ 25.130–25.139. Technical criteria required for a license is also described in depth in the FCC Regulations and include, for example, antenna size and performance standards, environmental impact standards, and power conservation standards.[52] Construction permits are not required for satellite Earth stations.[53] Spectrum is often

---

51. Amendment of the Commission's Regulatory Policies to Allow Non-U.S.-Licensed Space Stations to Provide Domestic and International Satellite Service in the United States, IB Docket No. 96-111, First Order on Reconsideration, FCC 99-325, 15 FCC Rcd. 7207(1999).
52. Details regarding the technical standards necessary for specific Earth station licenses are at 47 C.F.R. §§ 25.201–.261. *See also* FCC, Regulating Satellite Networks: Principles and Policies, http://transition.fcc.gov/connectglobe/sec8.html. Details regarding the technical standards necessary for specific Earth station licenses are at 47 C.F.R. §§ 25.201–.261.
53. 47 C.F.R. § 25.113(a).

shared between terrestrial and satellite operators, which requires coordination between Earth stations to prevent interference.[54]

GLOSSARY

Allocation: Entry in the Table of Frequency Allocations of a given frequency band for the purpose of its use by one or more terrestrial or space radiocommunication services or the radio astronomy service under specified conditions. (Radio Regulations)

Allotment: Entry of a designated frequency channel in an agreed plan, adopted by a competent conference, for use by one or more administrations for a terrestrial or space radiocommunication service in one or more identified countries or geographical areas and under specified conditions. (Radio Regulations)

Assignment: Authorization given by an administration for a radio station to use a radio frequency or radio frequency channel under specified conditions. (Radio Regulations)

Communication-satellite Earth station complex: Includes transmitters, receivers, and communications antennas at the Earth station site together with the interconnecting terrestrial facilities (cables, lines, or microwave facilities) and modulating and demodulating equipment necessary for processing of traffic received from the terrestrial distribution system(s) prior to transmission via satellite and of traffic received from the satellite prior to transfer of channels of communication to terrestrial distribution system(s). (47 C.F.R. 25.103)

Earth station: A station located either on the Earth's surface or within the major portion of the Earth's atmosphere intended for communication:

(a) With one or more space stations; or
(b) With one or more stations of the same kind by means of one or more reflecting satellites or other objects in space. (47 C.F.R. 25.201)

Electromagnetic spectrum: The range of wavelengths or frequencies over which electromagnetic radiation extends.

---

54. *See* 47 C.F.R. § 25.203 for more information on Earth station coordination with shared spectrum.

Harmful interference: Interference which endangers the functioning of a radionavigation service or of other safety services or seriously degrades, obstructs or repeatedly interrupts a radiocommunication service operating in accordance with the Radio Regulations. (ITU Annex)

Mutually exclusive: The FCC will consider applications to be mutually exclusive if their conflicts are such that the grant of one application would effectively preclude by reason of harmful electrical interference, or other practical reason, the grant of one or more other applications. (47 C.F.R. 25.155)

Radio-frequency spectrum: The spectrum of electromagnetic frequencies used for communications, including radio and radar and television.

Radio Regulations: The international treaty governing the use of the radio-frequency spectrum and the geostationary-satellite and non-geostationary-satellite orbits.

CHAPTER 8

Licensing Private Earth Remote Sensing Satellites

Space remote sensing is the act of sensing Earth's surface from outer space by making use of the properties of the electromagnetic waves emitted, reflected, or diffracted by the sensed objects. The National and Commercial Space Programs Act authorizes the U.S. Department of Commerce to license the operation of private U.S. space-based remote sensing systems.[1] Responsibility for implementing this authority has been delegated to the National Oceanic and Atmospheric Administration (NOAA). This delegation made sense because NOAA is the only agency within the Department of Commerce that has experience operating satellites. Under NOAA's National Environmental Satellite, Data, and Information Service (NESDIS), the Commercial Remote Sensing Regulatory Affairs Office (CRSRAO) is the primary office responsible for regulating private Earth remote sensing satellites.

This chapter discusses the licensing of U.S. private space-based remote sensing systems. Section I describes the NOAA licensing process. Section II discusses the 1986 United Nations resolution concerning space-based remote sensing that, while not binding in itself, is reflected in the NOAA licensing regulations.

I. NOAA LICENSING PROCESS

The NOAA licensing regulations address who must apply for a license, the license application process, the conditions for granting a license, and postgrant compliance and monitoring.[2] As with other laws

---

1. 51 U.S.C. § 60121(a).
2. 15 C.F.R. § 960.

addressing space-related activities, NOAA's licensing requirements reflect an attempt to balance commercial remote sensing needs and national security concerns.

A. Who Must Be Licensed

According to the NOAA Regulations, no "person" may operate a private remote sensing space system without a license.[3] "Person" includes

- any individual, regardless of citizenship, subject to U.S. jurisdiction
- a corporation, partnership, association, or other entity organized or existing under the laws of the United States
- a subsidiary (foreign or domestic) of a U.S. parent company
- an affiliate (foreign or domestic) of a U.S. company
- any other private remote sensing space system operator having substantial connections with the United States or deriving substantial benefits from the United States that support its international remote sensing operations sufficient to assert U.S. jurisdiction as a matter of common law.[4]

The Land Remote Sensing Policy Act defines "remote sensing system" broadly to include "any device, instrument, or combination thereof, the space-borne platform upon which it is carried, and any related facilities capable of actively or passively sensing the Earth's surface, including bodies of water, from space by making use of the properties of the electromagnetic waves emitted, reflected, or diffracted by the sensed objects."[5] The broad language used for the definition means that the licensing requirements apply to both large, commercial remote sensing satellites and small research satellites, such as CubeSats. Also note that the definition includes the entire surface of Earth, whether land or water, but does not include area above the surface, such as the atmosphere, or objects beyond the atmosphere, such as the Moon or other celestial bodies.

B. Application Process

The NOAA Regulations do not proscribe a formal application, but instead list information that must be provided to NOAA in order to

---

3. 15 C.F.R. § 960.4.
4. 15 C.F.R. § 960.3.
5. Id.

obtain a license. The application must include detailed information on the applicant's corporate structure and the plans for launch, space operation, and ground operation. This includes, in part, information on the system's operational and technical characteristics and expected dates of operation. The application must also include information describing the access to and distribution of unenhanced data obtained by the system and information on any agreements with foreign entities.[6] There is no application or license fee. Applicants are invited to confer with CRSRAO on their application before submission to ensure completeness.[7]

NOAA works in conjunction with other agencies to review an application, which it is statutorily required to complete within 120 days of submission. During review, NOAA consults with the Department of Defense on matters of national security, the Department of State on matters of foreign policy, the Department of the Interior on matters of storing acquired data in the National Land Remote Sensing Data Archive, and other federal agencies that are determined to have a substantial interest in the license application.[8]

C. Licensing Conditions

The NOAA Regulations require licensees to meet certain minimum licensing conditions. As discussed below, licensees are required to keep information current and must notify NOAA of any changes in a timely manner. The minimum requirements for operating a remote sensing system require that the licensee

- operate its system in a way that preserves national security and respects the foreign policy and international obligations of the United States
- maintain continuous operational control of spacecraft operations from within the United States
- maintain certain records for consistent monitoring and compliance purposes
- be prepared and willing to limit imaging during periods when national security, foreign policy, or international obligations are compromised as well as provide the U.S. government access to and use of data during these times upon request

---

6. For a complete list of the requirements, consult 15 C.F.R. § 960, app. 1.

7. CRSRAO may be contacted through its website at http://www.nesdis.noaa.gov/CRSRA.

8. 51 U.S.C. § 60121(c); 15 C.F.R. § 960.6(a).

- prior to entering into any significant or substantial foreign agreement, notify the U.S. government of its intent and obtain approval
- make unenhanced data available to a sensed state on reasonable cost terms and conditions
- make unenhanced data available to the U.S. National Satellite Land Remote Sensing Data Archive
- notify NOAA of any planned or unexpected deviations that violate conditions of the license
- obtain approval of all plans and procedures dealing with safe disposition of satellites
- submit a data protection plan that describes the process taken to protect data throughout operations
- will not mortgage, sell, or transfer the license.[9]

NOAA's most important finding during the licensing process is to assure that the operation will preserve national security.[10] Even after NOAA grants a license, if NOAA later discovers that the operation is adversely affecting national security, it may revoke the license.[11]

Another significant condition for a NOAA license is proper notification of significant or substantial foreign agreements.[12] Because NOAA monitors systems under U.S. jurisdiction, it must be aware of any new foreign agreements that could result in a foreign entity obtaining control over the system or unduly influencing its operations.

Data protection plans are also important, especially for highly capable systems that produce high-resolution images. These plans assure NOAA that the applicant has implemented, or will implement, appropriate measures to protect acquired data from end-to-end. The goal is to maintain data security to keep information out of the hands of national security threats.

All of these conditions are continuing obligations. Therefore, the secretary of commerce has the ability to seek an order from an appropriate federal district court to terminate, modify, or suspend a license

---

9. 51 U.S.C. § 60122(b); 15 C.F.R. § 960.11.
10. Id.
11. Glenn Tallia, *Licensing CubeSats: Introducing Universities to Space Law for the First Time* (ABA Section of Science & Technology Law, Space Law Comm., Jan. 20, 2012), http://www.americanbar.org/content/dam/aba/events/science_technology/jan_20_cubeaats_space.authcheckdam.pdf.
12. 15 C.F.R. § 960.3.

immediately for failing to comply with any of the above requirements. The secretary of commerce may also do so if the licensee fails to comply with international obligations or, as noted above, national security requirements of the United States.[13]

D. Monitoring and Compliance Requirements

The licensee of a remote sensing space system must continue to provide information to NOAA quarterly and annually under the Commercial Remote Sensing Compliance and Monitoring Program (CRSCMP). First, NOAA follows up on issued licenses through CRSCMP annual compliance audits. NOAA requests that licensees complete a form addressing changes to corporate structure, financial investments, and foreign agreements. The form is available on the NOAA website.[14] Second, NOAA requests CRSCMP quarterly reports to be completed by the program manager. The form requests information on system anomalies and any changes to the licensee's corporate structure or the system's operational and technical status.[15] Third, the CRSCMP includes an on-site audit conducted, at a minimum, on an annual basis.

E. Prohibition on Collection and Release of Satellite Imagery Relating to Israel

There is an additional condition imposed on applicants whose proposed operations will involve the remote sensing of the State of Israel. According to section 1064 of the Defense Authorization Act of 1997, known as the Kyl-Bingaman Amendment, "[a] department or agency of the United States may issue a license for the collection or dissemination by a non-Federal entity of satellite imagery with respect to Israel only if such imagery is no more detailed or precise than satellite imagery of Israel that is available from commercial sources."[16] Therefore, the application must include an explanation of how its system can restrict the dissemination of high resolution imagery collected of Israeli territory.

---

13. 51 U.S.C. § 60123(a)(2); 15 C.F.R. § 960.15.
14. NOAA, CRSCMP Annual Audit, http://www.nesdis.noaa.gov/CRSRA/files/Annual%20Audit.pdf.
15. NOAA, CRSCMP Quarterly Report. http://www.nesdis.noaa.gov/CRSRA/files/Quarterly%20Report.pdf.
16. National Defense Authorization Act for Fiscal Year 1997, Pub. L. No. 104-201, § 1064 (1997).

II. UNITED NATIONS RESOLUTION ON DATA DISSEMINATION

With the launch of ERTS-1 (Earth Resources Technology Satellite, later renamed Landsat-1) in the 1970s, satellite imagery became available for nonmilitary purposes for the first time. With this paradigm shift, the Legal Subcommittee of the U.N. Committee on the Peaceful Uses of Outer Space began considering the legal implications of the remote sensing of Earth from space. These deliberations resulted in the Principles Relating to Remote Sensing of Earth from Outer Space, which were adopted by the United Nations on December 3, 1986 (Appendix 3-7).[17]

The Principles were developed as recommended guidelines for States to develop their own national data policies. The Principles primarily concern data dissemination and what States should be allowed to do with the data generated from sensed States. They do not attempt to limit the right of States to conduct remote sensing operations, nor do the data-sharing requirements apply to military remote sensing systems. The Principles reflect a concern of those sensed States, especially developing countries, that observation of their country would be used against them to circumvent their economic sovereignty.

Considering this emphasis, Principle XII addresses nondiscriminatory access but not the generation of the data. It states:

> As soon as the primary data and the processed data concerning the territory under its jurisdiction are produced, the sensed State shall have access to them on a nondiscriminatory basis and on reasonable cost terms. The sensed State shall also have access to the available analyzed information concerning the territory under its jurisdiction in the possession of any State participating in remote sensing activities on the same basis and terms, particular regard being given to the needs and interests of the developing countries.[18]

In short, upon request, a sensing State should make available to the government of any sensed State data collected by its system concerning the territory under the jurisdiction of the sensed State. The United States incorporated these Principles into the first public law addressing commercial remote sensing by stressing that the "most beneficial use

---

17. G.A. Res. 41/65, Principles Relating to Remote Sensing of the Earth from Space, U.N. Doc. A/RES/41/65 (Dec. 3, 1986).
18. *Id.*

of land remote sensing data will result from maintaining a policy of non-discriminatory access."[19] The principle of nondiscriminatory access continued through to the current Act.[20] While the U.S. legislation refers to "unenhanced data" and the United Nations refers to "primary data," these terms are essentially the same. However, it is unclear how "processed data" in the Principles would correspond to the U.S. terminology.

GLOSSARY

Analyzed information: The information resulting from the interpretation of processed data, inputs of data and knowledge from other sources. (U.N. Res. 41/65)

Commercial Remote Sensing Regulatory Affairs Office (CRSRAO): The office within NOAA responsible for the licensing of private remote sensing space systems and subsequent compliance monitoring.

Data Protection Plan (DPP): The licensee's plan to protect data and information through the entire cycle of tasking, operations, processing, archiving and dissemination. At a minimum, this includes appropriate protection of communications links and/or delivery methods for tasking of the satellite, downlinking of data to a ground station (including relay stations), and delivery of data from the satellite to the licensee's central data storage facilities. (15 C.F.R. Part 960)

Kyl-Bingaman Amendment: Instituted a prohibition on the collection and release of high resolution satellite imagery relating to Israel.

National Environmental Satellite, Data, and Information Service (NESDIS): Entity within NOAA that provides access to global environmental data from satellites and other sources. It is responsible for acquiring and managing U.S. operational environmental satellites, operating the NOAA National Data Centers, providing data and information services including Earth system monitoring, performing official assessments of the environment, and conducting related research.

Primary data: The raw data that are acquired by remote sensors born by a space object and that are transmitted or delivered to the ground from space by telemetry in the form of electromagnetic signals, by photographic film, magnetic tape or any other means. (U.N. Res. 41/65)

---

19. *See* Land Remote Sensing Commercialization Act 1984, Pub. L. No. 98-365, § 101(5) (July 17, 1984).

20. 5 C.F.R. § 960.11(b)(10).

Processed data: The products resulting from the processing of the primary data, needed to make such data usable. (U.N. Res. 41/65)

Remote sensing system: Any device, instrument, or combination thereof, the space-borne platform upon which it is carried, and any related facilities capable of actively or passively sensing the Earth's surface, including bodies of water, from space by making use of the properties of the electromagnetic waves emitted, reflected, or diffracted by the sensed objects. (15 C.F.R. § 960.3)

Significant or substantial foreign agreement: An agreement with a foreign nation, entity, consortium, or person that provides for one or more of the following:

(1) Administrative control which may include distributorship arrangements involving the routine receipt of high volumes of the system's unenhanced data;
(2) Participation in the operations of the system; including direct access to the system's unenhanced data; or
(3) An equity interest in the licensee held by a foreign nation and/or person, if such interest equals or exceeds or will equal or exceed twenty (20) percent of total outstanding shares, or entitles the foreign person to a position on the licensee's Board of Directors. (15 C.F.R. §960.3)

Unenhanced data: Remote sensing signals or imagery products that are unprocessed or subject only to data preprocessing. Data preprocessing may include rectification of system and sensor distortions in remote sensing data as it is received directly from the satellite; registration of such data with respect to features of the Earth; and calibration of spectral response with respect to such data. It does not include conclusions, manipulations, or calculations derived from such data, or a combination of such data with other data. It also excludes phase history data for synthetic aperture radar systems or other space-based radar systems. (15 CFR Part 960.3)

CHAPTER 9

Compliance with Export Control Laws

The U.S. government regulates the foreign distribution of U.S. products, services, and technologies for national security purposes. The export of virtually all launch vehicles, spacecraft, and other space-related technologies are controlled in order to "reduce the possibility of missile-related and other technology spreading to foreign entities that could use it to threaten U.S. interests."[1]

Exports generally fall under one of two sets of regulations. The International Traffic in Arms Regulations (ITAR),[2] administered by Directorate of Defense Trade Controls (DDTC) of the U.S. Department of State, restrict the export of technologies and technical information classified as "defense articles" and "defense services," most of which are used for military applications. The Export Administration Regulations (EAR),[3] administered by the Bureau of Industry and Security (BIS) of the U.S. Department of Commerce, govern the export of "dual-use" items, which are items that are not primarily made for military applications but have both military and civilian uses. Other agencies also have jurisdiction over certain exports, such as the Office for Foreign Assets Control (OFAC) of the U.S. Department of the Treasury, which enforces economic sanctions and embargoes.

Prior to 1993, all satellite-related technologies were subject to the ITAR. In the mid-1990s, however, the Clinton administration began loosening the export restrictions on many dual-use technologies, and jurisdiction over communication satellites was transferred to the Department of Commerce. Following the failure of three Chinese Long

---

1. U.S. Dep't of Commerce & Fed. Aviation Admin., Introduction to U.S. Export Controls for the Commercial Space Industry 3 (2008), *available at* http://www.space.commerce.gov/library/reports/2008-10-intro2exportcontrols.pdf.
2. 22 C.F.R. §§ 120–130.
3. 15 C.F.R. §§ 730–774.

March rockets carrying U.S.-built satellites in 1992, 1995, and 1996, the Department of Commerce permitted two U.S. satellite manufacturers, Hughes Space and Communications International and Space Systems/Loral, to work with the Chinese government to analyze the launch failures. This transfer of U.S. spacecraft technology to China resulted in a major controversy and in 1998 Congress returned all satellites and related technologies to State Department jurisdiction, regardless of whether the satellites are used for military or civilian purposes.[4] In recent years, satellite manufacturers have argued that these strict controls have harmed U.S. competitiveness in the global commercial satellite market, and there have been numerous efforts to re-designate commercial satellites and related components as dual-use technologies under Commerce Department jurisdiction.

This chapter provides an overview of U.S. export control laws that apply to space technologies. Because almost all space technologies are subject to the ITAR, this chapter focuses on those regulations and only briefly covers the EAR and other export control regimes.

I. INTERNATIONAL TRAFFIC IN ARMS REGULATIONS

Section 38 of the Arms Export Control Act (AECA)[5] authorizes the president to control the export of defense articles and defense services. The ITAR are the regulations promulgated by the secretary of state to implement the AECA. These regulations comprise

- the United States Munitions List (USML), which specifies the controlled categories of defense articles, defense services, and related technical data
- the Missile Technology Control Regime, a list of controlled equipment, software, and technology related to missile technology

---

4. Strom Thurmond National Defense Authorization Act for Fiscal Year 1999, Pub. L. No. 105–261, § 1513, 112 Stat. 1920, 2174 (1998) ("Notwithstanding any other provision of law, all satellites and related items that are on the Commerce Control List of dual use items in the Export Administration Regulations (15 CFR part 730 et seq.) on the date of the enactment of this Act shall be transferred to the United States Munitions List and controlled under section 38 of the Arms Export Control Act (22 U.S.C. 2778)[.]"); Ryan Zelnio, *A Short History of Export Control Policy*, SPACE REVIEW, Jan. 9, 2006, http://www.thespacereview.com/article/528/1; REP. OF THE SELECT COMM. ON U.S. NAT'L SEC. & MILITARY/COMMERCIAL CONCERNS WITH THE PEOPLE'S REPUBLIC OF CHINA, H.R. REP. NO. 105-851 (1999) (commonly known as the Cox Report), http://www.house.gov/coxreport/pref/preface.html.

5. Pub. L. No. 90–629, 90 Stat. 744 (codified at 22 U.S.C. § 2778).

I. International Traffic in Arms Regulations

- supporting regulations that define terms, describe the procedures for requesting licenses, and specify various exemptions.

The ITAR regulates the export of defense articles, defense services, and technical data to destinations outside of the United States and to foreign persons both outside of and within the United States. Under the ITAR, the term "export" is defined broadly. It includes

- sending or taking a defense article out of the United States in any manner
- transferring registration, control, or ownership to a foreign person of any aircraft, vessel, or satellite covered by the USML
- disclosing (including oral or visual disclosure) or transferring in the United States any defense article to an embassy, any agency, or subdivision of a foreign government (e.g., diplomatic missions)
- disclosing (including oral or visual disclosure) or transferring technical data to a foreign person, whether in the United States or abroad
- performing a defense service on behalf of, or for the benefit of, a foreign person, whether in the United States or abroad.[6]

The delivery or transfer of any export-controlled product, service, or data not in the public domain to a foreign person is "deemed" to be an export, whether delivery takes place within the United States or abroad. The ITAR specifically states that a "launch vehicle or payload shall not, by reason of the launching of such vehicle, be considered an export."[7]

A "foreign person" is any natural or juridical person who is not a U.S. person.[8] A U.S. person is (1) a natural person (i.e., an individual), wherever located, who is a lawful permanent resident as defined in 8 U.S.C. § 1101(a)(20), meaning someone who is either a U.S. citizen, a foreign national with the visa status of legal permanent resident of the United States (a "green card" holder), or a "protected individual" as defined by 8 U.S.C. § 1324b(a)(3); or (2) a juridical person (i.e., a corporation, association, partnership, organization, trust, or other such entity) that is organized and entitled to do business under the laws of

---

6. 22 C.F.R. § 120.17.
7. *Id.*
8. 22 C.F.R. § 120.16.

the United States or any U.S. jurisdiction. Accordingly, the following are considered foreign persons:

- any natural person who is not a U.S. citizen, a lawful permanent resident, or a protected individual
- any foreign corporation, business association, partnership, trust, and the like, that is not incorporated or organized to do business in the United States
- any international organization (e.g., the United Nations or the World Bank), foreign government or governmental agency or division, foreign embassy, or other diplomatic mission.

A "defense article" is any item or technical data designated on the USML.[9] A "defense service" includes (1) the "furnishing of assistance (including training) to foreign persons, whether in the United States or abroad in the design, development, engineering, manufacture, production, assembly, testing, repair, maintenance, modification, operation, demilitarization, destruction, processing or use of defense articles"; (2) furnishing technical data to any foreign persons; and (3) providing military training to foreign forces.[10]

"Technical data" includes

- any information that is required for the design, development, production, manufacture, assembly, operation, repair, testing, maintenance, or modification of defense articles
- classified information relating to defense articles and defense services
- information covered by an invention secrecy order
- software that is directly related to defense articles.[11]

Technical data does not include (1) information concerning general scientific, mathematical, or engineering principles commonly taught in schools, colleges, and universities; (2) information in the public domain, as defined by the ITAR; or (3) basic marketing information on function or purpose or general system descriptions of defense articles.[12] For ITAR purposes, the term "public domain" is defined narrowly,

---

9. 22 C.F.R. § 120.6.
10. 22 C.F.R. § 120.9.
11. 22 C.F.R. § 120.10.
12. Id.

only encompassing information that is "generally accessible or available to the public" through one of eight specific mechanisms:

1. Sales at newsstands or bookstores.
2. Subscriptions available without restriction to any individual.
3. Second-class mailing privileges granted by the U.S. government.
4. Libraries open to the public.
5. Patents at any patent office.
6. Unlimited distribution at a conference, meeting, seminar, trade show, or exhibition in the United States that is generally accessible to the public.
7. Public release in any form after approval by the cognizant U.S. government department or agency.
8. Fundamental research in science and engineering at accredited institutions of higher learning in the United States where the resulting information is ordinarily published and shared broadly in the scientific community. University research is not considered "fundamental" if the researchers accept any publication restrictions or the research is funded by the U.S. government and the government imposes access and dissemination controls on the research results.[13]

The DDTC has issued guidance stating that information that would ordinarily be considered technical data but is freely available on the Internet is not automatically in the public domain for ITAR purposes. In order for information found on the Internet to be considered in the public domain, it must be available through one of the eight specific mechanisms identified in ITAR part 120.11.[14]

A. The United States Munitions List

The USML is enumerated in ITAR part 121.1 and consists of 21 product categories:

I. Firearms, Close Assault Weapons, and Combat Shotguns
II. Guns and Armament
III. Ammunition/Ordnance

---

13. 22 C.F.R. § 120.11.
14. Gibson, Dunn & Crutcher LLP, *ITAR Update: Technical Data Found on the Internet May Not Be Considered in the Public Domain*, May 9, 2008, http://www.gibsondunn.com/publications/pages/ITARUpdate-TechnicalDataOnInternet.aspx.

Checklist for ITAR-Controlled Technical Data

Materials related to a defense article or defense service may contain ITAR-controlled "technical data" if the materials include

- quantitative information (e.g., measurements and formulas) concerning the article's design, manufacture, or critical operational parameters
- data that do not consist wholly of information in the "public domain"
- data generated by private research and development for a military application
- data generated under a National Aeronautics and Space Administration, U.S. Department of Defense, or intelligence agency contract
- any graphic that depicts components at the engineering-drawing level of detail and is not merely an illustrative drawing, artist's impression, or rough sketch of the defense article
- information that provides understanding of or insight into the sensitive capabilities, limitations, or vulnerabilities of a defense article
- meaningful insight or practical instruction in the design or manufacture of a defense article
- details of critical design elements necessary for the proper function of a component of a defense article
- detailed and specific answers to questions involving "how to" and "why" relating to the design, manufacture, or operation of a defense article
- technical descriptions that go beyond general scientific, mathematical, or engineering principles commonly taught in schools, colleges, and universities
- descriptions of defense articles that go beyond basic marketing information on function or purpose

IV. *Launch Vehicles, Guided Missiles, Ballistic Missiles, Rockets, Torpedoes, Bombs, and Mines
V. *Explosives, Propellants, Incendiary Agents and Their Constituents
VI. Vessels of War and Special Naval Equipment
VII. Tanks and Military Vehicles
VIII. Aircraft and Associated Equipment
IX. Military Training Equipment
X. Protective Personnel Equipment
XI. Military Electronics
XII. *Fire Control, Range Finder, Optical and Guidance, and Control Equipment
XIII. *Auxiliary Military Equipment
XIV. Toxicological Agents and Equipment and Radiological Equipment
XV. *Spacecraft Systems and Associated Equipment
XVI. Nuclear Weapons Design and Related Equipment

XVII. Classified Articles, Technical Data, and Defense Services Not Otherwise Enumerated
XVIII. Directed Energy Weapons
XIX. Reserved
XX. Submersible Vessels, Oceanographic and Associated Equipment
XXI. Miscellaneous Articles

Defense articles in categories preceded by an asterisk (*) have been designated as "significant military equipment" (SME), to which special export controls apply "because of their capacity for substantial military utility or capability."[15] Any classified articles on the USML are also designated as SME. The SME designation extends to any technical data associated with a defense article that been designated as SME.

Category XXI—Miscellaneous Articles incorporates into the USML any "article not specifically enumerated in the other categories of the U.S. Munitions List which has substantial military applicability and which has been specifically designed or modified for military purposes."[16] This catch-all provision significantly expands the scope of items that may be export controlled.

The categories most relevant to the space industry are Category IV—Launch Vehicles, Guided Missiles, Ballistic Missiles, Rockets, Torpedoes, Bombs and Mines, which includes missile and space launch vehicle power plants, and Category XV—Spacecraft Systems and Associated Equipment, which includes communications satellites, remote sensing satellites, scientific satellites, research satellites, navigation satellites, experimental and multimission satellites, and other space technologies and related services and technical data. Unlike other USML categories, Category XV specifically includes items that were developed for civilian applications. The texts of Categories IV and XV that are current as of this writing are included in Appendix 9-1 for reference.

B. ITAR Export Authorizations

Exports and temporary imports of any defense article, defense services, or technical data require DDTC authorization, or, in cases where an

---

15. 22 C.F.R. § 120.7.
16. ITAR Category XXI—Miscellaneous, 22 C.F.R. § 121.1.

exemption applies, a written report to the DDTC. There are three basic types of DDTC authorizations:

- *License.* This is issued by the DDTC and permits the export or temporary import of a specific defense article or defense service.
- *Agreement.* This is an agreement, approved by the DDTC, for the performance of defense services or the disclosure of technical data to a foreign person, or granting a foreign person the right to manufacture or distribute defense articles.
- *Exemption.* This is an authorization granted under the ITAR for the export of certain defense articles without additional approval from the DDTC, although a report of any export transactions covered by the exemption may still be required.

Export authorizations must be requested from the DDTC in advance of an export transaction. The time required for the DDTC to approve authorizations can be as long as three or four months. With the recently expanded utilization of the D-Trade fully electronic licensing system, approvals may take less time.

As a precondition for the issuance of licenses and other export authorizations, all exporters and manufacturers of defense articles and defense services must register with the DDTC and renew the registration annually. Registration forms and related materials can be found on the DDTC website.[17] Once registered, most license requests may be submitted to DDTC through D-Trade.

The following subsections describe the types of licenses, agreements, and exemptions available under the ITAR. All exports of hardware controlled by the ITAR, regardless of the type of authorization (license, agreement, or exemption), require the filing of electronic export information (EEI) with the U.S. government's Automated Export System (AES).[18] The U.S. Census Bureau provides a free, Internet-based system for filing EEI, called AESDirect.[19]

---

17. U.S. Dep't of State, Directorate of Def. Trade Controls, Registration, http://www.pmddtc.state.gov/registration/index.html.
18. Foreign Trade Regulations, 15 C.F.R. §§ 30.2(a)(iv)(B) & (C).
19. http://www.aesdirect.gov.

1. Licenses

Export licenses are issued by the DDTC for the export or temporary import of specific defense articles, defense services, and related technical data, to specific end users for specific end uses. Once approved, a license remains valid until all the items identified on it have been shipped, or for a period of four years, whichever occurs first. The following are the forms that are most commonly used for license applications:

- *DSP-5*: Permanent export of unclassified defense articles.
- *DSP-61*: Temporary import or in-transit shipments of unclassified defense articles.
- *DSP-73*: Temporary export of unclassified defense articles that will be returned to the United States within four years and whose title will not transfer during that time.
- *DSP-83*: Non-Transfer and Use Certificate. This form is required for exports of SME and classified defense articles. The DDTC may also require that the foreign end user's government execute this certificate. The submission of any classified materials accompanying the license application must follow requirements of the National Industrial Security Program Operating Manual (NISPOM), which governs the protection of classified information by defense contractors.
- *DSP-85*: Permanent or temporary export or temporary import of classified defense articles including technical data. Form DSP-83 (Non-Transfer and Use Certificate) must also be attached when exporting classified technical data.

2. Agreements

Some ITAR-controlled export transactions involve an ongoing formal arrangement to provide defense services, technical data, manufacturing rights, distribution rights, or technical assistance to a foreign person for an extended period. Such transactions require the submission of a formal agreement to the DDTC for approval.

There are three basic types of agreements: manufacturing license agreements (MLAs), technical assistance agreements (TAAs), and warehousing and distribution agreements.[20] These agreements are binding

---

20. 22 C.F.R. § 120.21, .22, .23.

contracts between all parties involved in the transaction, but they may not be executed without written approval from DDTC. This approval is often granted by the DDTC subject to certain "provisos and limitations" that may place restrictions on the transactions covered by the agreement. A sample TAA/MLA is provided in Appendix 9-2.

3. Exemptions

There are a number of exemptions available under the ITAR that eliminate the need to obtain an export license or other prior approval from the DDTC, although certain reporting and record-keeping requirements will still apply. ITAR exemptions will not apply, however, if the export will be made to a destination on the list of countries proscribed or embargoed by the DDTC, or if a foreign person from one of these proscribed countries is involved.[21]

The following is a list of commonly used ITAR exemptions, along with the ITAR section reference where each exemption and its associated requirements are described.

For permanent exports:

- Shipments Between U.S. Possessions (123.12)
- Components and Parts Less Than $500 (123.16(b)(2))
- Exports of Models and Mock-Ups (123.16 (b)(4))
- Technical Data Pursuant to Written Request of DOD (125.4(b)(1))
- Technical Data in Furtherance of a TAA or MLA (125.4(b)(2))
- Technical Data in Furtherance of a U.S. Government Contract (125.4(b)(3))
- Technical Data Previously Authorized for Export (125.4(b)(4))
- Data on Basic Operations, Maintenance & Training (125.4 (b)(5))
- Technical Data Returned to Original Source of Import (125.4(b)(7))
- Technical Data for Use Only by U.S. Persons (125.4(b)(9))
- Technical Data Approved for Public Release (125.4(b)(13))
- Plant/Facility Visits (125.5(a)(b)(c))

---

21. 22 C.F.R. § 126.1.

- Shipments Conducted by U.S. Government Agencies (126.4(a))
- Shipments for End-Use by U.S. Government Agencies (126.4(c))
- Canadian Exemption (126.5)
- Foreign Military Sales Exemption (126.6(c))

For temporary imports:

- Imports for Overhaul Service or Repair (123.4(a)(1))
- Imports for Enhancement or Upgrade (123.4(a)(2))
- Imports for Exhibition, Demonstration, or Marketing (123.4(a)(3))

C. Special Export Controls for Foreign Satellite Launches

In addition to the controls described above, the ITAR applies "special export controls" (SECs) to the launching of U.S. satellites in, or by nationals of, a country that is not a member of the North Atlantic Treaty Organization (NATO) or a major non-NATO ally of the United States.[22] The SECs require U.S. licensees to implement government-approved technology control plans and to arrange and pay for the Department of Defense to monitor technical exchanges between U.S. and foreign personnel involved in launch activities. Export licenses and Department of Defense monitoring are also required for U.S. persons to participate in foreign launch failure investigations or analyses.

D. Violations and Penalties

The AECA and the ITAR impose both criminal and civil penalties on companies and individuals who violate their requirements.[23] ITAR violations can lead to fines and penalties with varying degrees of severity, depending on the nature and circumstances of the violation. Violations can result in civil penalties for corporations and individuals of $500,000 or more per violation, and the suspension or debarment from any future export of defense articles or defense services. Willful violations may result in criminal fines for exporters and/or the individual employees involved as high as $1 million per violation or, in some cases, twice the gross gain resulting from the violation, or the imprisonment of the individuals for up to 10 years, or both. ITAR violations

---

22. 22 C.F.R. § 124.15.
23. 22 C.F.R. § 127.

related to the handling of classified information or technical data obtained pursuant to U.S. government contracts may result in suspension or debarment from future government contracting.

II. EXPORT ADMINISTRATION REGULATIONS

While virtually all spacecraft-related technologies are controlled by the ITAR, it is important that attorneys in the space industry also be generally familiar with the EAR since space companies also utilize technologies that are not specifically designed for spacecraft. Moreover, if the export control reform efforts described at the end of this chapter are successful, certain commercial satellites and related components may eventually be removed from the USML and controlled as dual-use technologies under the EAR. Accordingly, a brief overview of the EAR is provided here.

The EAR generally covers "dual-use" items, which are items that have both commercial and military applications. In contrast to exports of defense articles and defense services under the ITAR, the majority of exports under the EAR do not require a formal export license. As described by the Department of Commerce:

> A fundamental difference between ITAR and EAR processes is their inherent presumptions regarding an applicant's rights to export. With the EAR, there is a "presumption of approval": The BIS specifically identifies only those items and the countries of destination for which an export license would be required, while still allowing for exceptions under certain circumstances. Conversely, the ITAR proceeds under a "presumption of denial": The DDTC makes determination of license applicability and approval, requiring the exporter to prove that their item or service does not pose significant risk to national security.[24]

Items that require an export license under the EAR are identified on the Commerce Control List (CCL), located in Supplement No. 1 to part 774 of the EAR. Each item on the CCL has a specific export control classification number (ECCN). The ECCN is an alpha-numeric code (e.g., 3A001) that describes the item and applicable licensing requirements. If an EAR-controlled item does not fit within any specific ECCN, it is designated "EAR99."

---

24. U.S. Dep't of Commerce & Fed. Aviation Admin., Introduction to U.S. Export Controls for the Commercial Space Industry 3 (2008), *available at* http://www.space.commerce.gov/library/reports/2008-10-intro2exportcontrols.pdf.

The first digit (e.g., 3A001) of the ECCN is one of the 10 broad CCL categories, which are numbered from 0 to 9:

0   Nuclear Materials, Facilities and Equipment and Miscellaneous
1   Materials, Chemicals, "Microorganisms," and Toxins
2   Materials Processing
3   Electronics
4   Computers
5   Telecommunications and Information Security
6   Lasers and Sensors
7   Navigation and Avionics
8   Marine
9   Propulsion Systems, Space Vehicles, and Related Equipment

Category 9 is the most relevant to the space industry, but almost all of the items listed there are simply referred to as being under the jurisdiction of the Department of State, by way of the ITAR.

Within each CCL category, products are arranged by the same five groups, identified by the letters A through E:

A   Equipment, Assemblies, and Components
B   Test, Inspection, and Production Equipment
C   Materials
D   Software
E   Technology

This group designator is the l second character the ECCN code (e.g., 3A001).

Finally, each product group is divided into three-digit codes (e.g., 3A001) corresponding to specific types of export controls:

000–099	National Security
100–199	Missile Technology
200–299	Nuclear Nonproliferation
300–399	Chemical & Biological Weapons
900–999	Foreign Policy
980–989	Short Supply/Crime Control
990–999	Anti-Terrorism/United Nations Embargo

Thus the example code 3A001 corresponds to electronic equipment, assemblies, and components that are primarily controlled for national security reasons.

Unlike the ITAR, which prohibits exports of controlled items to all foreign countries, the EAR is more nuanced and restricts exports only

154 COMPLIANCE WITH EXPORT CONTROL LAWS

to particular countries based on the designated reason(s) for control. For example, exports that are controlled because they contain missile technology require a license if they are being exported to any country other than Canada, whereas exports that are controlled for crime-control reasons may be exported without a license to Australia, Canada, most European countries, Japan, and New Zealand. Exports may be controlled for more than one reason. The Commerce Country Chart, located in Supplement No. 1 to part 738 of the EAR, contains licensing requirements based on the destination country and the reasons for control. The following is a list of all the possible reasons for control:

AT	Anti-Terrorism
CB	Chemical & Biological Weapons
CC	Crime Control
CW	Chemical Weapons Convention
EI	Encryption Items
FC	Firearms Convention
MT	Missile Technology
NP	Nuclear Nonproliferation
NS	National Security
RS	Regional Stability
SI	Significant Items
SL	Surreptitious Listening
SS	Short Supply
UN	United Nations Embargo

A chart explaining the basis for each of these reasons for control may be found on the BIS website.[25] The current official version of the EAR, including the CCL and the Commerce Country Chart, is also available online.[26]

As noted above, most non-USML exports do not require a formal export license. If an EAR-controlled product does require a license, however, the exporter will need to apply to BIS for an export license. If the license is approved, a license number will be issued for use on the export documents associated with the transaction. An EAR export

---

25. U.S. Dep't of Commerce, Bureau of Indus. & Sec., Basis of CCL Controls and Applicable EAR References, http://www.bis.doc.gov/policiesandregulations/basis_of_ccl_controls.htm.

26. U.S. Dep't of Commerce, Bureau of Indus. & Sec., Export Administration Regulation Downloadable Files, http://www.access.gpo.gov/bis/ear/ear_data.html.

III. Export Compliance and Monitoring

license is usually valid for two years. Detailed instructions on the license application process can be found on the BIS website.[27] All exports of hardware requiring a BIS license require the filing of electronic export information with the Automated Export System using AESDirect.[28]

Even if a license is generally required for an EAR-controlled export, a license exception that authorizes the export without a BIS license may be available. License exceptions and the conditions on their use are described in part 740 of the EAR. If an exception is available, the exporter must list the specific three-letter designation of that exception on the export documents for the transaction.

III. EXPORT COMPLIANCE AND MONITORING

All space companies that may export goods or services or interact with foreign nationals in the United States should have comprehensive internal policies and procedures for ensuring compliance with the ITAR, the EAR, and other export control laws. The full extent of these procedures is beyond the scope of this chapter and should be developed in consultation with experienced export counsel. Nevertheless, the following sections highlight some of the compliance practices and issues commonly addressed by companies in the space industry.

A. Overview of the Export Compliance Process

The export compliance process involves six basic steps:

1. Determine which U.S. government agency has jurisdiction over the product or technology to be exported and which export regulations it is subject to.
2. Determine whether current destination or end-use controls would prohibit the export transaction or impose restrictions on it.
3. Classify the export under the applicable regulations and determine what type of export authorization is required.
4. Obtain the appropriate export authorization.
5. Complete the export in accordance with terms of the authorization.

---

27. U.S. Dep't of Commerce, Bureau of Indus. & Sec., Applying for an Export License, http://www.bis.doc.gov/licensing/applying4lic.htm.
28. 15 C.F.R. § 30.2(a)(iv)(A).

6. Maintain records of the export transaction as required by the applicable regulations.

The type of export authorization required depends on the regulations governing a particular export transaction and the category or classification under which the product or technology falls. Authorizations generally fall under one of three broad categories:

1. No license is required for the export.
2. A license exemption (ITAR) or exception (EAR) applies to the export.
3. An export license or agreement is required for the export.

In order to know whether an export license will be required, and, if so, what kind of license, the exporter must determine whether the export is under Department of State (ITAR) or Department of Commerce (EAR) jurisdiction. If it is unclear whether an item is covered by the ITAR or the EAR or the exporter would like the Department of State to consider removing a particular item from the USML, the exporter may request that DDTC make an official determination by filing a Commodity Jurisdiction request.[29]

B. The Empowered Official

Each exporter must appoint an "empowered official" who (1) is directly employed by the exporter having authority for policy or management within the organization; (2) is legally empowered in writing to sign license applications or other requests for approval; and (3) understands the provisions and requirements of the various export control statutes and regulations, and the criminal liability, civil liability, and administrative penalties for violations. The ITAR requires that the empowered official have "independent authority" to (1) inquire into any aspect of a proposed export or temporary import, (2) verify the legality of the transaction and the accuracy of the information to be submitted, and (3) refuse to sign any license application or other request for approval without prejudice or other adverse recourse.[30]

---

29. More information about Commodity Jurisdiction requests may be found at U.S. Dep't of State, Directorate of Def. Trade Controls, Commodity Jurisdiction, http://www.pmddtc.state.gov/commodity_jurisdiction/index.html.
30. 22 C.F.R. § 120.25.

TABLE 9.1
DDTC Embargoed Countries (as of Jan. 2, 2012)

Afghanistan	Belarus
Burma	China (PR)
Côte d'Ivoire	Cuba
Cyprus	Congo
Eritrea	Fiji
Guinea	Haiti
Iran	Iraq
Kyrgyzstan	Lebanon
Liberia	Libya
Niger, Republic of	North Korea
Pakistan	Sierra Leone
Somalia	Sri Lanka
Sudan	Syria
Venezuela	Vietnam
Yemen	Zimbabwe

Source: U.S. Dep't of State, Directorate of Def. Trade Controls, Country Policies and Embargoes, http://www.pmddtc.state.gov/embargoed_countries/index.html

C. Screening for Restrictions and Diversion Risks
1. Destination and End Use Controls

In addition to the controls defined by the ITAR and the EAR, the U.S. government generally denies licenses for exports of all defense articles and defense services destined for certain countries.[31] Other U.S. products, including commercial and dual-use items, cannot be exported to certain countries without a license from the OFAC because of various trade embargoes. These country restrictions are known as "destination controls," because they determine which foreign countries can and cannot receive U.S. exports (see table 9.1).[32]

---

31. 22 C.F.R. § 126.1(a).
32. The current list of DDTC embargoed countries is available on the DDTC website: U.S. Dep't of State, Directorate of Def. Trade Controls, Country Policies and Embargoes, http://www.pmddtc.state.gov/embargoed_countries/index.html. A list of economically embargoed countries and various other associated lists and information about sanctions programs administered by the OFAC can be found at its website: U.S. Dep't of the Treasury, Office of Foreign Assets Control, http://www.treas.gov/offices/enforcement/ofac.

Although specific restrictions are stipulated for each country, the following are the basic guidelines for embargoed destinations:

- No exports of defense articles, technical data, or defense services.
- No proposals or brokering activities.
- No shipments on vessels, aircraft, or other means of conveyance that is owned, operated, or leased from a proscribed country or area.
- No use of ITAR exemptions or EAR exceptions without prior U.S. government approval.

"End-use controls" are regulations that govern how U.S. exports can be used. For example, U.S. companies are not permitted to export products or technology that could be used in the proliferation of weapons of mass destruction, including chemical/biological weapons, prohibited nuclear activities, and missile technology. This rule applies to all products "directly used" for these purposes, whether or not they are on the primary control lists.

Exporters are generally required to exercise due diligence and monitor transactions to screen for exports that may involve prohibited destinations or end uses and detect any red flags. The following are common red flags:

- The customer or purchasing agent is reluctant to offer information about the end use of the item.
- The product's capabilities do not fit the buyer's line of business, such as an order for sophisticated computers from a small bakery.
- The item ordered is incompatible with the technical level of the country to which it is being shipped, such as semiconductor manufacturing equipment being shipped to a country that has no electronics industry.
- The customer is willing to pay cash for a very expensive item when the terms of sale would normally call for financing.
- The customer has little or no business background.
- The customer is unfamiliar with the product's performance characteristics but still wants the product.
- The customer declines routine installation, training, or maintenance services.
- Delivery dates are vague, or deliveries are planned for out-of-the-way destinations.

- A freight-forwarding firm is listed as the product's final destination.
- The shipping route is abnormal for the product and destination.
- Packaging is inconsistent with the stated method of shipment or destination.
- When questioned, the buyer is evasive and especially unclear about whether the purchased product is for domestic use, for export, or for reexport.[33]

2. Denied Persons and Restricted Parties

U.S. companies are not permitted to transact business with persons and entities listed as denied persons and restricted parties unless the U.S. government approves the transaction. Generally, these persons and entities fall into one of the following categories:

- *Specially designated nationals and blocked persons list.* Organizations and persons located anywhere in the world that are owned or controlled by, or acting for or on behalf of, the government of a sanctioned country.[34]
- *Statutorily debarred parties list and administratively debarred parties list.* Persons and organizations that have been debarred by the DDTC.[35]
- *Nonproliferation sanctions.* The United States imposes sanctions under various legal authorities against foreign individuals, private entities, and governments that engage in proliferation activities. The official notifications for all nonproliferation sanctions determinations are published in the *Federal Register.*[36]

---

33. U.S. Dep't of Commerce, Bureau of Indus. & Sec., Red Flag Indicators, http://www.bis.doc.gov/complianceandenforcement/redflagindicators.htm.

34. For the current list, see OFFICE OF FOREIGN ASSETS CONTROL, SPECIALLY DESIGNATED NATIONALS AND BLOCKED PERSONS (May 8, 2012), http://www.treas.gov/offices/enforcement/ofac/sdn/t11sdn.pdf.

35. For the current list, see U.S. Dep't of State, Directorate of Def. Trade Controls, Lists of Parties Debarred for AECA Convictions, http://www.pmddtc.state.gov/compliance/debar_intro.html.

36. Links to those notices may be found on the Department of State website at http://www.state.gov/t/isn/c15231.htm.

- *Entity list.* Foreign end users whose presence in a transaction can trigger additional license requirements besides those required under the EAR.[37]

3. Boycott Requests

Export transactions must also be screened for the presence of any boycott requests. The antiboycott regulations prohibit U.S. companies from

- entering into agreements to refuse or actually refusing to do business with or in a certain nation (such as Israel) or with certain companies
- entering into agreements to discriminate or actually discriminating against other persons based on race, religion, sex, national origin, or nationality
- entering into agreements to furnish or actually furnishing information about business relationships with or in a certain nation (such as Israel) or with certain blacklisted companies
- entering into agreements to furnish or actually furnishing information about the race, religion, sex, or national origin of another person
- paying or implementing letters of credit containing prohibited boycott terms or conditions.[38]

D. Safeguarding Technical Data

Exporters and manufacturers of export-controlled items must undertake all reasonable steps to protect technical data associated with the manufacture of defense articles or defense services from unauthorized physical, visual, acoustical, or virtual access by employees, contractors, visitors, or guests who are not U.S. persons. Safeguards include physical security, information security, document marking, and verifying compliance by suppliers and contractors.

---

37. This list can be found at U.S. Dep't of Commerce, Bureau of Indus. & Sec., The Entity List, http://www.bis.doc.gov/Entities/.

38. *See* 15 C.F.R. § 760; U.S. Dep't of Commerce, Bureau of Indus. & Sec., Antiboycott Compliance, http://www.bis.doc.gov/complianceandenforcement/antiboycottcompliance.htm.

1. Physical Security

Physical security should be customized for each facility where ITAR-controlled technical data is stored, but at minimum should address the following common issues:

- visitor logs and screening
- badging
- secure document storage
- signage
- photography
- foreign visitor requests and escorts
- controls on outside contractors

2. Information Security

Companies should institute adequate information security controls to secure access to the company's network and computers, including any portable devices, such as laptop or notebook computers, where ITAR-controlled technical data may be stored in electronic form.

3. Document Marking

As a preventive measure against inadvertent ITAR violations, all documents containing ITAR-controlled technical data should be clearly labeled and identified as such before being transferred to any third party. Appropriate markings should appear, at a minimum, on the front page of any document containing ITAR-controlled technical data. Markings should include the authorizing license or agreement number (if any) and a statement similar to the following:

> **EXPORT-SENSITIVE—ITAR Restricted-Use Data**
>
> **Not to be copied or transferred to unauthorized third parties.**
>
> Information contained herein is subject to the International Traffic in Arms Regulations. This information may not be transferred to any person or entity outside of the United States, or transferred or made available to any foreign national within the United States, either in original form or after being incorporated through an intermediate process into another medium, without appropriate U.S. government authorization.

A legend, such as "ITAR Restricted Technical Data," should also be included on all subsequent pages where controlled technical data appears. Other labeling information may also be required in accordance with specific provisos or limitations of any related export license or agreement.

In the case of EAR-licensed exports, a destination control statement (DCS) must appear on the invoice and on the bill of lading, air waybill, or other export control document that accompanies each export shipment. The DCS must state:

> These commodities, technology or software were exported from the United States in accordance with the Export Administration Regulations. Diversion contrary to United States law prohibited.

This DCS is required for all exports of items on the CCL, except those categorized as EAR99.[39]

4. Verification of Export Compliance by Suppliers and Contractors

Violations involving the unauthorized export of controlled technical data can easily occur when negotiating for and procuring services or materials from a foreign supplier or contractor if technical data needs to be provided or discussed. Unintentional export violations can also occur when discussing export-controlled technical data with a domestic vendor where the vendor does not have adequate export controls in place. Consequently, whenever a project involves export-controlled technologies, it is important to make sure that suppliers and contractors are knowledgeable with regard to export control requirements, since their business practices will inevitably affect the company's own compliance efforts. Purchase orders and other contracts entered into in connection with export-controlled products or services should also expressly mandate compliance with the ITAR, the EAR, and other relevant export control laws.

E. Foreign National Employees

The employment of non-U.S. persons by businesses dealing with defense articles and defense services is allowed by the ITAR. However, depending on the job position, an export license may be required to allow the employee access to export-controlled hardware or technology. The DSP-5 export license is required before a foreign national is hired as a permanent or temporary employee who will be involved in any way with defense articles or defense services. This requirement also applies to foreign national consultants, contractors, or subcontractors located or performing services at a facility, as well as to foreign national

---

39. 15 C.F.R. § 758.6.

employees located outside of the United States who may have access to defense articles or defense services.

If foreign persons are employed at company facilities, a technology control plan (TCP) is required by section 126.13(c) of the ITAR and, for facilities cleared to receive classified information, by sections 10-509 and 2-307 of the NISPOM. The TCP will need to be approved by the DDTC and, in the case of DOD-related contracts, by the Defense Security Service (DSS) as well.

A TCP is an internal document stipulating how access to export-controlled information will be controlled in circumstances when foreign persons are present as employees or guests, or where a foreign entity owns or controls a U.S. company. It should describe the specific procedures that will be followed to safeguard sensitive technical information and data, as well as hardware and software, from use and observation by unauthorized foreign persons. The elements of a TCP generally include physical and information security measures, personnel screening and training, monitoring of international business travel by employees, custodial logging and tracking, and self-evaluation. The DSS template used for preparing a TCP is included in Appendix 9-3.

F. International Business Travel

Taking controlled defense articles and technical data on foreign business travel is technically considered an export of those items. Companies should ensure that their employees do not disclose export-controlled technical data, or hand-carry export-controlled items, during their foreign travel without the appropriate export authorization. This includes laptop computers or other data storage devices that may contain technical data; other export-controlled hardware items, such as test equipment or tool kits; models, mock-ups, exhibits, or displays for trade shows; sales presentations to be made at sales meetings; and technical papers to be presented at professional or academic conferences. However, an export license is not required for a U.S. corporation to send classified and unclassified technical data overseas to an employee who is a U.S. person or to a U.S. government agency, so long as (1) the technical data is to be used overseas solely by U.S. persons; (2) the U.S. person overseas is directly employed by the U.S. government or the U.S. corporation, and not by a foreign subsidiary; and (3) classified information is sent in accordance with the NISPOM.[40]

---

40. 22 C.F.R. § 125.4(b)(9).

G. Requesting Clearance for Information Release

All information designated for public release that may contain technical data should undergo an information release process in order to ensure that no export-controlled information is released without the proper authorization. "Public release" is the dissemination of information to an unrestricted audience, which is an audience that may be composed of foreign persons as well as U.S. persons. Examples of public release include publishing information in technical journals, making presentations at conference where the audience is not restricted to U.S. persons only, displaying information on public websites, providing information to the media, and exhibits at trade shows that non-U.S. persons may attend. The public release of ITAR-controlled technical data requires prior approval from the cognizant U.S. government department or agency or from the Office of Security Review of the Department of Defense.[41] Once cleared in this way, there is no required involvement by the DDTC for the public release of this information.

H. Record Keeping

Both the ITAR and the EAR contain specific record-keeping requirements. The mandatory period of retention for export documents is a minimum of five years, calculated from one of the following points in time:

Records relating to exports under the ITAR must be retained for a minimum of five years after

- the expiration of the license or other approval to which the documentation relates;
- the date the license or other authorization is exhausted or used completely; or
- the date the license or other authorization is suspended, revoked, or no longer valid.

Records relating to exports under the EAR must be retained for a minimum of five years after

- the date the export from the United States occurs;
- the date of any known reexport, transshipment, or diversion of such item;

---

41. 22 C.F.R. § 125.4(b)(13).

- the date of any termination of the transaction, whether contractual, legal, formally in writing, or by any other means; or
- in the case of records of, or pertaining to, transactions involving restrictive trade practices or boycotts, the date the regulated person receives the boycott-related request.

IV. EXPORT CONTROL REFORM

Since commercial satellite technology was transferred from Commerce Department jurisdiction to State Department jurisdiction in 1998, there has been considerable concern in the commercial satellite industry about the effect of these controls on U.S. competitiveness in the commercial satellite market. According to the Satellite Industry Association (SIA), the U.S. share of the worldwide satellite manufacturing market dropped precipitously from about 65 percent in 1999 to about 40 percent in 2001, and it remained roughly at this level through 2010.[42] While a number of factors contributed to this decline, including the growing capabilities of non-U.S. satellite manufacturers during this period, the president of SIA noted in 2009 that "it is increasingly clear that U.S. export controls have affected American firms' ability to compete globally." Specifically, she stated:

> One European manufacturer, Thales Alenia [S]pace, has begun to market an "ITAR-Free" satellite . . . [and] we are anecdotally observing increasing numbers of satellite operators from around the world preferring to purchase satellites that exclude U.S. technology and avoid the concomitant ITAR requirements. In some cases, proposals require delivery of technical data in timetables that are simply not feasible, given the needed time to secure the requisite ITAR approvals, licenses and Congressional notification. Some European Space Agency programs now have explicitly required "ITAR-Free" bids.[43]

---

42. SATELLITE INDUS. ASS'N, STATE OF THE SATELLITE INDUSTRY REPORT (June 2011); Satellite Indus. Ass'n, *Revenue Breakdown and U.S. Market Share of Commercial and Government Satellite Manufacturing, 1996–2008*, SPACE NEWS INT'L, June 16, 2011, http://www.hostedpayload.com/ppt/revenue-breakdown-commercial-and-government-satellite-manufacturing.

43. *Export Controls on Satellite Technology: Hearing Before the House Foreign Affairs Comm. (HFAC), Subcomm. on Terrorism, Non-Proliferation, and Trade* (written testimony of Patricia Cooper, President, Satellite Industry Association, Apr. 2, 2009, at 4–5, http://www.sia.org/PDF/HFAC-STNT_SIA_Written_Testimony__3_31_09_FINAL.pdf).

Reinforcing this point, former U.S. Secretary of Defense Robert Gates stated in a 2010 speech on export control reform that

> The [current export control] system has the effect of discouraging exporters from approaching the process as intended. Multinational companies can move production offshore, eroding our defense industrial base, undermining our control regimes in the process, not to mention losing American jobs. Some European satellite manufacturers even market their products as being not subject to U.S. export controls, thus drawing overseas not only potential customers, but some of the best scientists and engineers as well.[44]

Because of concerns such as these, "there have been considerable efforts to try and make it easier for American companies to export satellites and related components."[45] While many reform efforts have started and stalled over the last decade, the Obama administration and both Democratic and Republican members of Congress have placed renewed emphasis on this issue, believing that reforming export control laws will add jobs, make U.S. space companies more competitive internationally, and enhance national security by supporting the space and defense industrial base.[46]

On April 18, 2012, the Departments of Defense and State issued a joint report to Congress assessing the risks associated with removing satellites and related components from the USML, known as the "1248 Report" after the statutory provision that mandated the assessment.[47] The 1248 Report concluded that communications satellites that do not contain classified components, remote sensing satellites with performance parameters below certain thresholds, and certain other satel-

---

44. AEROSPACE INDUS. ASSOC., COMPETING FOR SPACE: SATELLITE EXPORT POLICY AND U.S. NATIONAL SECURITY 3 (2012), *available at* http://www.aia-aerospace.org/assets/CompetingForSpaceReport.pdf.

45. Jeff Foust, *The Sisyphean Task of Export Control Reform*, SPACE REVIEW, Nov. 7, 2011, http://www.thespacereview.com/article/1962/1.

46. *See, e.g.*, Brian H. Nilsson, Keynote Address at the Space Policy Institute Symposium on U.S. Export Controls and Space Science: The President's Export Control Reform Initiative: Implications for Space Science (Mar. 31, 2011) (describing the Obama administration's export control reform initiatives and its expected impact on space technologies), http://www.gwu.edu/~spi/assets/docs/Export_Controls_Nilsson.pdf.

47. U.S. DEP'TS OF DEFENSE AND STATE, REPORT TO CONGRESS, SECTION 1248 OF THE NATIONAL DEFENSE AUTHORIZATION ACT FOR FISCAL YEAR 2010 (PUBLIC LAW 111-84), RISK ASSESSMENT OF UNITED STATES SPACE EXPORT CONTROL POLICY (2012), *available at* http://www.defense.gov/home/features/2011/0111_nsss/docs/1248_Report_Space_Export_Control.pdf.

lite systems "do not contain technologies unique to the [U.S.] military industrial base nor are they critical to national security." Accordingly, the 1248 Report recommended that these items be removed from the USML and designated as dual-use items controlled under the EAR. As of this writing, the 1248 Report has been generally well-received on Capitol Hill and within the satellite industry, and legislation authorizing the president to remove commercial satellites and related components from the USML is moving through Congress.[48]

GLOSSARY

Arms Export Control Act (AECA): The statute that authorized the President to control the export of defense articles and defense services.

Automated Export System (AES): The system, including *AESDirect*, for collecting electronic export information from persons exporting goods from the United States, Puerto Rico, or the U.S. Virgin Islands; between Puerto Rico and the United States; and to the U.S. Virgin Islands from the United States or Puerto Rico, pursuant to the Foreign Trade Regulations.

Bureau of Industry and Security (BIS): The agency of the Department of Commerce responsible for implementing and enforcing the Export Administration Regulations (EAR).

Commerce Control List (CCL): The list of commodities controlled by the Department of Commerce's Bureau of Industry and Security (BIS) under the Export Administration Regulations (EAR). The CCL is found in Supplement 1 to Part 774 of the EAR.

Commerce Country Chart: This chart, located in Supplement No. 1 to Part 738 of the EAR, contains EAR licensing requirements based on destination and the Reason for Control.

Commodity Jurisdiction Ruling: In cases where an export is arguably covered by both the EAR and the ITAR, this is the procedure used by the Department of State's DDTC, upon request, to determine which agency will have jurisdiction.

Deemed Export: The term "deemed export" refers to the transfer of any covered goods or items, or the disclosure in written, oral, visual,

---

48. National Defense Authorization Act for Fiscal Year 2013, H.R. 4310, § 1241, 112th Cong. (2012).

or electronic form, or in some combination of these forms, of controlled information relating to any covered goods or items, to a foreign national or person in the United States, including the provision of certain types of services (such as training) to a foreign national or person in the United States.

Defense Article: Any item listed on the USML.

Defense Service: The furnishing of assistance (including training) to foreign persons, whether or not in the United States, with respect to defense articles, and the furnishing of any technical data associated with a defense article.

Directorate of Defense Trade Controls (DDTC): The agency of the Department of State responsible for implementing and enforcing the International Traffic in Arms Regulations (ITAR).

Dual-Use: Items, materials, or technologies not specifically military or defense-related that either contribute to the development of weapons systems of special concern (e.g., nuclear, chemical, and biological) or contribute to the development of military capabilities more generally (e.g., supercomputers, advanced telecommunications equipment, advanced materials used in armor or bullet-proof vests, advanced navigation equipment, satellite and space technology) and therefore might give an adversary a military advantage.

Empowered Official: An employee of an exporter who: (1) is directly employed by the exporter having authority for policy or management within the organization; (2) is legally empowered in writing to sign license applications or other requests for approval; and (3) understands the provisions and requirements of the various export control statutes and regulations, and the criminal liability, civil liability, and administrative penalties for violations.

End-use: How the ultimate consignee intends to use the commodities being exported. A detailed description of the end-use will generally be required to obtain authorization for the export of export-controlled items.

End-User: The person abroad who receives and ultimately uses the exported or re-exported items. The end-user may be the purchaser or ultimate consignee. A forwarding agent or intermediary is *not* an end-user.

Export: The transfer of anything by a U.S. person to a foreign person by any means, anytime, anywhere; or the transfer of anything to another U.S. person with the knowledge that it will be further transferred to a foreign person. The complete official definitions are contained in the ITAR (22 C.F.R. § 120.17) and the EAR (15 C.F.R. § 734.2), which should be consulted.

Export Administration Regulations (EAR): Regulations administered by the BIS that regulate the export and re-export of dual-use items and other commercial items.

Export Control Classification Number (ECCN): An alpha-numeric identifier assigned by the BIS to classify items on the Commerce Control List. This is the fundamental designation indicating the level of control for an item to be exported under the EAR.

Foreign Person: Any natural or juridical person who is not a U.S. person. That includes:

- Any natural person who is neither a U.S. citizen, nor a lawful permanent resident, nor a protected individual.
- Any foreign corporation, business association, partnership, society, trust, etc., that is not incorporated or organized to do business in the United States.
- Any international organization (e.g., the United Nations or the World Bank), foreign government or governmental agency or division, foreign embassy, or other diplomatic mission.

International Traffic in Arms Regulations (ITAR): Regulations issued pursuant to the AECA and administered by the DDTC to regulate the export of defense articles and defense services.

Manufacturing License Agreement (MLA): An authorization granted by a U.S. person to a foreign person or persons (government, industry, or individual) allowing the foreign person(s) to manufacture defense articles abroad for which technical data and defense services will be required. This may include the export of manufacturing know-how and/or the use by the foreign person of defense articles and technical data previously exported by the U.S. person.

Missile Technology Control Regime (MTCR): A list of controlled equipment, software, and technology related to missile technology.

National Industrial Security Program Operating Manual (NISPOM): A U.S. Government manual (DOD 5220.22-M) that provides baseline standards for the protection of classified information released or disclosed to industry in connection with classified contracts under the National Industrial Security Program (NISP).

Office for Foreign Assets Control (OFAC): The U.S. Department of the Treasury agency that enforces economic sanctions against certain countries and organizations linked with terrorism, narcotics trafficking, the proliferation of weapons of mass destruction, and other national security, foreign policy, or economic threats.

Public Domain: Information that is published and that is generally accessible or available to the public: (1) through sales at newsstands and bookstores; (2) through subscriptions that are available without restriction to any individual who desires to obtain or purchase the published information; (3) through second-class mailing privileges granted by the U.S. government; (4) at libraries open to the public or from which the public can obtain documents; (5) through patents available at any patent office; (6) through unlimited distribution at a conference, meeting, seminar, trade show, or exhibition, generally accessible to the public, in the U.S.; (7) through public release in any form after approval by the cognizant U.S. Government department or agency; and (8) through fundamental research in science and engineering at accredited institutions of higher learning in the U.S. where the resulting information is ordinarily published and shared broadly in the scientific community.

Public Release: The dissemination of information to an unrestricted audience, which is an audience that may be composed of foreign persons as well as U.S. persons.

"Red Flags": Warning signals or signs of potential compliance problems with an export transaction. "Red flags" are abnormal circumstances associated with a transaction that may indicate that the export is destined for an inappropriate end-use, end-user, or destination and that the exporter will therefore violate U.S. export laws and regulations by proceeding with the transaction.

Re-export or re-transfer: The transfer of any controlled export to an end-user, end-use, or destination that has not been previously authorized.

Significant Military Equipment (SME): Defense articles to which special export controls apply because of their capacity for substantial military utility or capability.

Special Export Controls (SECs): The ITAR requirement that the U.S. government monitor and review technical exchanges between U.S. and foreign personnel in association with the launch of a U.S. satellite in, or by nationals of, a country that is not a member of the North Atlantic Treaty Organization (NATO) or a major non-NATO ally of the United States.

Technical Assistance Agreement (TAA): An agreement between an U.S. person and a foreign person or persons (government, industry, or individual) involving the performance of defense services for the foreign person(s) by the U.S. person, or the disclosure of controlled technical data to the foreign person(s).

Technical Data: As defined by the ITAR, "technical data" includes the following:

- Information that is required for the design, development, production, manufacture, assembly, operation, repair, testing, maintenance, or modification of defense articles. This includes information in the form of blueprints, drawings, diagrams, photographs, plans, instructions, formulas, tables, manuals, or other documentation, whether written, printed, or recorded on other media or devices such as disk, tape, or read-only memories.
- Software that is directly related to defense articles.
- Classified information related to defense articles and defense services.
- Information covered by an invention secrecy order.
- Technical data does not include information concerning general scientific, mathematical, or engineering principles commonly taught in schools or universities, or information in the public domain, or basic marketing information on function or purpose or general system descriptions of defense articles.

United States Munitions List (USML): The part of the ITAR that specifies the controlled categories of defense articles, defense services, and related technical data.

U.S. Person: As defined by the ITAR, a U.S. person is:

- A natural person (i.e., an individual), wherever located, who is a lawful permanent resident as defined in 8 U.S.C. 1101(a)

(20)—meaning someone who is either a U.S. citizen, a foreign national with the visa status of legal permanent resident of the United States (a "green card holder"), or a "protected individual" as defined by 8 U.S.C. 1324b(a)(3); or
- A juridical person (i.e., a corporation, association, partnership, organization, trust, or other such entity) that is organized and entitled to do business under the laws of the United States or any U.S. jurisdiction.

Warehousing and Distribution Agreement (WDA): An agreement between a U.S. person and a foreign person for the warehousing and distribution of defense articles. This agreement must contain very specific conditions for distribution, including sales territory, end-use, and reporting.

CHAPTER 10

The U.S. Government as Your Customer

The U.S. government spends more than $40 billion each year on space-related activities and is the largest customer of many space companies. Selling products and services to the government is very different from selling products and services in the commercial marketplace (see table 10.1). Attorneys familiar only with commercial transactions face a steep learning curve when they first begin working with government contracts. This chapter will help attorneys new to government contracting become familiar with many of the unique aspects of this practice. Section I provides an overview of the different federal agencies that procure space-related products and services. Section II reviews the key federal contracting regulations. Section III discusses the most common types of government contracts. Section IV reviews the process for obtaining government contracts. Section V discusses the administrative issues that arise after a company has been awarded a government contract. Section VI outlines ethical and legal landmines that government contractors must avoid. Finally, Section VII describes the rights the government obtains to intellectual property developed under government contracts.

I. U.S. GOVERNMENT SPACE ACTIVITIES

In 2011, the U.S. government spent roughly $43 billion on space activities, representing about 61 percent of worldwide government space budgets.[1] To most people, the National Aeronautics and Space Administration (NASA) is synonymous with the U.S. space program.

---

1. EUROCONSULT, PROFILES OF GOVERNMENT SPACE PROGRAMS: ANALYSIS OF 60 COUNTRIES & AGENCIES (2012); Press Release, *Leading Government Space Programs Under Strong Budget Pressure*. EUROCONSULT, March 27, 2012, *available at* http://www.euroconsult-ec.com/news/press-release-33-1/55.html.

TABLE 10.1
Major Differences between Commercial and Government Contracting

Commercial Contracting	Government Contracting
Companies determine contracting practices, within certain legal boundaries	Contracting practices are proscribed by laws, regulations, directives, and policies
Profit is built into the total price	Government may negotiate a separate profit/fee, and that profit may be capped
Contract clauses are negotiated	Contract clauses are defined by law and generally not negotiable
Companies respond to social pressures, but social policies are usually not defined by law	Federal contracts must incorporate social policies enacted by Congress, such as policies favoring small businesses
Contracts often result from a fluid, organically developing business relationship	Government contracting process is formalistic and defined by laws and regulations

Nevertheless, as shown in table 10.2, many U.S. government agencies in addition to NASA are engaged in space-related activities. This section provides an overview of the space activities conducted across the federal government.

Three agencies design and operate space systems: the Department of Defense (DOD), NASA, and the National Oceanic and Atmospheric

TABLE 10.2
Estimated FY 2010 Federal Appropriations for Space Activities*

Agency	FY 2010 ($ millions)
Defense	26,463
NASA	18,179
Commerce	1,350
NSF	481
Energy	234
Interior	88
Agriculture	57
Smithsonian	24
FAA	15
Total	46,891

*Amounts estimated based on an analysis of various federal budget documents. For authoritative budget figures for FY 2008, see Aeronautics and Space Report of the President: FY 2008 Activities, *available at* http://history.nasa.gov/presrep2008.pdf

Administration (NOAA). DOD has the largest space budget of any government agency. DOD develops, builds, launches, and operates spacecraft for a wide variety of military purposes, including communications, reconnaissance, early warning, and meteorology. The National Reconnaissance Office, which is part of DOD but works closely with the intelligence community, operates the nation's spy satellites and coordinates the collection and analysis of satellite imagery.

The National Aeronautics and Space Act of 1958 created NASA to control the civil space activities sponsored by the United States.[2] NASA conducts its space activities through two mission directorates: (1) Human Exploration and Operations, which is responsible for the International Space Station and human exploration beyond low Earth orbit, and (2) Science, which conducts Earth and space science research. NASA also supports the civil space activities of other government agencies and coordinates with DOD on space activities of common interest.

The Department of Commerce (DOC), through NOAA's National Environmental Satellite, Data, and Information Service, operates the U.S. civil environmental and weather satellite programs. The DOC Office of Space Commercialization coordinates space commerce policy and assists commercial space companies in their efforts to do business with the U.S. government.

Three agencies have responsibility for regulating aspects of nongovernmental space activities. The Federal Aviation Administration licenses nongovernmental spaceports and the launch of reentry of nongovernmental spacecraft. NOAA is responsible for licensing commercial Earth remote sensing satellites. The Federal Communications Commission licenses all nongovernment satellites and coordinates usage of the radio frequency spectrum by satellites and ground stations.

Several other agencies fund space-related activities, but generally do not have a direct role in procuring, operating, or regulating space systems. The Department of the Interior, through the U.S. Geological Survey, works with NASA to operate the Landsat remote sensing satellites and uses remote sensing data for land management purposes. The

---

2. 51 U.S.C. § 20102(b) (declaring that NASA shall exercise "control over aeronautical and space activities sponsored by the United States, except that activities peculiar to or primarily associated with the development of weapons systems, military operations, or the defense of the United States . . . shall be the responsibility of, and shall be directed by, the Department of Defense").

Department of Agriculture uses remote sensing data to solve problems related to food production. The National Science Foundation is the lead federal agency supporting ground-based astronomy. The Department of Energy works with NASA on research on space physics and space nuclear propulsion and manages the infrastructure that supplies nuclear materials for radioisotope power systems on spacecraft. The Smithsonian Astrophysical Observatory conducts research in astronomy, astrophysics, and space science and operates several ground- and space-based observatories. Finally, the State Department coordinates international diplomacy in support of U.S. space activities.

Many of these government activities, particularly those of DOD and NASA, are conducted by private companies working under government contracts. In 2010, for example, private contractors received about $16 billion of the $18.1 billion appropriated to NASA for space programs.[3] Because so much government space activity is performed by private contractors, both government and private sector attorneys in the space industry must be familiar with the fundamental principles of government contracting.

II. KEY GOVERNMENT CONTRACTING LAWS AND REGULATIONS

The principal body of law governing government contracts is the Federal Acquisition Regulations (FAR).[4] The FAR was promulgated pursuant to the Office of Federal Procurement Policy Act of 1974[5] and was "established for the codification and publication of uniform policies and procedures for acquisition by all executive agencies."[6] Prior to the adoption of the FAR, each government agency had its own acquisition regulations, often dating back to World War II. The FAR is maintained by the General Services Administration (GSA), DOD, and NASA.

The FAR covers every aspect of federal government contracting, from acquisition planning through contract termination. The Decem-

---

3. USASpending.gov.
4. 48 C.F.R. §§ 1.000 et seq.
5. 41 U.S.C. §§ 401 et seq.
6. FAR 1.101.

ber 2011 version of the FAR's table of contents provides a good overview of the entire contracting process:

- Subchapter A: General
  - Part 1. Federal Acquisition Regulations System
  - Part 2. Definitions of words and terms
  - Part 3. Improper business practices and personal conflicts of interest
  - Part 4. Administrative matters
- Subchapter B: Acquisition Planning
  - Part 5. Publicizing contract actions
  - Part 6. Competition requirements
  - Part 7. Acquisition planning
  - Part 8. Required sources of supplies and services
  - Part 9. Contractor qualifications
  - Part 10. Market research
  - Part 11. Describing agency needs
  - Part 12. Acquisition of commercial items
- Subchapter C: Contracting Methods and Contract Types
  - Part 13. Simplified acquisition procedures
  - Part 14. Sealed bidding
  - Part 15. Contracting by negotiation
  - Part 16. Types of contracts
  - Part 17. Special contracting methods
  - Part 18. Emergency acquisitions
- Subchapter D: Socioeconomic Programs
  - Part 19. Small business programs
  - Parts 20–21. *Reserved*
  - Part 22. Application of labor laws to government acquisitions
  - Part 23. Environment, energy and water efficiency, renewable energy technologies, occupational safety, and drug-free workplace
  - Part 24. Protection of privacy and freedom of information
  - Part 25. Foreign acquisition
  - Part 26. Other socioeconomic programs
- Subchapter E: General Contracting Requirements
  - Part 27. Patents, data, and copyrights
  - Part 28. Bonds and insurance
  - Part 29. Taxes
  - Part 30. Cost accounting standards administration

- Part 31. Contract cost principles and procedures
- Part 32. Contract financing
- Part 33. Protests, disputes, and appeals
- Subchapter F: Special Categories of Contracting
  - Part 34. Major system acquisition
  - Part 35. Research and development contracting
  - Part 36. Construction and architect-engineer contracts
  - Part 37. Service contracting
  - Part 38. Federal supply schedule contracting
  - Part 39. Acquisition of information technology
  - Part 40. *Reserved*
  - Part 41. Acquisition of utility services
- Subchapter G: Contract Management
  - Part 42. Contract administration and audit services
  - Part 43. Contract modifications
  - Part 44. Subcontracting policies and procedures
  - Part 45. Government property
  - Part 46. Quality assurance
  - Part 47. Transportation
  - Part 48. Value engineering
  - Part 49. Termination of contracts
  - Part 50. Extraordinary contractual actions and the safety act
  - Part 51. Use of government sources by contractors
- Subchapter H: Clauses and Forms
  - Part 52. Solicitation provisions and contract clauses
  - Part 53. Forms
  - Parts 54–99. *Reserved*

Perhaps the most commonly referenced section of the FAR is Part 52, which contains the required contract clauses and solicitation provisions. The matrix found at FAR 52.301 outlines the FAR clauses that are required for each type of contract. If a required FAR clause is omitted from a contract without proper authorization, the courts will interpret the contract as though the clause were still included. According to the *Christian* doctrine, which is derived from the court opinion in *G. L. Christian & Associates v. United States*,[7] because government regulations have the force of law and government personnel may not devi-

---

7. 312 F.2d 418 (Ct. Cl.), *cert. denied*, 375 U.S. 954 (1963).

ate from the law without authorization, government contractors are presumed to be familiar with the FAR, and FAR clauses are treated as included in every contract, even if inadvertently omitted.

In addition to the standard FAR clauses, each government agency is permitted to issue its own supplemental contracting regulations. These supplements generally parallel the FAR in numbering, scope, and format, but they may not conflict with or supersede relevant FAR provisions. For space companies, the most relevant supplements are the Defense Federal Acquisition Regulation Supplement (DFARS)[8] and the NASA FAR Supplement (NFS).[9]

Numerous other statutes and regulations apply to government contracting. Many of these legal requirements are discussed in this chapter. The statutes that are considered a part of the statutory foundation of government contracts law are the Armed Services Procurement Act of 1947,[10] which governs the acquisition of all property, construction, and services by defense agencies; the Federal Property and Administrative Services Act of 1949,[11] which governs the acquisition of all property, construction, and services by civilian agencies; and the Competition in Contracting Act of 1984 (CICA),[12] which requires both defense and civilian agencies to seek and obtain "full and open competition" wherever possible in the contract award process, except in limited circumstances.

III. TYPES OF GOVERNMENT CONTRACTS

A. FAR-Based Contracts

The FAR provides for several types of contract vehicles. The two most common basic contract types are fixed-price contracts and cost-reimbursement contracts. Fixed-price contracts (FAR 16.2) provide for a firm price or a price that may be adjusted, but only in limited, stated circumstances. These contracts are used for acquiring commercial items or other supplies or services on the basis of reasonably definite specifications. In a fixed-price contract, the contractor bears the financial risk of being unable to perform the contract at the contract price.

---

8. FAR 200–299.
9. FAR 1800–1899.
10. 10 U.S.C. §§ 2301 *et seq.*
11. 40 U.S.C. §§ 471 *et seq.* & 41 U.S.C. §§ 251 *et seq.*
12. 10 U.S.C. §§ 2301 *et seq.* & 41 U.S.C. § 403.

Cost-reimbursement contracts (FAR 16.3) provide for payment of allowable incurred costs, and often a negotiated profit, or fee. These contracts establish an estimate of total cost for the purpose of obligating funds and establishing a ceiling that the contractor may not exceed without government approval. Cost-reimbursement contracts are used when the agency is unable to define its requirements sufficiently to allow for a fixed-price contract, such as in an early-stage research and development effort. In a cost-reimbursement contract, the government bears the financial risk of unanticipated cost overruns.

Other commonly used contract types include incentive contracts, indefinite-delivery contracts, time-and-materials contracts, and letter contracts. Incentive contracts (FAR 16.4) are used in connection with both fixed-price and cost-reimbursement contracts and reward a contractor for efficiently managing costs or achieving improved delivery or technical performance. Indefinite-delivery contracts (FAR 16.5) are used to acquire goods or services when the exact times or quantities of future deliveries are not known at the time of contract award. There are three types of indefinite-delivery contracts: definite-quantity contracts, requirements contracts, and indefinite-quantity contracts. Time-and-materials (T&M) contracts (FAR 16.6) are used when it is not possible at the time of awarding the contract to estimate accurately the extent or duration of the work or to anticipate costs with any reasonable degree of confidence. A T&M contract provides for acquiring supplies or services on the basis of a fixed hourly rate for labor, which includes indirect expenses and profit, and the actual cost of materials. A letter contract (FAR 16.6) is a written preliminary contractual instrument that authorizes the contractor to begin immediately manufacturing supplies or performing services. Letter contracts are used when performance must start immediately and negotiating a definitive contract is not possible in sufficient time to meet the government's requirements.

B. CRADAs and Space Act Agreements

Not all government contracts are governed by the FAR. Two of the most common non-FAR contracts used in the space industry are cooperative research and development agreements and Space Act Agreements. A cooperative research and development agreement (CRADA) is an agreement between a private company and a government laboratory to work together on a mutually beneficial research and development project.[13] Private sector funds may be used to fund portions of the government's effort on the project, but the government may not use federal funds to support the private sector participant. Under a CRADA, research results may be kept confidential for up to five years, and the private company may retain ownership of and apply for patents on inventions developed under the CRADA, with the government receiving a license to the patents.

A Space Act Agreement (SAA) is a type of legal agreement specified in the National Aeronautics and Space Act, which permits NASA to enter into "other transactions as may be necessary in the conduct of its work and on such terms as it may deem appropriate, with any agency or instrumentality of the United States, or with any state, territory, or possession, or with any political subdivision thereof, or with any person, firm, association, corporation, or educational institution."[14] Because SAAs are not governed by the FAR, they are more flexible than typical government procurement contracts. SAAs may be (1) reimbursable, where NASA's costs are reimbursed by the contractor; (2) nonreimbursable, where each party bears the cost of its own participation; or (3) funded, where NASA pays the contractor to perform its activities. Funded SAAs may only be used when NASA's objectives cannot be accomplished through the use of a procurement contract, grant, or cooperative agreement.[15]

IV. FEDERAL CONTRACTING PROCESS

One of the major differences between government and commercial contracting is that the government contracting process is highly regimented. This section will describe the principal actors involved in government contracting and the government contracting process.

A. Government Contracting Personnel and Agencies

Authority and responsibility to contract for supplies and services are vested in the head of the agency that is procuring such supplies and

---

13. 15 U.S.C. § 3710a.
14. 51 U.S.C. § 20113(e).
15. NASA Office of Gen. Counsel, NASA Policy Directive (NPD) 1050.1I, Authority to Enter into Space Act Agreements (Dec. 23, 2008); *see also* NASA Office of Gen. Counsel, NASA Advisory Implementing Instruction (NAII) 1050-1B, Space Act Agreement Guide (June 10, 2011).

services. The agency head may establish contracting activities and delegate broad authority to manage the agency's contracting functions.[16]

Contracts may be entered into and signed on behalf of the government only by contracting officers (COs) appointed pursuant to FAR 1.603. COs have authority to enter into, administer, and terminate contracts on behalf of the government and make related determinations and findings. COs may bind the government only to the extent of the authority delegated to them. This authority is stated in the CO's certificate of appointment, which is formally called a "warrant." For example, a CO's warrant may limit the CO's purchasing authority to supplies and services not exceeding $500,000. There is no concept of apparent authority in government contract law, so any action taken by a CO that exceeds the authority described in the CO's warrant is not binding on the government.

The CO may be assisted by several other individuals. The administrative contracting officer administers day-to-day contract activities following contract award as a delegate of the CO. For DOD contracts, this function is performed by the Defense Contract Management Agency. The contracting officer representative or contracting officer technical representative is usually a government employee with experience in the technical area relevant to the contract, who monitors the contractor's progress in fulfilling the technical requirements of the contract, approves invoices, and performs the final inspection and acceptance of the contract deliverables.[17] Unlike the CO, these other officers are not authorized to make any commitments on behalf of the government, grant the contractor permission to deviate from the contract requirements, or direct the contractor to perform any work outside the scope of the contract. The Defense Contract Audit Agency is responsible for performing all contract audits for the DOD, NASA, and several other agencies.

B. Solicitation Process

The contracting process begins when a government agency identifies a requirement and decides to proceed with an acquisition. With limited exceptions, CICA requires that COs provide for full and open competition in soliciting offers and awarding government contracts. To fulfill the full and open competition requirement, the government generally

---

16. FAR 1.601(a).
17. FAR 1.604.

uses one of two basic contracting methods: sealed bidding or competitive negotiation.

Sealed bidding is the more rigid of the two methods and is used when (1) time permits the solicitation, submission, and evaluation of sealed bids; (2) the award will be made on the basis of price and other price-related factors; (3) it is not necessary to conduct discussions with the responding offerors about their bids; and (4) there is a reasonable expectation of receiving more than one sealed bid.[18] This contracting method is most often used for purchasing commodities and nontechnical services. The process for sealed bidding is governed by FAR 14.

Competitive negotiation is used either when the source selection decision must take into account factors other than price, such as technical expertise and experience, or when discussions with the prospective contractors are required.[19] Because of the technical complexity of most aerospace projects, competitive negotiation is more common than sealed bidding in the space industry. The process for competitive negotiation is governed by FAR 15.

The competitive negotiation process begins with the CO issuing a request for proposal (RFP) that indicates the government's requirements, the anticipated contract terms, information that must be included in the prospective contractor's proposal, and the factors that the agency will consider in evaluating proposals. RFPs are advertised in a variety of public forums, and RFPs for procurements with an estimated value over $25,000 must be advertised on the Federal Business Opportunities website.[20]

Any interested party may respond to an RFP. Ultimately, the CO is required to select the proposal that represents the "best value" to the government based on a variety of factors that will differ depending on the nature of the contract.[21] To help make a decision, the CO may hold "discussions" with a subset of prospective contractors who fall within a certain competitive range. Once the contract is awarded, the award decision must contain an analysis of why the awardee's proposal represents the best value to the government. Unsuccessful offerors may request a debriefing, where the CO will explain the bases for the selection decision.

---

18. FAR 6.401(a).
19. FAR 6.401(a).
20. http://www.FedBizOpps.gov.
21. FAR 15.1.

Despite CICA's broad applicability, there are numerous exceptions to its competition requirements. In particular, full and open competition is not required under the following circumstances:

- the supplies or services required by the agency are available from only one responsible source, or, for DOD, NASA, and the Coast Guard, from only one or a limited number of responsible sources
- the agency's need for the supplies or services is of such an unusual and compelling urgency that the government would be seriously injured unless the agency is permitted to limit the number of sources from which it solicits bids or proposals
- to maintain a facility, producer, manufacturer, or other supplier available for furnishing supplies or services in case of a national emergency or to achieve industrial mobilization
- to establish or maintain an essential engineering, research, or development capability to be provided by an educational or other nonprofit institution or a federally funded research and development center
- to acquire the services of an expert or neutral person for any current or anticipated litigation or dispute
- when authorized or required by statute or an international agreement
- when disclosure of the agency's needs would compromise national security
- when the agency head determines that it is not in the public interest in the particular acquisition concerned.[22]

Full and open competition is also not required for contract modifications within the scope and terms of an existing contract, for orders placed under previously awarded indefinite-delivery contracts, and for procurements under set-aside programs, such as those for small businesses, HUBZones, and businesses owned by women and disabled veterans.[23] Agencies also have broad discretion to use simplified acquisition procedures for goods and services not exceeding the simplified acquisition threshold ($100,000) or the micro-purchase thresh-

---

22. FAR 6.302.
23. FAR 6.2. The HUBZone (Historically Underutilized Business Zone) program is administered by the U.S. Small Business Administration; see the program's home page at http://www.sba.gov/hubzone.

old ($3,000).[24] Finally, agencies may acquire commercial goods and services through the Federal Supply Schedule (FSS) program, which is maintained by the GSA.[25] The FSS program offers an online shopping service through which agencies may place orders from contractors on the GSA Schedule using prenegotiated pricing and terms and conditions.

C. Protests

An interested party that believes the government has made an improper contracting decision may challenge, or "protest," the decision with the contracting agency, the Government Accountability Office, the U.S. Court of Federal Claims, or the local federal district court. An "interested party" is an "actual or prospective offeror whose direct economic interest would be affected by the award of a contract or by the failure to award a contract."[26] The procedures for protests are described in FAR 33.

V. CONTRACT ADMINISTRATION

The award of a contract is only the first step in the government contracting process. After contract award, the contract must be administered. Contract administration is a complex, rule-driven exercise. This section discusses the following aspects of contract administration: cost accounting principles, contract changes, contract termination, contract disputes, subcontracting, liability issues, socioeconomic obligations, and the protection of classified information.

A. Accounting

In order to receive payment under a government contract for non-commercial items, the contractor must comply with the Allowable Cost and Payment Clause (FAR 52.166-7) if the payment is based on costs incurred. If the payment is based on milestones, the contract provisions on payment will dictate the billing terms. The requirements for cost-based payments require the contractor to capture the cost of each contract separately and have systems in place that can be audited by the government. Such contractual obligations may

---

24. FAR 13.
25. FAR 8.4.
26. FAR 33.101.

require the implementation of specialized accounting systems for work performed under government contracts that are separate from a company's standard accounting system for private sector transactions. Except in limited circumstances, the government may audit the contractor's compliance with these requirements for up to three years after the date of the final contract payment.[27]

The FAR cost principles are defined in FAR 31. Generally, for contracts where the government agrees to reimburse a contractor's costs, the contractor may recover direct and indirect costs that are (1) allowable, (2) allocable, and (3) reasonable. FAR 31.205 describes more than 50 types of costs that may or may not be allowable. Recovery of costs associated with many common business practices, such as advertising and marketing, entertainment, lobbying, and executive compensation above a certain threshold, is either restricted or completely prohibited under these cost principles.

Most negotiated contracts for more than $700,000 in goods and services are also subject to the Cost Accounting Standards (CAS), which instruct the contractor how to maintain its accounting system and account for certain types of costs.[28] Sealed-bid contracts, negotiated contracts not in excess of $700,000, contracts with small businesses, firm fixed-price contracts awarded without the submission of certified cost or pricing data, and contracts with foreign entities are not subject to CAS.[29]

Contractors that receive a single CAS-covered contract award of $50 million or more or received $50 million or more in net CAS-covered awards during its preceding cost accounting period are subject to full CAS coverage and must comply with all of the CAS specified in 48 C.F.R. part 9904. Contractors who meet the $50 million threshold may also be required to submit a description of their cost accounting practices in a disclosure statement. Contractors that do not meet the thresholds for full CAS coverage are subject to modified CAS coverage, which only requires compliance with a subset of CAS requirements, and do not have to submit a disclosure statement.

---

27. FAR 52.215-2.
28. FAR 52.230-2.
29. 48 C.F.R. § 9903.201-1.

B. Changes

The government may, at any time and without notice, make changes within the general scope of the contract to any one of the following: (1) drawings, designs, or specifications when the supplies to be furnished are to be specially manufactured for the government in accordance with the drawings, designs, or specifications; (2) method of shipment or packing; and (3) place of delivery.[30] If the change would result in an increase in cost for or time of performance, the contractor may request an equitable adjustment to the contract price and/or delivery schedule, which must be filed within 30 days of receiving the change notice. Changes outside the general scope of the contract will usually require the initiation of a new contract solicitation subject to full and open competition, unless one of the CICA exceptions discussed above apply to the circumstances.

C. Termination

The government may terminate a contract, in whole or in part, for its convenience or for default. The CO may terminate a contract for convenience if he determines that it is in the government's interest to do so.[31] The contractor and the government then negotiate a termination settlement, which will include reimbursement of the contractor's allowable costs through the effective date of termination, plus a reasonable profit.

The CO may terminate a contract for default if the contractor fails to perform its obligations under the contract and does not cure the failure within 10 days of receiving a notice of default from the government.[32] In a termination for default, the contractor is only entitled to recover costs for goods or services actually accepted by the government. If the government must repurchase the contracted goods or services from a third party, then the contractor may be responsible for reimbursing the government for any excess repurchasing costs.

---

30. FAR 52.243-1, -2.
31. *See, e.g.*, FAR 52.249-6(a)(1) (termination for convenience of cost-reimbursement contracts).
32. *See, e.g.*, FAR 52.249-6(a)(2) (termination for default of cost-reimbursement contracts).

D. Disputes

All disputes arising under a government contract are resolved pursuant to the Contract Disputes Act of 1978,[33] which is implemented in FAR 52.233-1. A contractor initiates the dispute process by presenting a "claim" to the CO. If the contractor and government are unable to negotiate a resolution to the dispute, the CO must issue a "final decision." If the contractor disagrees with the CO's final decision or the CO does not issue the final decision in the required time frame, the contractor may commence litigation against the government in an agency Board of Contract Appeals or the U.S. Court of Federal Claims. An adverse decision in either forum may be appealed to the U.S. Court of Appeals for the Federal Circuit.

E. Subcontracting

On large government programs such as those common in the space industry, there are often multiple tiers of private companies performing work on any particular government contract. The company with the direct contractual relationship with the government is the prime contractor. Companies that receive subcontracts from the prime contractor are subcontractors. Subcontractors do not have privity of contract with the government, so the contract between the prime contractor and each subcontractor is fundamentally a commercial contract between two private entities, although certain important differences exist between purely commercial contracts and government subcontracts. Most notably, government subcontractors are subject to many of the same FAR obligations as the prime contractor, which are required to be incorporated into, or flowed down to, each subcontract.

Prime contractors are ultimately responsible for ensuring that the goods or services the government contracted for are delivered on time and in conformance with the contract specifications. This means that the prime contractor is liable for any actions or omissions of the subcontractor that cause the prime contractor to fail to meet its prime contract obligations. For example, if the prime contract requires the prime contractor to deliver certain reports to the government, but the prime contractor is unable to deliver the reports because a subcontractor failed to provide required information, then, from the govern-

---

33. 41 U.S.C. §§ 601 *et seq.*

ment's perspective, it is the prime contractor, not the subcontractor, who is in default.

Because of this, prime contractors usually flow down most, if not all, of the prime contract clauses to which they are subject, meaning that each subcontractor has the same substantive contract obligations as the prime contractor. In fact, the FAR requires that certain clauses, such as those relating to accounting, socioeconomic, ethical, and intellectual property requirements, be flowed down to each subcontractor that meets the applicable contract value threshold.[34] This way, if a subcontractor breaches a FAR requirement and the government penalizes the prime contractor for that breach, the prime contractor can in turn seek recourse against the subcontractor for breaching its subcontract obligations.

F. Liability

The allocation of liability risk is a crucial function of any contract. This is particularly true for inherently risky activities like spaceflight. The risk analysis is somewhat different in government contracts than in commercial contracts because liability risk is often dictated by public policy and established in the relevant FAR clauses, rather than subject to negotiation between the parties. In some cases, the government may assume liability for loss of or damage to government property after the government accepts the contracted-for goods or services.[35] In other cases, the contractor may be asked to provide a warranty for the goods or services it is providing to the government.[36]

In certain circumstances, the government may relieve the contractor of liability to third parties that arises out of the performance of the contract. For example, FAR 52.227-1 (Authorization and Consent) provides that the government assumes responsibility for third-party patent infringement liability for the use and manufacture of an invention covered by a U.S. patent in performing a contract when the

---

34. *See, e.g.*, FAR 52.203-7 (Anti-Kickback Procedures); FAR 52.222-36 (Affirmative Action for Workers with Disabilities); FAR 52.222.50 (Combating Trafficking in Persons); DFARS 252.204-7008 (Export-Controlled Items).

35. *See, e.g.*, FAR 52.546-23 (Limitation of Liability); FAR 52.546-24 (Limitation of Liability—High-Value Items); FAR 52.246-25 (Limitation of Liability—Services).

36. *See, e.g.*, FAR 52.246-17 (Warranty of Supplies of a Noncomplex Nature); FAR 52.246-18 (Warranty of Supplies of a Complex Nature); FAR 52.246-19 (Warranty of Systems and Equipment under Performance Specifications or Design Criteria); FAR 52.246-20 (Warranty of Services); FAR 52.246-21 (Warranty of Construction).

infringing item was (1) accepted by the government under the contract or (2) used in accordance with contract specifications.

Another example of the government protecting contractors can be found in FAR 52.228.7 (Insurance—Liability to Third Persons), which provides that the government will reimburse the contractor for third-party personal injury and property damage liability to the extent not covered by the contractor's insurance, so long as (1) appropriated funds are available, (2) the contractor obtained any insurance coverage required by the contract, and (3) the liability did not result from the contractor's willful misconduct or lack of good faith. FAR 52.228.7 is only available for cost-reimbursement contracts and, unlike many FAR clauses, this clause is not a required flow-down for subcontracts. Indeed, since a prime contractor is not authorized to impose a binding indemnification obligation on the government, the prime contractor cannot flow this clause down to a subcontractor without the CO's express consent. If the clause is flowed down to a subcontractor without permission, the practical consequence in the event of liability is that the prime contractor, not the government, will be responsible for reimbursing the subcontractor for liability that is reimbursable under this clause.

FAR 52.250-1 (Indemnification under Public Law 85-804) provides the broadest third-party liability protection available from the government. Under this provision, the government agrees to indemnify the contractor against (1) third-party claims for death, personal injury, and property damage; (2) loss of or damage to the contractor's property; and (3) loss of or damage to the government's property. This clause is available only when the "approving official determines that the contractor shall be indemnified against unusually hazardous or nuclear risks."[37]

In addition to statutory and contractual protections, for more than two decades government contractors have relied on the "government contractor defense" to protect them from third-party liability resulting from defective products and services delivered to the government. The government contractor defense was established by the U.S. Supreme Court in *Boyle v. United Technologies Corp.*[38] It shields government contractors from liability in state and federal tort actions brought by those injured by defective equipment manufactured or supplied by

---

37. FAR 50.104-4.
38. 487 U.S. 500 (1988).

the contractor if the contractor can demonstrate that (1) the government approved reasonably precise specifications, (2) the equipment conformed to those specifications, and (3) the contractor warned the government about any dangers in the use of the equipment that were known to the contractor but not to the government. Although the Court in *Boyle* only addressed the defense in the context of a procurement contract, most, but not all, courts have extended the defense to service contracts as well.[39]

G. Socioeconomic Obligations

Most contractors and subcontractors are subject to numerous socioeconomic obligations. For example, a contractor or subcontractor with a contract over $10,000 must take affirmative action to ensure that applicants and employees are treated without regard to race, color, religion, sex, or national origin.[40] Government contractors must also comply with certain labor standard laws[41] and drug-free workplace requirements.[42] A contractor that fails to comply with these socioeconomic obligations and their associated reporting requirements may be disqualified from obtaining future government contracts.

H. Protecting Classified Information

Many contractors in the space industry must access classified information in order to perform their contracts. Executive Order 12,829 established the National Industrial Security Program (NISP) to safeguard classified information that is released to government contractors,

---

39. *See* Hudgens v. Bell Helicopters/Textron, 328 F.3d 1329 (11th Cir. 2003) (applying the government contractor defense to prohibit state tort claims against a service contractor); *but see* Filippi v. Sullivan, 866 A.2d 599 (Conn. 2005) (holding that the government contractor defense applies only to military procurement contracts).

40. FAR 52.222-26.

41. *See, e.g.*, Walsh-Healey Act, 41 U.S.C. §§ 35–45 (governing supply contracts, implemented in FAR 52.222-20); Davis-Bacon Act, 40 U.S.C. 276a (governing construction contracts, implemented in FAR 52.222-6); Service Contract Act of 1965, 41 U.S.C. §§ 351 *et seq.* (governing service contracts, implemented in FAR 52.222-41); Contract Work Hours and Safety Standards Act, 40 U.S.C. §§ 327–333 (requiring payment at 1.5 times the basic rate of pay for all hours worked by laborers or mechanics in excess of 40 hours per week, implemented in FAR 52.222-4).

42. Drug-Free Workplace Act of 1988, Pub. L. No. 200-690 (implemented in FAR 52.223-6).

licensees, and grantees.[43] The NISP requirements are implemented through the National Industrial Security Program Operating Manual (NISPOM).[44] The NISPOM is issued and maintained by the secretary of defense, with the concurrence of the secretary of energy, the Nuclear Regulatory Commission, and the director of Central Intelligence. A contractor that must access classified information must agree to the Department of Defense Security Agreement,[45] which incorporates the NISPOM requirements into its contract.[46] The Defense Security Service administers the NISP on behalf of the DOD and 22 non-DOD federal agencies, and is the primary security interface for cleared defense contractors.

VI. ETHICAL OBLIGATIONS

The ethical practices of government contractors are heavily regulated. For example, contractors and COs are required to identify, avoid, and mitigate organizational and personal conflicts of interest in government contracting. An organizational conflict of interest (OCI) exists when a contractor has past, present, or currently planned interests that relate to the work to be performed under the government contract and such interests may diminish the contractor's capacity to give impartial, technically sound, objective performance, or result in the contractor having an unfair competitive advantage. OCIs may arise in situations where (1) a company has set the ground rules for another government contract by, for example, writing the statement of work or the specifications; (2) a company's work under one government contract involves evaluating itself, either through an assessment of performance under another contract or the evaluation of proposals; or (3) a company has access to nonpublic information as part of its performance of a government contract where that information may provide a competitive advantage in a later competition for a government contract.[47] A personal conflict of interest exists when a covered employee has a financial interest, personal activity, or relationship that could impair

---

43. Exec. Order No. 12,829, Jan. 6, 1993, 58 Fed. Reg. 3479, as amended by Exec. Order No. 12,885, Dec. 14, 1993, 58 Fed. Reg. 65,863.
44. DoD 5220.22-M.
45. From DD 441.
46. FAR 52.204-2.
47. FAR 9.5.

VI. Ethical Obligations

the employee's ability to act impartially and in the best interest of the government when performing under the contract. "Covered employees" are contractor employees who are performing acquisition functions closely associated with inherently governmental functions for or on behalf of federal agencies.[48]

Other significant ethical restrictions and obligations include:

- With limited exceptions, contractors are prohibited from giving gratuities to government employees for the purpose of obtaining a contract or favorable treatment under a contract.[49]
- The False Claims Act imposes sanctions on the submission to the government of fraudulent claims for payment.[50] Noncompliance with the FAR cost principles, the CAS, and the Truth in Negotiations Act can also result in false claim liability.
- Contractors are prohibited from making false, fictitious, or fraudulent statements or representations to the government.[51]
- The Truth in Negotiations Act requires a government contractor or subcontractor to submit certified "cost or pricing data" for any negotiated contract, subcontract, or modification that is expected to exceed $700,000 (subject to adjustment for inflation).[52] The contract price may be reduced if it is found after award that the contractor or subcontractor submitted cost and pricing data that were not accurate, current, and complete, as certified. The government will also look at failure to supply the required information as potential fraud.
- The Anti-Kickback Act of 1986 prohibits making or accepting payments for the purpose of improperly obtaining favorable treatment in connection with a prime contract or a subcontract.[53]
- The Procurement Integrity Act prohibits prospective contractors from obtaining contractor "bid or proposal information" or "agency source selection information" prior to the award of a federal contract.[54]

---

48. FAR 3.11.
49. FAR 52.203-3.
50. 31 U.S.C. §§ 3729–3733.
51. 18 U.S.C. § 1001(a).
52. Codified at 10 U.S.C. § 2306A, 41 U.S.C. § 254B, implemented in FAR 15.403-4.
53. 41 U.S.C. §§ 51–58 (implemented in FAR 3.502).
54. 41 U.S.C. § 423 (implemented in FAR 3.104).

- Contractors are prohibited from discussing employment with certain federal employees and former federal employees are restricted from performing certain services once they leave the government.[55]
- Contractors are prohibited from using appropriated government funds for lobbying purposes.[56]

Many of these regulations impose affirmative disclosure obligations on contractors in the event a contractor discovers a violation. In addition, contractors and subcontractors with contracts worth at least $5 million and with a performance period of at least 120 days must have a written code of business ethics and compliance, a business ethics awareness and compliance program, and an internal ethics control system.[57] Failure to comply with these regulations may result in suspension or debarment from government contracting, contract termination, contract damages, or civil and criminal penalties.

VII. INTELLECTUAL PROPERTY RIGHTS

When contracting for products or services that involve innovative research and development and technically advanced systems, such as space systems, the government needs to obtain rights to inventions, technical data, and computer software developed in the performance of government contracts in order to (1) use and maintain systems it purchases, (2) promote full and open competition, and (3) insure that the results of federally funded research are commercialized and available to the public.[58] Yet intellectual property is one of a technology company's most valuable assets. Companies are often leery of contracting with the government because they fear that proprietary information provided to the government will be given to their competitors.[59] The FAR recognizes these concerns and that the protection of proprietary data is "necessary to encourage qualified contractors to participate in and apply innovative concepts to Government programs." Accordingly, the

---

55. 41 U.S.C. § 423(c) & (d).
56. 31 U.S.C. § 1352 (implemented in FAR 3.8).
57. FAR 52.203-13.
58. FAR 27.302(a) & 27.402(a).
59. GEN. ACCOUNTING OFFICE, GAO-02-723T, INDUSTRY AND AGENCY CONCERNS OVER INTELLECTUAL PROPERTY RIGHTS (May 10, 2002) [hereinafter GAO REPORT], *available at* http://www.gao.gov/new.items/d02723t.pdf.

VII. Intellectual Property Rights

FAR's intellectual property provisions attempt to "balance the Government's needs and the contractor's legitimate proprietary interests."[60] This section describes how this balance is struck.

A. Inventions

The FAR divides intellectual property into two broad categories: inventions and data. Inventions include any "invention or discovery that is or may be patentable or otherwise protectable under title 35 of the U.S. Code, or any variety of plant that is or may be protectable under the Plant Variety Protection Act."[61] Prior to 1980, the government generally retained title to any inventions created under federal research grants and contracts, with each agency establishing its own intellectual property policies. This policy led to concerns that the results of government-owned research were not being properly used because contractors had little incentive to advance government-owned technologies.[62] In 1980, Congress enacted the Bayh-Dole Act,[63] which created a uniform patent policy for federally funded inventions and permitted recipients of federal contracts and grants to retain ownership of their inventions. In 2002, *The Economist* magazine called the Bayh-Dole Act "possibly the most inspired piece of legislation to be enacted in America over the past half-century."[64]

The Bayh-Dole Act requirements are implemented in government contracts through FAR 52.227-11. Under this clause, contractors must disclose inventions made in the performance of the contract, known as "subject inventions," to the agency within two months after the inventor first discloses the invention in writing to the contractor personnel responsible for patent matters. The contractor may elect to retain title to the subject invention within two years of disclosure to the agency. If the contractor elects to retain title, the contractor must file a patent application on the invention within one year of the election.

---

60. FAR 27.402(b).
61. FAR 52.227-11(a).
62. GAO Report, *supra* note 59, at 5.
63. The Bayh-Dole Act is the common name for the Patent and Trademark Law Amendments of 1980, Pub. L. No. 96-517 (Dec. 12, 1980), codified at 35 U.S.C. §§ 200–212, and implemented by 37 C.F.R. § 401; Presidential Memorandum on Government Patent Policy to the Heads of Executive Departments and Agencies (Feb. 18, 1983); and Exec. Order No. 12,591, Facilitating Access to Science and Technology (Apr. 10, 1987).
64. Opinion, *Innovation's Golden Goose*, Economist, Dec. 12, 2002, *available at* http://www.economist.com/node/1476653.

The government may, upon request, require the contractor to assign the subject invention to the government if the contractor fails to follow the disclosure and election requirements required by the clause.

If the contractor retains title to the subject invention, the government receives (1) a "nonexclusive, nontransferable, irrevocable, paid-up license to practice, or have practiced for or on its behalf, the subject invention throughout the world;" and (2) "march-in rights," which is the right of the government to compel a contractor to license the invention under limited circumstances, such as the contractor's failure to use the subject invention within a reasonable period of time or for health and safety needs that are not being met by the contractor.[65] Although march-in rights are in theory very powerful, to date no government agency has exercised the march-in rights provision.[66]

An important caveat to the Bayh-Dole Act framework is that NASA treats inventions differently than other agencies. The National Aeronautics and Space Act provides that NASA will own any invention made in the performance of any NASA contract.[67] Contractors may request that NASA waive its rights to inventions under the NASA Patent Waiver Regulations, which require that the waiver be granted, except in limited circumstances.[68] Once waived, NASA inventions are generally subject to the standard Bayh-Dole Act requirements. If NASA's rights are not waived and NASA retains title to the invention, the contractor will be granted a nonexclusive, royalty-free license to practice the invention.

B. Data Rights

Unlike patent rights, there is no unifying policy governing data rights. Each agency is generally free to adopt its own policies, and the rights granted to the government may be negotiable. Nevertheless, most civilian agencies rely on FAR 52.227-14. DOD has its own data rights requirements, located at DFARS 252.227-7013 and -7014.

Both the FAR and the DFARS have separate rules for technical data and computer software. Technical data is "recorded information, regardless of the form or method of the recording, of a scientific or technical nature (including computer software documentation)."

---

65. FAR 52.227-11(h).
66. GAO REPORT, *supra* note 59, at 10.
67. 51 U.S.C. § 20135(b).
68. 14 C.F.R. ¶ 1245.

TABLE 10.3
FAR and DFARS Data Rights Provisions

FAR		
Type of Rights	**Applies to**	**Scope of Rights**
Unlimited Rights (FAR 27.404-1)	Default for technical data and computer software first produced in the performance of a contract	Government may use in any manner and for any purpose, and to have or permit others to do so
Limited Rights (FAR 27.404-2(c))	Technical data that embody trade secrets or are commercial or financial and confidential or privileged, pertaining to items, components, or processes developed at private expense, and that is delivered with the appropriate limited rights notice	Government may reproduce and use, but may not be used for manufacturing purposes or disclosed outside the government, except as agreed to by the contractor
Restricted Rights (FAR 27.404-2(d))	Noncommercial computer software developed at private expense that is a trade secret, is commercial or financial and confidential or privileged, or is copyrighted computer software, and that is delivered with the appropriate restricted rights notice	Government may • use on the computer for which it was acquired • use on a backup computer • archive and backup copies • modify, adapt, or combine with other software • disclose to service support contractors • use on replacement computer
Government Purpose Rights (FAR 27.408)	Co-sponsored R&D where the contractor is required to make substantial contributions of funds or resources and the contractor's and the government's respective contributions are not segregable—most common in CRADAs	Negotiated, but at a minimum the government must be able to use the data for agreed-to government purposes

DFARS		
Type of Rights	**Applies to**	**Scope of Rights**
Unlimited Rights (-7013(b)(1) & -7014(b)(1))	Technical data and computer software developed exclusively with government funds	Government may use in any manner and for any purpose, and to have or permit others to do so
Limited Rights (-7013(b)(3))	Technical data developed exclusively at private expense and marked with the limited rights legend	Government may use internally, but cannot disclose outside the government

	DFARS (continued)	
Type of Rights	Applies to	Scope of Rights
Restricted Rights (-7014(b)(3))	Noncommercial computer software developed exclusively at private expense	Restricts the government's use to a single computer per copy of the software, and prohibits all but backup or archival copies
Government Purpose Rights (-7013(b)(2) & -7014(b)(2))	Technical data and computer software developed with mixed funding	Government may use internally without restriction and disclose to third parties for government purposes; coverts to Unlimited Rights after five years unless otherwise negotiated

Technical data does not include "computer software" or information that is "incidental to contract administration," such as financial and management information.[69] Computer software includes computer programs, source code, source code listings, object code listings, design details, algorithms, processes, flow charts, formulae, and related material that would enable the software to be reproduced, recreated, or recompiled. Computer software does not include computer databases or computer software documentation.[70]

In general, the contractor retains ownership of all technical data and computer software, subject to the rights granted to the government and other restrictions on the disclosure and use of classified or export-controlled information. The scope of the government's rights depends on whether the technical data or computer software was developed with government or private funding, as described in table 10.3.

Commercial software is treated differently from noncommercial software. Commercial software is any software that is customarily used for nongovernmental purposes and that is sold, leased, or licensed to the general public.[71] The government may acquire commercial software under the license customarily provided to the public, so long as such license is "consistent with Federal law and otherwise satisfies the Government's needs."[72]

---

69. FAR 52.227-14(a); DFARS 252.227-7013(a)(14).
70. FAR 52.227-14(a); DFARS 252.227-7013(a)(3).
71. FAR 2.101.
72. FAR 27.405-3(a).

GLOSSARY

Administrative Contracting Officer (ACO): A government employee who administers day-to-day contract activities following contract award as a delegate of the Contracting Officer.

Anti-Kickback Act of 1986: Statute that prohibits making or accepting payments for the purpose of improperly obtaining favorable treatment in connection with a prime contract or a subcontract. 41 U.S.C. §§ 51-58.

Armed Services Procurement Act of 1947 (ASPA): The statute that governs the acquisition of all property, construction, and services by defense agencies. 10 U.S.C. § 2301 et seq.

Bayh-Dole Act: Statute that created a uniform patent policy for federally funded inventions and permitted recipients of federal contracts and grants to retain ownership of their inventions. 35 U.S.C. §§ 200-212.

Christian Doctrine: The doctrine that states that that because government regulations have the force of law and government personnel may not deviate from the law without authorization, government contractors are presumed to be familiar with the FAR, and FAR clauses are treated as included in every contract, even if inadvertently omitted.

Competition in Contracting Act of 1984 (CICA): Statute that requires both defense and civilian agencies to seek and obtain "full and open competition" wherever possible in the contract award process, except in limited circumstances. 10 U.S.C. § 2301 et seq. & 41 U.S.C. § 403.

Competitive negotiation: One of two basic government contracting methods that awards the contract to the proposer that represents the "best value" to the government based on a variety of factors that will differ depending on the nature of the contract.

Computer Software: As used in government contract data rights provisions, includes computer programs, source code, source code listings, object code listings, design details, algorithms, processes, flow charts, formulae and related material that would enable the software to be reproduced, recreated, or recompiled. The term does not include computer databases or computer software documentation.

Contract Disputes Act of 1978 (CDA): Statute governing the resolution of disputes arising under a government contract. 41 U.S.C. §§ 601, et seq.

Contracting Officer (CO): A government employee authorized to enter into, administer, and terminate contracts on behalf of the Government and make related determinations and findings.

Contracting Officer Representative (COR) or Contracting Officer Technical Representative (COTR): A government employee with experience in the technical area relevant to the contract, who monitors the contractor's progress in fulfilling the technical requirements of the contract, approves invoices and performs the final inspection and acceptance of the contract deliverables.

Cooperative Research and Development Agreement (CRADA): An agreement between a private company and a government laboratory to work together on a mutually beneficial research and development project.

Cost Accounting Standards (CAS): Rules that instruct the contractor how to maintain its accounting system and account for certain types of costs.

Cost-reimbursement contracts: Government contract type that provides for payment of allowable incurred costs, and often a negotiated profit, or fee.

Defense Contract Audit Agency (DCAA): Agency that is responsible for performing all contract audits for the DOD, NASA, and several other agencies.

Defense Contract Management Agency (DCMA): Agency that administers day-to-day contract activities on behalf of the DOD.

False Claims Act: Statute that imposes sanctions on the submission to the government of fraudulent claims for payment. 31 U.S.C. §§ 3729-3733.

FAR cost principles: Rules defined in FAR 31 that govern which costs contractors may recover from the Government under cost-reimbursement contracts.

Federal Acquisition Regulations (FAR): The principle body of law governing government contracts. 48 C.F.R. §§ 1.000 et. seq.

Federal Property and Administrative Services Act of 1949 (FPASA): The statute that governs the acquisition of all property, construction, and services by civilian agencies. 40 U.S.C. § 471, et seq. & 41 U.S.C. § 251 et seq.

Fixed-price contracts: Government contract type that provides for a firm price or a price that may be adjusted, but only in limited, stated circumstances.

Government contractor defense: Established by the U.S. Supreme Court in *Boyle v. United Technologies Corp.*, this litigation defense shields government contractors from liability in state and federal tort actions brought by those injured by defective equipment manufactured or supplied by the contractor under a government contract, if the contractor can demonstrate that: (1) the government approved reasonably precise specifications; (2) the equipment conformed to those specifications; and (3) the contractor warned the government about any dangers in the use of the equipment that were known to the contractor but not to the government.

Incentive contracts: Government contract type that is used in connection with both fixed-price and cost-reimbursement contracts and reward a contractor for efficiently managing costs or achieving improved delivery or technical performance.

Indefinite-delivery contracts: Government contract type that is used to acquire goods or services when the exact times or quantities of future deliveries are not known at the time of contract award. There are three types of indefinite-delivery contracts: definite-quantity contracts, requirements contracts, and indefinite-quantity contracts.

Invention: Any invention or discovery that is or may be patentable or otherwise protectable under title 35 of the U.S. Code, or any variety of plant that is or may be protectable under the Plant Variety Protection Act.

Letter contract: A written preliminary contractual instrument that authorizes the contractor to begin immediately manufacturing supplies or performing services.

March-in Rights: The right of the government under the Bayh-Dole Act to compel a contractor to license a government-funded invention under limited circumstances, such as the contractor's failure to use the invention within a reasonable period of time or for health and safety needs that are not being met by the contractor.

National Industrial Security Program (NISP): Program established by Executive Order 12,829 to safeguard classified information that is released to government, contractors, licensees, and grantees.

National Industrial Security Program Operating Manual (NISPOM): The manual that is issued and maintained by the Secretary of Defense, with the concurrence of the Secretary of Energy, the Nuclear Regulatory Commission, and the Director of Central Intelligence, to implement the NISP requirements.

Office of Federal Procurement Policy Act of 1974: The statute that, among other things, authorized the creation of the FAR. 41 U.S.C. §§ 401 et. seq.

Organizational Conflict of Interest (OCI): An OCI exists when a contractor has past, present, or currently planned interests that relate to the work to be performed under the government contract and such interests may diminish the contractor's capacity to give impartial, technically sound, objective performance, or result in the contractor having an unfair competitive advantage.

Personal Conflict of Interest (PCI): A PCI exists when an employee of a contractor who is performing certain acquisition functions for the government has a financial interest, personal activity, or relationship that could impair the employee's ability to act impartially and in the best interest of the government when performing under the contract.

Prime Contractor: The company with the direct contractual relationship with the government.

Procurement Integrity Act: Statute that prohibits prospective contractors from obtaining contractor "bid or proposal information" or "agency source selection information" prior to the award of a federal contract. 41 U.S.C. § 423.

Sealed bidding: The more rigid of the two basic government contracting methods that awards the contract to the lowest responsive and responsible bidder.

Space Act Agreement (SAA): A type of non-FAR based legal agreement that NASA is permitted to enter into pursuant to the National Aeronautics and Space Act of 1958.

Subcontractor: A company that receives a subcontract from a prime contractor and does not have privity of contract with the government.

Subject Invention: Any invention made in the performance of a government contract.

Technical Data: As used in government contract data rights provisions, recorded information, regardless of the form or method of the recording, of a scientific or technical nature (including computer software documentation).

Time-and-materials (T&M) contracts: Government contract type provides for acquiring supplies or services on the basis of a fixed hourly rate for labor, which includes indirect expenses and profit, and the actual cost of materials.

Truth in Negotiations Act (TINA): Statute that requires a government contractor or subcontractor to submit certified "cost or pricing data" for any negotiated contract, subcontract, or modification that is expected to exceed $700,000 (subject to adjustment for inflation). 10 U.S.C. § 2306A, 41 U.S.C. § 254B.

Warrant: The certificate of appoint for a Contracting Officer (CO) that states the extent of the CO's authority to contractually bind the United States government.

CHAPTER 11

Space Operations and the Environment

From Earth, outer space seems like an endless expanse. It can be hard to believe that we need to worry about contaminating the outer space environment. Many people once held similar views about the oceans, yet we now know that human activity can have a very detrimental impact on the marine environment. The same is true for outer space. While outer space is vast, the usable regions of outer space are relatively small. Certain regions are already becoming overcrowded. Space operations can also have a detrimental impact on terrestrial environments, both on Earth and on other planetary bodies.

This chapter discusses the environmental impact of human activities in outer space and the laws, regulations, and standards that attempt to mitigate environmental harm. Section I discusses the international laws that apply to outer space activities. Section II addresses contamination of Earth and other planetary bodies by space operations. Section III covers the use of nuclear power sources in outer space. Section IV discusses the problem of orbital space debris.

I. INTERNATIONAL ENVIRONMENTAL LAW

The foundational principle of modern international environmental law, that the "polluter pays," was established in 1941 in what has become known as the Trail Smelter Arbitration. In the late 1920s and early 1930s, a smelter in Trail, British Columbia, was generating smoke that traveled south to pollute farmlands in the United States. The United States demanded that Canada compensate it for this damage and in 1935 the two countries agreed to arbitration of the dispute. In 1941, after several interim decisions and extensive expert testimony, the arbitration tribunal decided in favor of the United States, reasoning that "no State has the right to use or permit the use of its territory in such a manner as to cause injury by fumes in or to the territory of another or the properties or persons therein, when the case is of serious consequence and the injury is establish by clear and convincing evidence."[1]

This principle was formally recognized by the United Nations in 1972 when it adopted the Declaration of the United Nations Conference on the Human Environment, commonly known as the Stockholm Declaration. Principle 21 of the Stockholm Declaration provides that "States have, in accordance with the Charter of the United Nations and the principles of international law, the sovereign right to exploit their own resources pursuant to their own environmental policies, and the responsibility to ensure that activities within their jurisdiction or control do not cause damage to the environment of other States or of areas beyond the limits of national jurisdiction." As subsequently restated by the U.N. General Assembly, "In the exploration, exploitation and development of their natural resources, States must not produce significant harmful effects in zones situated outside their national jurisdiction."[2]

Article IX of the Outer Space Treaty established the principal framework governing the protection of the outer space environment.[3] Article IX generally extends the principles established in the Trail Smelter Arbitration and the Stockholm Declaration to outer space. However, the scope of Article IX is broader than environmental contamination; it addresses the "harmful contamination" of the outer space environment and also "harmful interference" with the peaceful activities of other nations in outer space.

The first sentence of Article IX is a general mandate that States conduct their activities in outer space "with due regard to the corresponding interests of all other States Parties to the Treaty." The next two sentences expand upon the principle of "due regard." First, States must conduct their studies and exploration of outer space, includ-

---

1. Decision of the Tribunal, Trail Smelter Arbitration (Mar. 11, 1941), *reprinted in* 3 UNITED NATIONS REPORTS OF INTERNATIONAL ARBITRAL AWARDS 1905–1982, at 1065 (2006), *available at* http://untreaty.un.org/cod/riaa/cases/vol_III/1905-1982.pdf.

2. Declaration of the United Nations Conference on the Human Environment, June 5–16, 1972, Principle 21; G.A. Res. 2995 (XXVII), Co-operation between States in the Field of the Environment, U.N. GAOR, 2112th Sess. U.N. Doc. A/RES/2995, at 1 (Dec. 15, 1972).

3. Treaty on Principles Governing the Activities of States in the Exploration and Use of Outer Space, Including the Moon and Other Celestial Bodies, art. IX, Oct. 10, 1967, 18 U.S.T. 2410, T.I.A.S. No. 6347.

ing the Moon and other celestial bodies, "so as to avoid their harmful contamination." Second, States must "adopt appropriate measures" to avoid "adverse changes in the environment of the Earth resulting from the introduction of extraterrestrial matter." Third, States must "undertake appropriate international consultations before proceeding with" any activity or experiment that "would cause potentially harmful interference with activities of other States Parties in the peaceful exploration and use of outer space."

The last sentences of Article IX give States the right to request consultations if they believe that "an activity or experiment planned by another State . . . would cause potentially harmful interference with activities in the peaceful exploration and use of outer space." Article IX does not provide a procedure for these consultations, or for the prospective consultations required by the previous sentence. Despite many instances of States directly (e.g., jamming) and indirectly (e.g., anti-satellite missile tests resulting in debris) interfering with the peaceful outer space activities of other countries, Article IX's consultation mechanisms have never been formally invoked by any party to the Outer Space Treaty.[4]

With these general principles as background, we can now delve into the specific problems of environmental contamination, nuclear power, and space debris.

II. CONTAMINATION

There are generally three types of contamination that may result from space-related activities: (1) contamination of Earth's biosphere from the operation of launch vehicles; (2) back contamination, which is the contamination of Earth from outer space by the return of astronauts or space objects; and (3) forward contamination, which is the contamination of outer space or celestial bodies by Earth through human space activities.

---

4. *See* Michael C. Mineiro, *FY-1C and USA-193 ASAT Intercepts: An Assessment of Legal Obligations under Article IX of the Outer Space Treaty*, 35(2) J. SPACE L. 321, 354 (Winter 2008) (China's "lack of consultation prior to [its 2007] ASAT experiment is consistent with the Cold War practices of the United States and the Soviet Union.").

A. Launch Vehicles

The operation of launch vehicles can harm Earth's biosphere in several ways. The exhaust produced by a launch vehicle can introduce harmful pollutants into the atmosphere. The improper handling or storage of chemicals used at launch sites can contaminate the local environment. Hazardous materials carried by the launch vehicle can fall back to Earth in the event of an accident.

There have been surprisingly few scientific studies on the effects of the operation of launch vehicles on Earth's atmosphere. Those studies that have been conducted have generally concluded that at the current launch rate, launch vehicles have an insignificant impact on the atmosphere compared with other industrial activities, but atmospheric effects may become problematic if launch rates increase significantly without any corresponding reduction in the pollution generated from rockets on a per-launch basis.

The Earth's atmosphere consists of four main layers: the troposphere, where weather phenomena occur (surface to 8–16 km); the stratosphere, which contains the ozone layer (from troposphere to 50 km); the mesosphere (from stratosphere to 80 km), and the ionosphere (from stratosphere to 2,000 km). The primary environmental impacts from the operation of launch vehicles are pollution from particulate matter (PM) released into the troposphere and the release of ozone-depleting and global warming-inducing gases into the stratosphere. Depending on the type of engine that is used, launch vehicles may release chlorine (Cl), hydrogen chloride (HCl), aluminum oxide (Al_2O_3), and nitrogen dioxide (NO_2) directly into the stratosphere, all of which can lead to ozone depletion. Engines can also emit greenhouse gases, including water vapor, carbon dioxide (CO_2), methane (CH_4), ozone (O_3), chlorofluorocarbons (CFCs), hydrofluorocarbons, and perfluorinated carbons. Table 11.1 lists the substances emitted by various types of engines.

TABLE 11.1
Exhaust Products from Launch Vehicle Propulsion Systems

Jet Engines	Solid	Liquid Hydrocarbon	Cryogenic	Hybrid Systems
HCL, CO, N_2 CO_2, NO_X, CL, H_2O, PM	HCL, CO, N_2 CO_2, NO_X, CL, H_2O, PM	HCL, CO, N_2 CO_2, NO_X, CL, H_2O, PM	HCL, CO, N_2 CO_2, NO_X, CL, H_2O, PM	HCL, CO, N_2 CO_2, NO_X, CL, H_2O, PM

Source: FAA

II. Contamination

The Federal Aviation Administration (FAA) has analyzed the impact of launch-related activities in environmental impact statements (EIS) issued in connection with the licensing of the launch and reentry of spacecraft and the operation of spaceports. A 2012 FAA environmental assessment of the operation of Virgin Galactic's SpaceShipTwo at Mojave Air and Space Port estimated that the CO_2 emissions generated by 30 launches and reentries per year would "represent about one hundred-thousandth of one percent of U.S. [greenhouse gas] emissions and two millionths of one percent of global [greenhouse gas] emissions."[5] Impact statements prepared in 2001, 2005, and 2008 similarly concluded that the amount of ozone-depleting chemicals produced by launch vehicles would be significantly lower than those produced by industrial sources. For example, a 2001 EIS for FAA-licensed launches estimated that the total amount of chlorine emitted from launch vehicles would be less than 0.064 percent of the total chlorine that was emitted from industrial sources in 1994. The FAA also estimated that CO_2 emissions for launch vehicles between 2000 and 2010 would be about 25,000 tons annually, which is negligible compared to the estimated 6.6 million tons of CO_2 that was emitted by all sources in the United States in 2009. The FAA further concluded that water vapor produced from launch vehicles would have an "insignificant impact on global warming."[6]

Several academic studies have also been published on the impact of space activities on Earth's atmosphere. A 2009 study noted that rocket emissions have the potential to become a significant contributor to ozone depletion because of three unique characteristics: (1) rocket emissions "are the only human-produced source of ozone-destroying

---

5. Fed. Aviation Admin., Office of Comm. Space Trans., Final Environmental Assessment for the Launch and Reentry of SpaceShipTwo Reusable Suborbital Rockets at the Mojave Air and Space Port 31 (2012), *available at* http://www.faa.gov/about/office_org/headquarters_offices/ast/media/20120502_Mojave_SS2_Final_EAandFONSI.pdf.

6. *See* Fed. Aviation Admin., Office of Comm. Space Trans., Final Programmatic Environmental Impact Statement for Licensing Launches (2001), *available at* http://www.faa.gov/about/office_org/headquarters_offices/ast/licenses_permits/media/Volume1-PEIS.pdf. *See also* Fed. Aviation Admin., Office of Comm. Space Trans., Final Programmatic Environmental Impact Statement for the Spaceport America Commercial Launch Site, Sierra Country, New Mexico (2008), *available at* http://www.faa.gov/about/office_org/headquarters_offices/ast/media/Spaceport%20America-FINAL%20EIS%20Vol%201.pdf; Fed. Aviation Admin., Office of Comm. Space Trans., Final Programmatic Environmental Impact Statement for Horizontal Launch and Reentry of Reentry Vehicles (2005), *available at* http://www.faa.gov/about/office_org/headquarters_offices/ast/licenses_permits/media/Final_FAA_PEIS_Dec_05.pdf.

compounds injected directly" into the stratosphere; (2) ozone levels are controlled by trace amounts of reactive gases and particles, so "relatively small absolute amounts of these reactive compounds can significantly modify ozone levels;" and (3) all rocket engines emit some type of ozone-depleting compounds, regardless of the propellant used. The study concluded that annual ozone loss attributable to current rocket launches is "insignificant" at 0.03 percent. However, if the Space Shuttle had launched weekly, as originally planned, the annual ozone loss attributable to rocket launches would have reached 0.2 percent, which the authors characterized as the "possible upper bound that defines the limit of 'acceptable' ozone loss."[7]

In 2010, the American Geophysical Union published a study on the possible effects on global warming of black carbon emissions from suborbital launch vehicles.[8] This study looked specifically at the type of hybrid rocket engine that is planned to be used on SpaceShipTwo. The authors developed a model that showed that if 1,000 suborbital flights launched each year from a single spaceport, the resulting soot emissions could cool the latitude band around the spaceport by 0.7°C, while warming Antarctica by 0.8°C.[9] Industry trade groups were critical of the study's assumptions and methodology, and Virgin Galactic's former president has said that the "carbon output of a SpaceShipTwo flight into space will be slightly less than a business class ticket on Virgin Atlantic from London to New York," and that "when you compare that to something like the Shuttle or Soyuz, the carbon output is negligible."[10]

B. Planetary Protection

The scientific community has long recognized forward and back contamination as a possible unintended consequence of space exploration. In 1956, before the launch of Sputnik 1, the International Astronautical Federation discussed the topic at its Seventh Congress in Rome.

---

7. Martin Ross, Darin Toohey, Manfred Peinemann & Patrick Ross, *Limits on the Space Launch Market Related to Stratospheric Ozone Depletion*, 7 ASTROPOLITICS 1 (2009), available at http://www.tandfonline.com/doi/pdf/10.1080/14777620902768867.

8. Martin Ross, Michael Mills & Darin Toohey, *Potential Climate Impact of Black Carbon Emitted by Rockets*, 37 GEOPHYSICAL RESEARCH LETTER (2010), available at http://nldr.library.ucar.edu/repository/assets/osgc/2010GL044548.pdf.

9. Press Release, Am. Geophyiscal Union, Soot from Space Tourism Rockets Could Spur Climate Change (Oct. 22, 2010), http://www.agu.org/news/press/pr_archives/2010/2010-34.shtml.

10. Jeff Foust, *Climate Change and Suborbital Spaceflight*, SPACE REVIEW, Nov. 8, 2010, http://www.thespacereview.com/article/1723/1.

II. *Contamination* 211

FIGURE 11.1
President Nixon visits Apollo 11 crew in quarantine. The practice of quarantining astronauts was discontinued after the Apollo 14 mission, as it was deemed unnecessary based on studies from the Apollo 11 and 12 missions

Credit: David Compton, NASA Special Publication-4214: Where No Man Has Gone Before: A History of Apollo Lunar Exploration Missions, Ch. 12-9 (NASA History Series, 1989), *available at* http://www.hq.nasa.gov/office/pao/History/SP-4214/contents.html

In 1958, the International Council of Scientific Unions formed the international Committee on Space Research (COSPAR), "in part to co-ordinate anti-contamination action."[11] COSPAR continues this work today, and in 2002 it adopted the COSPAR Planetary Protection Policy (CPPP), based on the principle that, "Although the existence of life elsewhere in the solar system may be unlikely, the conduct of scientific investigations of possible extraterrestrial life forms, precursors, and remnants must not be jeopardized. In addition, the Earth must be protected from the potential hazard posed by extraterrestrial matter carried by a spacecraft returning from another planet."[12]

---

11. Francis Lyall & Paul B. Larsen, Space Law: A Treatise 285 (Ashgate 2009).
12. Comm. on Space Research, Planetary Protection Policy (Oct. 20, 2002, amended Mar. 24, 2005), http://cosparhq.cnes.fr/Scistr/Pppolicy.htm.

TABLE 11.2
COSPAR Planetary Protection Policy Categories and Requirements

Category	Mission Type	Protection Requirements
I	No biological interest (e.g., Sun, Mercury)	No protection requirements
II	Significant biological interest, but insignificant probability of contamination (e.g., Earth's Moon, Venus, comets, Jupiter, Pluto)	Simple documentation
III	Flyby and orbiter missions to locations with the potential to host life and for which there is a possibility of contamination (e.g., Mars, Europa, Titan)	More extensive documentation than Category II; some implementing procedures, including trajectory biasing, the use of cleanrooms during spacecraft assembly and testing, and possibly bioburden reduction
IV	Lander or probe missions to locations with the potential to host life and for which there is a possibility of contamination (e.g., Mars, Europa, Titan)	More extensive documentation than Category III; implementing procedures may include trajectory biasing, cleanrooms, bioload reduction, possible partial sterilization of the direct contact hardware, and a bioshield for that hardware
V	Earth-return mission	If unrestricted Earth return, none. If restricted Earth return, considerable documentation requirements; hardware must be sterile; no destructive impact of probe on return to Earth to avoid escape of contaminants

The CPPP establishes controls on contamination for certain space mission/target body combinations. The restrictions for forward contamination are generally based on whether the target body may be of interest for understanding the process of chemical evolution and/or the origin of life, and the likelihood that the mission will contaminate the body in a manner that will jeopardize future biological experiments. For missions where samples will be returned to Earth, the primary question is whether the target body has the potential to support life. If it does not, the mission is an "unrestricted Earth return" mission and there are no requirements other than those imposed based upon the category for the outbound journey. If the target body could potentially support life, the mission is a "restricted Earth return" mission with stringent containment requirements, in addition to the requirements imposed on the outbound leg. Table 11.2 describes CPPP protection requirements for the five mission categories.

III. NUCLEAR POWER SOURCES

Nuclear power has been an important source of spacecraft power since the dawn of the space age. Radioisotope thermoelectric generators (RTGs) and radioisotope heater units (RHUs), which generate power and warm sensitive instruments using the heat generated by the decay of radioactive material such as plutonium-238, have been used on many U.S. space missions since 1961. As of 2011, 45 RTGs have powered 25 U.S. space vehicles, including Pioneer, Viking, Voyager, Galileo, Cassini, and the Mars Science Laboratory.[13] Fission nuclear reactors have been used primarily by Russia, but new research on these systems is being conducted in the United States. The only U.S. spacecraft to be powered by a fission reactor, an experimental satellite launched in 1965 called SNAP-10A, was shut down after 43 days due to a malfunction unrelated to its nuclear reactor and it remains in orbit today.

The Soviet Union launched 31 fission-powered Radar Ocean Reconnaissance Satellites (RORSATs) between 1967 and 1988. The reactor core of one of the RORSATs, Cosmos 954, failed to separate and boost into a nuclear-safe orbit after it launched on September 18, 1977. The satellite's orbit decayed until it reentered the atmosphere on January 24, 1978, depositing debris over an uninhabited area of northwest Canada. The Soviet Union initially claimed that the satellite was destroyed during reentry. However, a joint U.S.-Canadian team recovered 12 large pieces of the satellite, all but two of which were radioactive. The Canadian government brought a claim against the U.S.S.R. pursuant to the 1972 Convention on International Liability for Damage Caused by Space Objects for over C$6 million as reimbursement for cleanup expenses and future unpredicted expenses. The Canadian and Soviet governments eventually settled the claim for C$3 million in 1981.[14]

In response to the Cosmos 954 incident, the United Nations' Committee on the Peaceful Uses of Outer Space (COPUOS) formed a working group to study nuclear-powered satellites. This effort eventually led to the United Nations General Assembly's unanimous adoption in 1992 of the Principles Relevant to the Use of Nuclear Power Sources

---

13. World Nuclear Ass'n, Nuclear Reactors for Space (updated Nov. 2011), http://world-nuclear.org/info/inf82.html.

14. Protocol Between the Government of Canada and the Government of the Union of Soviet Socialist Republics, Apr. 2, 1981, *available at* http://www.jaxa.jp/library/space_law/chapter_3/3-2-2-1_e.html.

in Outer Space (Appendix 3-8).[15] The Principles establish nonbinding operating guidelines and criteria for the safe use of nuclear power sources in outer space and notification and assistance requirements in the event of the accidental reentry of space objects containing radioactive materials. They apply only to "nuclear power sources in outer space devoted to the generation of electric power on board space objects for non-propulsive purposes."

The operational guidelines described in the Principles are predicated on the idea that "In order to minimize the quantity of radioactive material in space and the risks involved, the use of nuclear power sources in outer space shall be restricted to those space missions which cannot be operated by non-nuclear energy sources in a reasonable way." In particular, nuclear reactors may only be used (1) on interplanetary missions, (2) in high orbits, and (3) in low Earth orbits if they are stored in high orbits after the operational part of their missions is completed. RTGs may only be used for "interplanetary missions and other missions leaving the gravity field of the Earth" and in Earth orbit if they are stored in a high orbit after their operational mission is completed. The Principles also reiterate that States "shall bear international responsibility for national activities involving the use of nuclear power sources in outer space" and that "each State which launches or procures the launching of a space object and each State from whose territory or facility a space object is launched shall be internationally liable for damage caused by such space objects or their component parts."

In 2009, COPUOS and the International Atomic Energy Agency (IAEA) jointly issued the Safety Framework for Nuclear Power Source Applications in Outer Space.[16] The framework combines the expertise of both organizations to provide high-level guidance to the operators of space nuclear power systems in the form of a model safety framework that allows for flexibility in national implementation. The framework includes recommendations on governmental responsibilities, management responsibilities, and technical guidance for missions using nuclear power sources.

The U.S. government's use of nuclear space power sources is subject to two separate policy reviews. The National Environmental Policy Act

---

15. U.N. Doc. RES/47/68 (Dec. 14, 1992).
16. U.N. COMMITTEE ON THE PEACEFUL USES OF OUTER SPACE & INT'L ATOMIC ENERGY AGENCY, SAFETY FRAMEWORK FOR NUCLEAR POWER SOURCE APPLICATIONS IN OUTER SPACE (2009), *available at* http://www.fas.org/nuke/space/iaea-space.pdf.

(NEPA) requires a federal agency anticipating taking action that significantly affects the quality of the human environment to prepare an EIS, which must review the environmental impact of the proposed action, any adverse environmental effects that cannot be avoided, and any alternatives to the proposed action, including no action. If the agency does not comply with NEPA, a plaintiff may challenge the agency's action under the Administrative Procedure Act. The decision of the National Aeronautics and Space Administration (NASA) to launch spacecraft powered by RTGs has been challenged under NEPA on at least three occasions, each time without success.[17]

In addition to NEPA, Presidential Directive/National Security Council Memorandum No. 25 requires that prior to the launch of a nuclear space system, an ad hoc Interagency Nuclear Safety Review Panel (INSRP), composed of experts from NASA, the Department of Energy, the Department of Defense, and the Environmental Protection Agency, conduct an independent evaluation of the safety of the proposed mission. Based on the INSRP's findings, the agency sponsoring the launch must then request the president's approval for the flight through the Office of Science and Technology Policy.[18]

IV. SPACE DEBRIS

Many useful regions of Earth orbit are crowded. The U.S. government currently tracks about 22,000 man-made objects in Earth orbit. It is estimated, however, that more than 500,000 man-made objects larger than a centimeter, and millions of objects smaller than a centimeter, are currently circling Earth.[19] Of these objects, only about 1,000 are operational spacecraft. The rest are dead satellites, discarded equipment, spent rocket boosters, fragments from collisions and explosions,

---

17. Fla. Coal. for Peace & Justice v. Bush, Civil Action No. 89-2682-OG (D.D.C. 1989) (Galileo); Fla. Coal. for Peace & Justice v. Bush, Civil Action No. 89-2682-Og (D.D.C. 1990) (Ulysses); Haw. Cnty. Green Party v. Clinton, 980 F. Supp. 1160 (D. Haw. 1997) (Cassini).

18. Presidential Directive/National Security Council Memorandum No. 25 (PD/NSC-25), Scientific or Technological Experiments with Possible Large-Scale Adverse Environmental Effects and Launch of Nuclear Systems into Space, ¶ 9 (1977, amended May 8, 1996), *available at* http://www.marshall.org/pdf/materials/837.pdf; *see also* NASA Office of Safety and Mission Assurance, Coordinate Nuclear Launch Safety Approval (NLSA) Process, HOWI 8710-GD000014, Jan. 17, 2006, *available at* http://nodis3.gsfc.nasa.gov/iso_docs/pdf/H_OWI_8710_GD000_014_D_.pdf.

19. NASA, Frequently Asked Questions: Orbital Debris, http://www.nasa.gov/news/debris_faq.html.

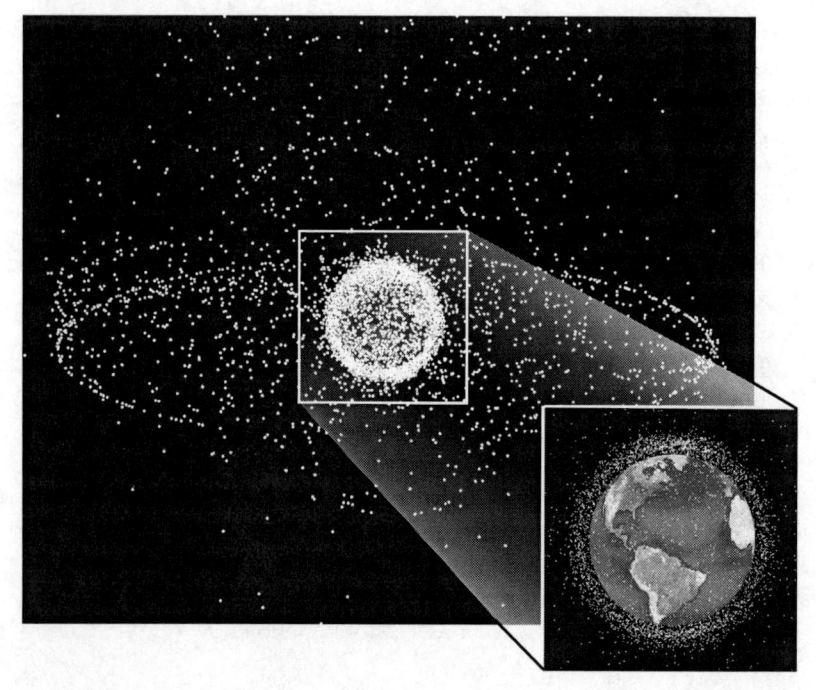

FIGURE 11.2
Computer-generated images of tracked space debris in Earth orbit
Credit: NASA

paint chips, and other by-products of human space activities, collectively known as space debris. Space debris in low orbits will eventually succumb to atmospheric drag and reenter Earth's atmosphere. Most will burn up, but some fragments will survive reentry and strike the ground, like the debris from Cosmos 954. Debris in higher orbits will remain in space for hundreds or thousands of years, or even forever.

Space debris was first recognized as a serious problem in the late 1970s and early 1980s.[20] However, several events in recent years have significantly increased the amount of debris in orbit. In 2007, the Chinese government used a dead Chinese weather satellite, Fengyun-1C, as a target for the test of an anti-satellite missile system. As of 2011, 3,135 pieces of debris from the Fengyun-1C satellite have been cataloged in low Earth orbit (LEO), and NASA's Orbital Debris Program

---

20. Jim Schefter, *The Growing Peril of Space Debris*, POPULAR SCI., July 1982.

IV. *Space Debris* 217

Office estimates that more than 150,000 pieces of debris larger than one centimeter in diameter were generated by this impact.[21] In 2009, a satellite operated by Iridium Communications, Inc., was destroyed in a collision with a defunct Russian military satellite, the first major collision between an operational satellite and space debris. The Iridium collision generated about 1,000 pieces of debris larger than 10 centimeters, and many more smaller pieces.[22] Experts fear that if the amount of debris in orbit reaches a critical mass, a few major debris-creating episodes (whether intentional or accidental) could set off a sequence of ever more frequent collisions—a chain reaction that would expand until, within decades, certain portions of Earth orbit would be rendered virtually unusable. This "collisional cascading" is known as the Kessler syndrome, named after NASA scientist Donald Kessler, who first proposed this possibility in 1978.[23] Some studies indicate that Earth orbit is already unstable at certain altitudes, in the sense that absent debris mitigation efforts, a collisional cascade could already be inevitable.[24]

Efforts to address the space debris problem fall into three categories: (1) debris tracking, (2) growth mitigation, and (3) removal.

A. Tracking

In order for operational spacecraft to avoid space debris, their operators must have comprehensive knowledge of the population of space objects, particularly of any objects that might cross their orbital paths. Knowledge of the space environment is referred to as space situational awareness (SSA). SSA is obtained through space surveillance, which is

---

21. CelesTrak.com, Chinese ASAT Test (updated May 20, 2011), http://celestrak.com/events/asat.asp; NASA, *Fengyun-1C Debris: One Year Later*, 12 ORBITAL DEBRIS QUARTERLY NEWS 1, 3 (2008), *available at* http://www.orbitaldebris.jsc.nasa.gov/newsletter/pdfs/ODQNv12i1.pdf.

22. Veronika Oleksyn, *What a Mess! Experts Ponder Space Junk Problem*, USA TODAY, Feb. 19, 2009, *available at* http://www.usatoday.com/tech/science/space/2009-02-19-space-junk_N.htm; Philip Hattis, *The Growing Menace of Orbital Debris*, LIVEBETTER eMAGAZINE, Jan. 2012, http://www.centerforabetterlife.com/eng/magazine/article_detail.lasso?id=265.

23. Donald J. Kessler & Burton G. Cour-Palais, *Collision Frequency of Artificial Satellites: The Creation of a Debris Belt*, 83 J. GEOPHYSICAL RESEARCH 2637–46 (1978).

24. Donald J. Kessler et al., *The Kessler Syndrome: Implications to Future Space Operations*, 33d Annual AAS Guidance and Control Conference, Paper AAS 10-016 (Feb. 2010), http://webpages.charter.net/dkessler/files/Kessler%20Syndrome-AAS%20Paper.pdf; Donald J. Kessler & Phillip D. Anz-Meador, *Critical Number of Spacecraft in Low Earth Orbit: Using Fragmentation Data to Evaluate the Stability of the Orbital Debris Environment*, Third European Conference on Space Debris (Mar. 2001), http://webpages.charter.net/dkessler/files/CriticalNumberofSpacecraftinLow.pdf.

the detection, correlation, characterization, and orbit determination of space objects. The U.S. Department of Defense's Space Surveillance Network (SSN) maintains the "world's most capable space surveillance network, consisting of ground-based radars, optical telescopes, and satellites."[25] The SSN is capable of tracking objects as small as five centimeters in diameter in LEO and about one meter in diameter in geosynchronous Earth orbit (GEO).[26] The United States makes much of this data publicly available through the SSA Sharing Program,[27] which is administered by the U.S. Strategic Command.

In addition to the SSA Sharing Program, several non-U.S. government organizations also provide SSA tracking data. The European Space Agency is developing its own SSA capability under the European SSA Preparatory Programme.[28] In 2009, the three largest commercial satellite companies, Intelsat, Inmarsat, and SES, formed the nonprofit Space Data Association to share SSA data. The Center for Space Standards and Innovation (CSSI) also provides SSA services. CSSI operates the satellite tracking website CelesTrak, which redistributes two-line element sets and information from the SSA Sharing Program, supplemented with its own data and analysis. CSSI also offers a service called Satellite Orbital Conjunction Reports Assessing Threatening Encounters in Space (SOCRATES), which provides information on pending orbital conjunctions during the coming week. Finally, the International Scientific Optical Network (ISON) describes itself as a "scientific project" initiated by the Keldysh Institute of Applied Mathematics of the Russian Academy of Sciences in 2001. ISON receives its data from 25 optical telescopes located at 18 observation facilities.[29]

B. Mitigation

Debris mitigation efforts first began in the early 1980s. One early solution implemented by McDonnell Douglas, for example, was to modify

---

25. James D. Randleman & Robert E. Ryals, *Spacecraft Operator Duty of Care*, AIAA Space 2011 Conference & Exposition 3 (2011).
26. NASA, Space Debris and Human Spacecraft, http://www.nasa.gov/mission_pages/station/news/orbital_debris.html.
27. Available at Space-Track, http://www.space-track.org.
28. European Space Agency, Space Situational Awareness, http://www.esa.int/esaMI/SSA.
29. Tiffany Chow, *Space Situational Awareness Sharing Program: An SWF Issue Brief*, SECURE WORLD FOUNDATION (Sept. 22, 2011), http://swfound.org/media/3584/ssa_sharing_program_issue_brief_nov2011.pdf. *See generally* Space Data Association, http://www.space-data.org/sda/; CelesTrak, http://celestrak.com; Satellite Orbital Conjunction Reports Assessing Threatening Encounters in Space (SOCRATES), http://celestrak.com/SOCRATES/; International Scientific Optical Network (ISON), www.isonteam.com.

its Delta booster so that it did not explode in orbit after completing its mission.[30] In 1995, NASA became the first space agency to issue comprehensive orbital debris mitigation guidelines.[31] Two years later, an interagency working group led by NASA and the Department of Defense developed Orbital Debris Mitigation Standard Practices (Appendix 11-1) for U.S. government satellites and launch vehicles. The Standard Practices provide guidelines for achieving four key objectives:

1. Control of debris released during normal operations.
2. Minimizing debris generated by accidental explosions.
3. Selection of safe flight profiles and operational configurations.
4. Post-mission disposal of space structures.[32]

The FAA, the Federal Communications Commission (FCC), and the National Oceanographic and Atmospheric Administration (NOAA) have also incorporated debris mitigation requirements into their respective licensing regimes for nongovernmental spacecraft.[33]

Internationally, other countries and organizations, including the European Space Agency (ESA), France, Japan, and Russia, have followed the United States' lead by adopting their own debris mitigation guidelines. In 2002, the Inter-Agency Space Debris Coordination Committee (IADC), which is composed of the space agencies of 11 countries and ESA, adopted a set of common guidelines designed to mitigate the growth of the orbital debris population (Appendix 11-2). In 2007, after a multiyear effort, the COPUOS developed and adopted a consensus set of space debris mitigation guidelines similar to the IADC guidelines, which were subsequently endorsed by the full U.N. General Assembly.[34] The COPUOS guidelines include seven recommendations:

1. Limit debris released during normal operations.
2. Minimize the potential for break-ups during operational phases.
3. Limit the probability of accidental collision in orbit.
4. Avoid intentional destruction and other harmful activities.

---

30. Schefter, *supra* note 20.
31. NASA Orbital Debris Program Office, Orbital Debris Mitigation, http://orbitaldebris.jsc.nasa.gov/mitigate/mitigation.html.
32. U.S. GOVERNMENT ORBITAL DEBRIS MITIGATION STANDARD PRACTICES, *available at* http://orbitaldebris.jsc.nasa.gov/library/USG_OD_Standard_Practices.pdf.
33. 14 CFR §§ 417.129 & 431.43 (c)(3) (FAA); 47 C.F.R. §§ 5.63(3), 25.114(d)(14) & 97.207(g) (FCC); 15 C.F.R. § 960.11(b)(12) (NOAA).
34. G.A. Res. 62/217, International Cooperation in the Peaceful Uses of Outer Space, U.N. GAOR, 62d Sess. U.N. Doc. A/RES/62/217, at ¶ 26 (Dec. 22, 2007).

5. Minimize potential for postmission break-ups resulting from stored energy.
6. Limit the long-term presence of spacecraft and launch vehicle orbital stages in LEO region after the end of their mission.
7. Limit the long-term interference of spacecraft and launch vehicle orbital stages with the GEO region after the end of their mission.[35]

In addition to the COPUOS, IADC, and national orbital debris mitigation guidelines, there have been several efforts to develop more comprehensive "rules of the road" for outer space operations. In December 2008, the European Union proposed a draft Code of Conduct for Outer Space Activities, which would require, among other things, that participating States "refrain from intentional destruction of any on-orbit space object or other harmful activities which may generate long-lived space debris," and "adopt, in accordance with their national legislative process, the appropriate policies and procedures in order to implement" the COPUOS debris mitigation guidelines.[36] Although initially responding favorably to this effort, the United States announced in January 2012 that it would not support the European Code of Conduct because the U.S. government found it to be "too restrictive."[37] Instead, U.S. Secretary of State Hillary Rodham Clinton announced that the United States would "join with the European Union and other nations to develop an International Code of Conduct for Outer Space Activities" that "will help maintain the long-term sustainability, safety, stability, and security of space by establishing guidelines for the responsible

---

35. UNITED NATIONAL OFFICE FOR OUTER SPACE AFFAIRS, SPACE DEBRIS MITIGATION GUIDELINES OF THE COMMITTEE ON THE PEACEFUL USES OF OUTER SPACE, Official Records of the General Assembly, 62d Session, Supp. No. 20 (A/62/20), paras. 117 & 118 and annex., available at http://orbitaldebris.jsc.nasa.gov/library/Space%20Debris%20Mitigation%20 Guidelines_COPUOS.pdf.

36. COUNCIL OF THE EUROPEAN UNION, DRAFT CODE OF CONDUCT FOR OUTER SPACE ACTIVITIES, at art. II, § 5 (2008), available at http://www.eu2008.fr/webdav/site/PFUE/ shared/import/1209_CAGRE_resultats/Code%20of%20Conduct%20for%20outer%20space% 20activities_EN.pdf.

37. Marcus Weisberger, *U.S. Won't Accept EU Code of Conduct for Space*, SPACE NEWS, Jan. 12, 2012, http://www.spacenews.com/policy/120112-wont-adopt-code-conduct-space .html (quoting U.S. Undersecretary of State for Arms Control and International Security Ellen Tauscher).

use of space" and "reverse the troubling trends that are damaging our space environment."[38]

C. Removal

There are generally two mechanisms for removing debris from orbit: self-removal and external removal. For self-removal, a spacecraft at the end of its useful life either enters into a decaying orbit that will reenter the atmosphere in a reasonable time frame or moves itself into a graveyard orbit out of the way of other spacecraft. The U.S. government's Standard Practices provide that spacecraft using the atmospheric reentry method should be left in an orbit where atmospheric drag will limit the lifetime of the spacecraft to no longer than 25 years after the completion of its mission, and that the risk of human casualties upon reentry should be less than 1 in 10,000.[39] The 25-year postmission lifetime limit has become the generally accepted standard in the space industry for spacecraft in LEO. A spacecraft achieves this by either performing its mission in a low orbit that will naturally decay within the 25-year window or by using on-board propellant to move itself into a lower orbit at the end of its mission. Technologies are currently being developed as alternatives to using valuable propellant to deorbit spacecraft, such as electrodynamic tethers, inflatable balloons, and sail-like attachments that would be installed on a spacecraft prior to launch and, when deployed at the end of the spacecraft's useful life, would create drag that would quickly deorbit the spacecraft.

Spacecraft in high Earth orbits cannot realistically carry enough extra propellant to descend into an atmospheric disposal orbit at the end of their missions. Because there is little or no atmospheric drag, spacecraft at these altitudes will remain in orbit almost indefinitely. To reduce the threat of collision with operational spacecraft, they must be moved out of commonly used orbits into graveyard orbits. To this end, the International Telecommunications Union (ITU) requires that member States ensure that geostationary satellites are transferred to a "supersynchronous graveyard orbit" that does not intersect with

---

38. Press Release, U.S. Sec'y of State Hillary Rodham Clinton, International Code of Conduct for Outer Space Activities (Jan. 17, 2012), http://www.state.gov/secretary/rm/2012/01/180969.htm.

39. U.S. GOVERNMENT ORBITAL DEBRIS MITIGATION STANDARD PRACTICES ¶ 4-1(a), *available at* http://orbitaldebris.jsc.nasa.gov/library/USG_OD_Standard_Practices.pdf.

the geostationary orbit (GSO) at the end of their useful lives.[40] This requirement has been implemented in the United States in section 25.283(a) of the FCC's rules for satellite operations, which provides that "a space station authorized to operate in the geostationary satellite orbit under this part shall be relocated, at the end of its useful life, barring catastrophic failure of satellite components, to an orbit with a perigee with an altitude of no less than: 36,021 km. [(22,382 miles)]."[41] Unfortunately, spacecraft operators do not always comply with the ITU requirement. In 2009, for example, only 11 of 21 satellites in GSO that reached their end of life were disposed of in proper graveyard orbits.[42]

While self-removal is the preferred mechanism for removing debris from orbit, this is not possible for the thousands of pieces of debris currently in orbit without any built-in means of self-removal. Proposed mechanisms for external removal of debris from orbit include, for example, building spacecraft to rendezvous with and capture debris, and ground-based "laser brooms" to sweep debris from orbit.

There are two problems with currently proposed external removal mechanisms. First, none of the proposed mechanisms are economically feasible with today's technology. Second, since the Outer Space Treaty provides that launching States retain ownership, jurisdiction, and control of their space objects and there is no law of salvage for outer space similar to the law of salvage under maritime law, one State cannot remove another State's nonfunctioning spacecraft from orbit without permission. States would likely be willing to consent to the removal of their debris in most cases, but obtaining consent for each item of debris would be time-consuming and inefficient and States would be less willing to allow another State to interact with debris that may contain national security secrets, such as derelict reconnaissance satellites. Accordingly, these solutions to the space debris problem would need to be accompanied by an international legal framework authorizing the removal efforts.

---

40. LYALL & LARSEN, *supra* note 11, at 246 (citing International Telecommunications Union Radiocommunication Sector Recommendation S.1003-1 (01/04), "Environmental Protection of the Geostationary Orbit").

41. 47 C.F.R. § 25.283(a).

42. EUROPEAN SPACE AGENCY, CLASSIFICATION OF GEOSYNCHRONOUS OBJECTS 126 (Feb. 2010), *available at* http://lfvn.astronomer.ru/files/COGO-issue12.pdf.

GLOSSARY

Back contamination: The contamination of Earth from space by the return of astronauts or space objects.

Committee on Space Research (COSPAR): Committee formed in 1958 by the International Council of Scientific Unions in part to coordinate anti-contamination policies.

COSPAR Planetary Protection Policy (CPPP): Committee on Space Research policy that establishes controls on contamination for certain space mission/target planet combinations.

Forward contamination: The contamination of space or celestial bodies by Earth through human space activities.

Kessler Syndrome: A scenario where the amount of space debris in Earth orbit reaches a critical mass where one collision sets off a chain reaction of other collisions that renders Earth orbit virtually unusable. Also referred to as "collisional cascading."

Principles Relevant to the Use of Nuclear Power Sources in Outer Space: United Nations resolution adopted in 1992 that establishes operating guidelines and criteria for the safe use of nuclear power sources and notification and assistance requirements in the event of the accidental re-entry of space objects containing radioactive materials.

Radioisotope heater unit (RHU): A small device that uses the heat generated from the decay of a small amount of plutonium-238 or other radioactive material to keep instruments warm in outer space.

Radioisotope thermoelectric generator (RTG): A spacecraft power system that produces electricity from the decay of radioactive materials, such as plutonium-238.

Safety Framework for Nuclear Power Source Applications in Outer Space: A 2007 joint publication of the U.N. Committee on the Peaceful Uses of Outer Space and the International Atomic Energy Agency that provides high level guidance on the use of nuclear power sources in outer space.

Space debris: Dead satellites, discarded equipment, spent rocket boosters, fragments from collisions and explosions, paint chips, and other by-products of human space activities.

Space Situation Awareness (SSA): The comprehensive knowledge of the space environment, including the population of space objects and existing threats to spacecraft.

Space Surveillance: The detection, correlation, characterization, and orbit determination of space objects.

SSA Sharing Program: The program administered by the U.S. Strategic Command that makes U.S. Government space tracking data available to the public through the website www.space-track.org.

Stockholm Declaration: 1972 Declaration of the United Nations Conference on the Human Environment, which stated that States have the "responsibility to ensure that activities within their jurisdiction or control do not cause damage to the environment of other States or of areas beyond the limits of national jurisdiction."

Trail Smelter Arbitration: 1941 arbitration between the United States and Canada that established the "polluter pays" principle in international environmental law.

CHAPTER 12

Property Rights

Property rights are vital to any commercial endeavor. "Property" can consist of tangible goods and real estate or intangible inventions, ideas, and creative works. Both forms of property can be important in the outer space context. Section I of this chapter addresses rights to tangible property in outer space. Section II discusses rights to intangible intellectual property in space technologies.

I. TANGIBLE PROPERTY

Tangible property includes both physical items and real estate. The drafters of the Outer Space Treaty recognized that the ownership of physical items launched into or manufactured in outer space would be important to government and private space operators alike. Accordingly, Article VIII of the Outer Space Treaty clearly states that "the ownership of objects launched into outer space and items manufactured in outer space does not change based on their being in outer space." Thus, the entity that owns an object that is placed in orbit or on the Moon or other celestial body retains ownership of that object and enjoys all of the rights and privileges of ownership that it would enjoy on Earth. In 1993, British video game developer and private space tourist Richard Garriott purchased Lunokhod-2, a Soviet rover that landed on the Moon in 1973, from the Russian government for $68,500.[1] Garriott then proclaimed that he is proud to be "the world's only private owner of an object on a celestial body."

The ownership of real estate on a celestial body is treated differently. The Outer Space Treaty provides that "[o]uter space, including the moon and other celestial bodies, is not subject to national appropriation by claim of sovereignty, by means of use or occupation, or

---

1. Sean Blair, *Space Property: Who Owns It?*, Focus Magazine, Dec. 25, 2011, http://sciencefocus.com/feature/health/whoowns-space.

by other means."[2] The Agreement Governing the Activities of States on the Moon and Other Celestial Bodies (known as the Moon Agreement) goes even further, stating that lunar property cannot become the property of "any State, international intergovernmental or non-governmental organization, national organization or non-governmental entity or of any natural person."[3] No major space power has signed the Moon Agreement, however, so it is generally treated as dormant.

Some individuals have attempted to claim private property on the Moon and asteroids, arguing that since the nonappropriation provision of the Outer Space Treaty bans only "national" appropriation, the Outer Space Treaty implicitly allows private parties to claim property ownership.[4] This argument is generally not accepted in the space law community because the appropriation of unoccupied land requires a government to either grant express authority for the appropriation or give its endorsement and approval to the appropriation, both actions that would be in violation of the Outer Space Treaty.[5]

Ownership of resources extracted from the Moon or other celestial bodies is more complicated. Article I of the Outer Space Treaty provides that "[o]uter space, including the moon and other celestial bodies, shall be free for exploration and use by all States without discrimination of any kind, on a basis of equality and in accordance with international law, and there shall be free access to all areas of celestial bodies." Some commentators argue that extracting resources from celestial bodies is a "use" permitted by the Outer Space Treaty, and that once extracted, these materials may become the property of the entity

---

2. Treaty on Principles Governing the Activities of States in the Exploration and Use of Outer Space, Including the Moon and Other Celestial Bodies, 18 U.S.T. 2410, art. II (1967).

3. Agreement Governing the Activities of States on the Moon and Other Celestial Bodies, art 11, para. 3, 18 I.L.M. 1434 (1979).

4. *See, e.g.,* Ronald E. Roel, *On a Full Moon, Empty Lots are $19.99,* Newsday, Oct. 2, 2003, http://www.newsday.com/columnists/glenn-gamboa/2.1091/on-a-full-moon-empty-lots-are-19-99-1.384996 (discussing the claims made by Dennis Hope, founder of Lunar Embassy Corp.); Robert Kelly, Note, *Nemitz v. United States, a Case of First Impression: Appropriation, Private Property Rights and Space Law Before the Federal Courts of the United States,* 30 J. Space L. 297 (2004) (describing the claim made by Gregory W. Nemitz over the asteroid Eros).

5. Virgiliu Pop, Who Owns the Moon?: Extraterrestrial Aspects of Land and Mineral Ownership 68 (2010); *see also* Kelly, *supra* note 4, at 305–08 (arguing that the Outer Space Treaty prohibits property ownership by natural persons, corporations, and other nongovernmental entities, in addition to States).

that performed the extraction.[6] Indeed, even the Moon Agreement permits the extraction of resources from the Moon for private purposes after an international authority is formed to regulate resource extraction and as long as the "benefits derived from those resources" are equitably shared by all parties to the Moon Agreement. This equitable sharing requirement was one reason that the Moon Agreement was not accepted by the United States and other space powers. It was feared that this would be a disincentive to the development of a commercial space economy based on the use of lunar resources.[7]

The United States has been opaque in its official interpretation of the status of resources extracted from the Moon. Nevertheless, the positions taken by the United States in court cases concerning Moon rocks brought back to Earth by the Apollo missions weigh in favor of an interpretation that resources extracted from the Moon may be owned by the party that extracts them. For example, in a 2003 case concerning the recovery of an Apollo Moon rock that was given to the government of Honduras as a goodwill gesture and was stolen during a period of political upheaval and later purchased by a private individual in the United States, the U.S. government argued that while a private citizen could not own Moon rocks, the United States government could, presumably because the U.S. government went to the Moon to retrieve it.[8]

Devising mechanisms for allocating property rights to outer space territory and resources is a rich area of academic debate. Yet this issue is not particularly relevant to the current crop of commercial space companies, largely because the technology to economically mine and use outer space resources does not yet exist. Once technology advances to the point where companies are able to close a business case built on extracting, using, and selling these resources, however, an equitable system for allocating property rights and settling disputes will become necessary both to incentivize private investment in these businesses and prevent conflict over outer space resources.

---

6. *See, e.g.,* Richard B. Bilder, *A Legal Regime for the Mining of Helium-3 on the Moon: U.S. Policy Options,* 33 FORDHAM INT'L L.J. 243 (2010); Eric Husby, Comment, *Sovereignty and Property Rights in Outer Space,* 3 J. INT'L L. & PRAC. 359, 366, 370 (1994).

7. *See* Rosanna Sattler, *Transporting a Legal System for Property Rights: From the Earth to the Stars,* 6 CHIC. J. INT'L L. 23 (2005) (noting that the Moon Agreement "was not widely accepted, and no major space power has signed it because it further restricts ownership and prohibits and property rights until an international body is created").

8. United States v. One Lucite Ball Containing Lunar Material (One Moon Rock) and One Ten Inch by Fourteen Inch Wooden Plaque, 252 F. Supp. 2d 1367 (S.D. Fla. 2003).

II. INTELLECTUAL PROPERTY

In addition to rights to tangible property in outer space, rights to intangible property, particularly intellectual property, are also important to space companies. The four basic types of intellectual property are (1) patents, which protect inventions; (2) copyrights, which protect works of authorship, such as books, music, movies, and software code; (3) trademarks, which protect brands; and (4) trade secrets, which protect information of economic value that is maintained in secrecy. In most cases, these property rights apply to space companies no differently than they do to companies in other industries, so a full discussion of each is beyond the scope of this book. This section focuses on two types of intellectual property protection that are of particular importance in the space industry: patents and trade secrets. It also briefly discusses the principle international treaties that attempt to harmonize national intellectual property laws.[9]

A. Patents

A patent is an exclusive right granted by a national government to an inventor to exclude others from making, using, selling, or importing an invention for a limited period of time, usually 20 years. In exchange for this monopoly, the inventor must disclose the patented invention to the public. To receive a patent, the invention must be new, useful, and nonobvious. Patents generally cannot be obtained for inventions that have previously been disclosed to the public, either by the inventor or a third party, used in a public setting, or sold or offered for sale to a third party.

Because patents are granted by national governments, they are inherently territorial and may only be enforced within the jurisdiction of the granting government.[10] The holder of a U.S. patent, for example, may enforce the patent only against someone who makes, uses, or sells

---

9. For a more detailed discussion of the application of intellectual property rights to space-related activities, see WORLD INTELLECTUAL PROP. ORG., INTELLECTUAL PROPERTY AND SPACE ACTIVITIES: ISSUE PAPER PREPARED BY THE INTERNATIONAL BUREAU (Apr., 2004), *available at* http://www.wipo.int/patent-law/en/developments/pdf/ip_space.pdf.

10. 35 U.S.C. § 154 (a patent grants the patentee "the right to exclude others from making, using, offering for sale, or selling the invention *throughout the United States*") (emphasis added).

the patented invention within the United States.[11] For this reason, an inventor must file a separate patent application in each jurisdiction where he wishes to obtain exclusive rights to an invention. An inventor who wishes to apply for patent protection in multiple countries may file an application under the Patent Cooperation Treaty (PCT), which provides a unified process for filing patent applications in each of the PCT's 145 member countries. While filing a PCT application allows the inventor to "reserve" his international patent rights, the inventor must still file a domestic application in each jurisdiction where he wishes to obtain patent protection.

None of the major international space treaties specifically addresses the applicability of national patent laws to activities in outer space. Nonetheless, the Outer Space Treaty provides that a space object's State of registration "shall retain jurisdiction and control over such object, and over any personnel thereof, while in outer space or on a celestial body."[12] This principle is analogous to the "floating island" principle that exists in maritime law with respect to ships in international waters.[13] Under this principle, States are permitted to extend their laws, including their patent laws, to their registered space objects.[14]

Consistent with the framework established by the Outer Space Treaty, in 1990 the United States extended the reach of its patent laws to U.S.-registered spacecraft by enacting 35 U.S.C. § 105.[15] The purpose of this legislation was "to clarify U.S. patent law with respect to its extraterritorial application aboard U.S.-flag spacecraft, in order to

---

11. The U.S. Patent Act defines patent infringement as follows: "Except as otherwise provided in this title, whoever without authority makes, uses, offers to sell, or sells any patented invention, within the United States, or imports into the United States any patented invention during the term of the patent therefor, infringes the patent." 35 U.S.C. § 271(a) (emphasis added). For a discussion of the limited circumstances under which a country may enforce its patents outside its borders, see Kurt G. Hammerle and Theodore U. Ro, *The Extra-Territorial Reach of U.S. Patent Law on Space-Related Activities: Does the "International Shoe" Fit as We Reach for the Stars?*, 34 J. SPACE L. 241 (2008).

12. Treaty on Principles Governing the Activities of States in the Exploration and Use of Outer Space, Including the Moon and Other Celestial Bodies, art. VIII, Oct. 10, 1967, 18 U.S.T. 2410, T.I.A.S. No. 6347.

13. Glenn H. Reynolds, *Legislative Comment: The Patents in Space Act*, 3 HARV. J. L. & TECH. 14, 19 (1990).

14. *See* FRANCIS LYALL & PAUL B. LARSEN, SPACE LAW: A TREATISE 124–27 (Ashgate 2009) (proposing that the jurisdictional control that states exert over their own space objects enables them to issue patent rights to inventors whose inventions are created within those space objects).

15. Pub. L. No. 101-580, § 1(a), 104 Stat. 2863 (Nov. 15, 1990).

encourage private investment in research and manufacture conducted in outer space."[16] Section 105(a) provides that "any invention made, used, or sold in outer space on a space object or component thereof under the jurisdiction or control of the United States shall be considered to be made, used or sold within the United States for the purposes of [U.S. patent laws]." An invention conceived or first reduced to practice on a U.S.-registered spacecraft is therefore deemed to have been made in the United States. Further, an infringement lawsuit based on a U.S. patent for activities concerning the making, use, or selling of an invention in outer space on a U.S.-registered spacecraft must be brought in a U.S. court and would succeed only if the activity is covered by the claims of the U.S. patent.

There are two exceptions to the application of U.S. jurisdiction under section 105(a): (1) if an international agreement to which the United States is a party specifies a different jurisdiction or (2) if the space object or component thereof is carried on a non-U.S. registry in accordance with the 1975 Convention on Registration of Objects Launched into Outer Space.[17] In addition, section 105(b) provides that a space object or component thereof registered on a non-U.S. registry will be subject to U.S. jurisdiction if U.S. jurisdiction is specified in an international agreement to which the United States is a party.[18]

The concept of national patent jurisdiction based on flag registration was incorporated into the 1998 intergovernmental agreement concerning cooperation on the International Space Station (ISS).[19] Under

---

16. SENATE COMM. ON THE JUDICIARY, INVENTIONS IN OUTER SPACE, S. REP. NO. 101-266, at 2 (1990).

17. 35 U.S.C. § 105(a). For a more in-depth discussion of 35 U.S.C. § 105, including a loophole that it created in U.S. patent law that could allow infringers on patents on space technologies to avoid patent infringement liability in the United States, see Theodore U. Ro, Matthew J. Kleiman & Kurt G. Hammerle, *Patent Infringement in Outer Space in Light of 35 U.S.C. § 105: Following the White Rabbit Down the Rabbit Loophole*, 17 B.U. J. SCI. & TECH. L. 202 (2011), *available at* http://www.bu.edu/law/central/jd/organizations/journals/scitech/volume172/documents/Kleiman_Web.pdf; Matthew J. Kleiman, *Patent Rights and Flags of Convenience in Outer Space*, 23 AIR & SPACE L. 4 (2011).

18. 35 U.S.C. § 105(b).

19. *See* Agreement Among the Government of Canada, Governments of the Member States of the European Space Agency, the Government of Japan, the Government of the Russian Federation, and the Government of the United States of America Concerning Cooperation on the Civil International Space Station, art. 21, ¶ 2, Jan. 28, 1998 ("[F]or the purposes of intellectual property law, an activity occurring in or on a Space Station flight element shall be deemed to have occurred only in the territory of the [country] of that element's registry, except that for [European Space Agency]-registered elements any European Partner State may deem the activity to have occurred within its territory").

this agreement, patent jurisdiction over an activity on the ISS belongs to the country of registration of the space station module wherein that activity occured. Consequently, Japan, Russia, and the United States each have exclusive patent jurisdiction over activities conducted in their respective ISS module(s), and any European partner state may claim patent jurisdiction over activities conducted in the ISS modules registered to the European Space Agency.

B. Trade Secrets

Many companies do not apply for patent protection on their inventions because they do not wish for the invention to be made public. This is usually because they expect the invention to remain valuable for longer than the 20-year monopoly granted by patent protection. Instead, these companies maintain their inventions as trade secrets. Trade secrets include "all forms and types of financial, business, scientific, technical, economic, or engineering information" that (1) the owner has taken reasonable measures to keep secret and (2) derives independent economic value from not being generally known to the public.[20] Probably the most famous trade secret is the Coca-Cola formula. In the space industry, trade secrets might include manufacturing processes and the source code to proprietary software.

Trade secrets are protected by both U.S. federal and state law. Under U.S. federal law, trade secrets are protected by the Economic Espionage Act of 1996, which makes the theft or misappropriation of a trade secret to benefit a foreign power or for commercial or economic purposes a federal crime.[21] Most U.S. states have also criminalized the theft of trade secrets by adopting the Uniform Trade Secrets Act.[22] It is also a federal crime for a U.S. government employee to disclose trade secrets or other private confidential information that he learns in the course of his official duties.[23] Trade secrets and other forms of privileged or confidential commercial or financial information are also exempt from disclosure under the U.S. Freedom of Information Act, and most similar state laws.[24]

---

20. 18 U.S.C. § 1839(3).
21. 18 U.S.C. §§ 1831–1839.
22. The Uniform Trade Secrets Act is available at http://www.law.upenn.edu/bll/archives/ulc/fnact99/1980s/utsa85.pdf.
23. 18 U.S.C. § 1905.
24. 5 U.S.C. § 552(b)(4).

In order to maintain trade secret protection the owner must have taken "reasonable measures" to keep the information secret. The most common "reasonable measures" are notifying employees, contractors, and other business partners of the existence and nature of trade secrets, and only disclosing such information under the protection of a written nondisclosure agreement. In addition, disclosing information to the U.S. government or a prime contractor on a government contract or subcontract with "unlimited rights" (see chapter 10) can strip trade secrets of their protected status.[25] Accordingly, government contractors who wish to use trade secrets on government contracts or subcontracts should ensure that information delivered to the government or prime contractor is labeled with appropriate proprietary markings.

C. International Treaties

Because intellectual property law is inherently territorial, the protection of intellectual property is generally left to national laws and enforcement mechanisms. Nonetheless, there are several international treaties that attempt to harmonize national intellectual property laws, although none specifically address the application of these rights in outer space. The PCT (discussed in section II.A above), which streamlined the process for filing patent applications internationally, is perhaps the most relevant intellectual property treaty for technology-based companies.

The Paris Convention for the Protection of Industrial Property, signed on March 20, 1883, was one of the first treaties addressing intellectual property, and is still in force today. The Paris Convention established two important principles. The "national treatment principle" provides that when a foreign applicant applies for patent or trademark protection in a country, the application must receive the same treatment as if it was filed by a domestic applicant. The "priority right" principle provides that an applicant from one country may use as its first filing date the date of filing in another country, if the application is filed within six months (trademarks) or 12 months (patents) of the first filing date. Of note to lawyers in the space industry, the Convention also provides that the use of a patented invention on board a vessel

---

25. *See* Plainville Elec. Prods. Co. v. Bechtel Bettis, Inc., No. 3:06cv920 (SRU), 2009 WL 801639, at 17 (D. Conn. Mar. 26, 2009) ("Because the 2003 subcontract gave [the prime contractor] the unlimited right to use, distribute, or disclose that data, as discussed above, there can be no 'misappropriation' of the Type I data, and therefore no trade secret claim[.]").

or aircraft that enters a country temporarily or accidentally does not constitute infringement of a patent in that country.

The Berne Convention for the Protection of Literary and Artistic Works is the primary treaty in the field of copyright. The Berne Convention establishes basic principles such as the national treatment principle, which is similar to the national treatment principle under the Paris Convention, and the principle of automatic protection, which means a nation's laws will provide copyright protection without the formality of registration. The Berne Convention was first accepted in 1886, but the United States did not become a party until 1988. The World Intellectual Property Organization Copyright Treaty (WCT) was adopted in 1996 to cover information technology matters not addressed in the Berne Convention. In particular, the WCT provides for the protection of computer programs and databases, to the extent the selection or arrangement of a database's content constitutes an intellectual creation.

The Agreement on Trade-Related Aspects of Intellectual Property Rights (TRIPS) was adopted in 1994. TRIPS contains minimum standards that nations' laws must meet for copyrights, industrial designs, integrated circuit layout designs, patents, new plant varieties, trademarks, trade dress, and confidential information. TRIPS also provides that any advantage, favor, privilege, or immunity granted by a party to the nationals of any other party must be accorded to the nationals of all other parties, which is referred to as "most-favored-nation treatment." Finally, TRIPS provides that patents must be available and patents rights enjoyable without discrimination as to the place of invention.

APPENDICES

APPENDIX 3-1

Treaty on Principles Governing the Activities of States in the Exploration and Use of Outer Space, including the Moon and Other Celestial Bodies

The States Parties to this Treaty,

Inspired by the great prospects opening up before mankind as a result of man's entry into outer space,

Recognizing the common interest of all mankind in the progress of the exploration and use of outer space for peaceful purposes,

Believing that the exploration and use of outer space should be carried on for the benefit of all peoples irrespective of the degree of their economic or scientific development,

Desiring to contribute to broad international co-operation in the scientific as well as the legal aspects of the exploration and use of outer space for peaceful purposes,

Believing that such co-operation will contribute to the development of mutual understanding and to the strengthening of friendly relations between States and peoples,

Recalling resolution 1962 (XVIII), entitled "Declaration of Legal Principles Governing the Activities of States in the Exploration and Use of Outer Space", which was adopted unanimously by the United Nations General Assembly on 13 December 1963,

Recalling resolution 1884 (XVIII), calling upon States to refrain from placing in orbit around the earth any objects carrying nuclear weapons or any other kinds of weapons of mass destruction or from installing such weapons on celestial bodies, which was adopted unanimously by the United Nations General Assembly on 17 October 1963,

Taking account of United Nations General Assembly resolution 110 (II) of 3 November 1947, which condemned propaganda designed or likely to provoke or encourage any threat to the peace, breach of the peace or act of aggression, and considering that the aforementioned resolution is applicable to outer space,

Convinced that a Treaty on Principles Governing the Activities of States in the Exploration and Use of Outer Space, including the Moon and Other Celestial Bodies, will further the purposes and principles of the Charter of the United Nations,

Have agreed on the following:

ARTICLE I

The exploration and use of outer space, including the moon and other celestial bodies, shall be carried out for the benefit and in the interests of all countries, irrespective of their degree of economic or scientific development, and shall be the province of all mankind.

Outer space, including the moon and other celestial bodies, shall be free for exploration and use by all States without discrimination of any kind, on a basis of equality and in accordance with international law, and there shall be free access to all areas of celestial bodies.

There shall be freedom of scientific investigation in outer space, including the moon and other celestial bodies, and States shall facilitate and encourage international co-operation in such investigation.

ARTICLE II

Outer space, including the moon and other celestial bodies, is not subject to national appropriation by claim of sovereignty, by means of use or occupation, or by any other means.

ARTICLE III

States Parties to the Treaty shall carry on activities in the exploration and use of outer space, including the moon and other celestial bodies, in accordance with international law, including the Charter of the United Nations, in the interest of maintaining international peace and security and promoting international co-operation and understanding.

ARTICLE IV

States Parties to the Treaty undertake not to place in orbit around the earth any objects carrying nuclear weapons or any other kinds of weapons of mass destruction, install such weapons on celestial bodies, or station such weapons in outer space in any other manner.

The moon and other celestial bodies shall be used by all States Parties to the Treaty exclusively for peaceful purposes. The establishment of military bases, installations and fortifications, the testing of any type of weapons and the conduct of military manoeuvres on celestial bodies shall be forbidden. The use of military personnel for scientific research or for any other peaceful purposes shall not be prohibited. The use of any equipment or facility necessary for peaceful exploration of the moon and other celestial bodies shall also not be prohibited.

ARTICLE V

States Parties to the Treaty shall regard astronauts as envoys of mankind in outer space and shall render to them all possible assistance in the event of accident, distress, or emergency landing on the territory of another State Party or on the high seas. When astronauts make such a landing, they shall be safely and promptly returned to the State of registry of their space vehicle.

In carrying on activities in outer space and on celestial bodies, the astronauts of one State Party shall render all possible assistance to the astronauts of other States Parties.

States Parties to the Treaty shall immediately inform the other States Parties to the Treaty or the Secretary-General of the United Nations of any phenomena they discover in outer space, including the moon and other celestial bodies, which could constitute a danger to the life or health of astronauts.

ARTICLE VI

States Parties to the Treaty shall bear international responsibility for national activities in outer space, including the moon and other celestial bodies, whether such activities are carried on by governmental agencies or by non-governmental entities, and for assuring that national activities are carried out in conformity with the provisions set forth in the present Treaty. The activities of non-governmental entities in outer space, including the moon and other celestial bodies, shall require authorization and continuing supervision by the appropriate State Party to the Treaty. When activities are carried on in outer space, including the moon and other celestial bodies, by an international organization, responsibility for compliance with this Treaty shall be borne both by the international organization and by the States Parties to the Treaty participating in such organization.

ARTICLE VII

Each State Party to the Treaty that launches or procures the launching of an object into outer space, including the moon and other celestial bodies, and each State Party from whose territory or facility an object is launched, is internationally liable for damage to another State Party to the Treaty or to its natural or juridical persons by such object or its component parts on the Earth, in air or in outer space, including the moon and other celestial bodies.

ARTICLE VIII

A State Party to the Treaty on whose registry an object launched into outer space is carried shall retain jurisdiction and control over such object, and over any personnel thereof, while in outer space or on a celestial body. Ownership of objects launched into outer space, including objects landed or constructed on a celestial body, and of their component parts, is not affected by their presence in outer space or on a celestial body or by their return to the Earth. Such objects or component parts found beyond the limits of the State Party to the Treaty on whose registry they are carried shall be returned to that State Party, which shall, upon request, furnish identifying data prior to their return.

ARTICLE IX

In the exploration and use of outer space, including the moon and other celestial bodies, States Parties to the Treaty shall be guided by the principle of co-operation and mutual assistance and shall conduct all their activities in outer space, including the moon and other celestial bodies, with due regard to the corresponding interests of all other States Parties to the Treaty. States Parties to the Treaty shall pursue studies of outer space, including the moon and other celestial bodies, and conduct exploration of them so as to avoid their harmful contamination and also adverse changes in the environment of the Earth resulting from the introduction of extraterrestrial matter and, where necessary, shall adopt appropriate measures for this purpose. If a State Party to the Treaty has reason to believe that an activity or experiment planned by it or its nationals in outer space, including the moon and other celestial bodies, would cause potentially harmful interference with activities of other States Parties in the peaceful exploration and use of outer space, including the moon and other celestial bodies, it shall undertake appropriate international consultations before proceeding with any such activity or experiment. A State Party to the Treaty which has reason to believe that an activity or experiment planned by another State Party in outer space, including the moon and other celestial bodies, would cause potentially harmful interference with activities in the peaceful exploration and use of outer space, including the moon and other celestial bodies, may request consultation concerning the activity or experiment.

ARTICLE X

In order to promote international co-operation in the exploration and use of outer space, including the moon and other celestial bodies, in conformity with the purposes of this Treaty, the States Parties to the Treaty shall consider on a basis of equality any requests by other States Parties to the Treaty to be afforded an opportunity to observe the flight of space objects launched by those States. The nature of such an opportunity for observation and the conditions under which it could be afforded shall be determined by agreement between the States concerned.

ARTICLE XI

In order to promote international co-operation in the peaceful exploration and use of outer space, States Parties to the Treaty conducting activities in outer space, including the moon and other celestial bodies, agree to inform the Secretary-General of the United Nations as well as the public and the international scientific community, to the greatest extent feasible and practicable, of the nature, conduct, locations and results of such activities. On receiving the said information, the Secretary-General of the United Nations should be prepared to disseminate it immediately and effectively.

ARTICLE XII

All stations, installations, equipment and space vehicles on the moon and other celestial bodies shall be open to representatives of other States Parties to the Treaty on a basis of reciprocity. Such representatives shall give reasonable advance notice of a projected visit, in order that appropriate consultations may be held and that maximum precautions may betaken to assure safety and to avoid interference with normal operations in the facility to be visited.

ARTICLE XIII

The provisions of this Treaty shall apply to the activities of States Parties to the Treaty in the exploration and use of outer space, including the moon and other celestial bodies, whether such activities are carried on by a single State Party to the Treaty or jointly with other States, including cases where they are carried on within the framework of international intergovernmental organizations.

Any practical questions arising in connection with activities carried on by international intergovernmental organizations in the exploration and use of outer space, including the moon and other celestial bodies, shall be resolved by the States Parties to the Treaty either with the appropriate international organization or with one or more States members of that international organization, which are Parties to this Treaty.

ARTICLE XIV

1. This Treaty shall be open to all States for signature. Any State which does not sign this Treaty before its entry into force in accordance with paragraph 3 of this article may accede to it at anytime.

2. This Treaty shall be subject to ratification by signatory States. Instruments of ratification and instruments of accession shall be

deposited with the Governments of the United Kingdom of Great Britain and Northern Ireland, the Union of Soviet Socialist Republics and the United States of America, which are hereby designated the Depositary Governments.

3. This Treaty shall enter into force upon the deposit of instruments of ratification by five Governments including the Governments designated as Depositary Governments under this Treaty.

4. For States whose instruments of ratification or accession are deposited subsequent to the entry into force of this Treaty, it shall enter into force on the date of the deposit of their instruments of ratification or accession.

5. The Depositary Governments shall promptly inform all signatory and acceding States of the date of each signature, the date of deposit of each instrument of ratification of and accession to this Treaty, the date of its entry into force and other notices.

6. This Treaty shall be registered by the Depositary Governments pursuant to Article 102 of the Charter of the United Nations.

ARTICLE XV

Any State Party to the Treaty may propose amendments to this Treaty. Amendments shall enter into force for each State Party to the Treaty accepting the amendments upon their acceptance by a majority of the States Parties to the Treaty and thereafter for each remaining State Party to the Treaty on the date of acceptance by it.

ARTICLE XVI

Any State Party to the Treaty may give notice of its withdrawal from the Treaty one year after its entry into force by written notification to the Depositary Governments. Such withdrawal shall take effect one year from the date of receipt of this notification.

ARTICLE XVII

This Treaty, of which the English, Russian, French, Spanish and Chinese texts are equally authentic, shall be deposited in the archives of the Depositary Governments. Duly certified copies of this Treaty shall be transmitted by the Depositary Governments to the Governments of the signatory and acceding States.

IN WITNESS WHEREOF the undersigned, duly authorized, have signed this Treaty.

DONE in triplicate, at the cities of London, Moscow and Washington, the twenty-seventh day of January, one thousand nine hundred and sixty-seven.

APPENDIX 3-2

Agreement on the Rescue of Astronauts, the Return of Astronauts and the Return of Objects Launched into Outer Space

The Contracting Parties,
Noting the great importance of the Treaty on Principles Governing the Activities of States in the Exploration and Use of Outer Space, including the Moon and Other Celestial Bodies, which calls for the rendering of all possible assistance to astronauts in the event of accident, distress or emergency landing, the prompt and safe return of astronauts, and the return of objects launched into outer space,
Desiring to develop and give further concrete expression to these duties,
Wishing to promote international co-operation in the peaceful exploration and use of outer space,
Prompted by sentiments of humanity,
Have agreed on the following:

ARTICLE 1

Each Contracting Party which receives information or discovers that the personnel of a spacecraft have suffered accident or are experiencing conditions of distress or have made an emergency or unintended landing in territory under its jurisdiction or on the high seas or in any other place not under the jurisdiction of any State shall immediately:

(A) notify the launching authority or, if it cannot identify and immediately communicate with the launching authority, immediately make a public announcement by all appropriate means of communication at its disposal;
(b) notify the Secretary-General of the United Nations, who should disseminate the information without delay by all appropriate means of communication at his disposal.

ARTICLE 2

If, owing to accident, distress, emergency or unintended landing, the personnel of a spacecraft land in territory under the jurisdiction of a Contracting Party, it shall immediately take all possible steps to rescue them and render them all necessary assistance. It shall inform the launching authority and also the Secretary-General of the United Nations of the steps it is taking and of their progress. If assistance by the launching authority would help to effect a prompt rescue or would contribute substantially to the effectiveness of search and rescue operations, the launching authority shall co-operate with the Contracting Party with a view to the effective conduct of search and rescue operations. Such operations shall be subject to the direction and control of the Contracting Party, which shall act in close and continuing consultation with the launching authority.

ARTICLE 3

If information is received or it is discovered that the personnel of a spacecraft have alighted on the high seas or in any other place not under the jurisdiction of any State, those Contracting Parties which are in a position to do so shall, if necessary, extend assistance in search and rescue operations for such personnel to assure their speedy rescue. They shall inform the launching authority and the Secretary-General of the United Nations of the steps they are taking and of their progress.

ARTICLE 4

If, owing to accident, distress, emergency or unintended landing, the personnel of a spacecraft land in territory under the jurisdiction of a

Contracting Party or have been found on the high seas or in any other place not under the jurisdiction of any State, they shall be safely and promptly returned to representatives of the launching authority.

ARTICLE 5

1. Each Contracting Party which receives information or discovers that a space object or its component parts has returned to Earth in territory under its jurisdiction or on the high seas or in any other place not under the jurisdiction of any State, shall notify the launching authority and the Secretary-General of the United Nations.

2. Each Contracting Party having jurisdiction over the territory on which a space object or its component parts has been discovered shall, upon the request of the launching authority and with assistance from that authority if requested, take such steps as it finds practicable to recover the object or component parts.

3. Upon request of the launching authority, objects launched into outer space or their component parts found beyond the territorial limits of the launching authority shall be returned to or held at the disposal of representatives of the launching authority, which shall, upon request, furnish identifying data prior to their return.

4. Notwithstanding paragraphs 2 and 3 of this article, a Contracting Party which has reason to believe that a space object or its component parts discovered in territory under its jurisdiction, or recovered by it elsewhere, is of a hazardous or deleterious nature may so notify the launching authority, which shall immediately take effective steps, under the direction and control of the said Contracting Party, to eliminate possible danger of harm.

5. Expenses incurred in fulfilling obligations to recover and return a space object or its component parts under paragraphs 2 and 3 of this article shall be borne by the launching authority.

ARTICLE 6

For the purposes of this Agreement, the term "launching authority" shall refer to the State responsible for launching, or, where an international intergovernmental organization is responsible for launching, that organization, provided that organization declares its acceptance of the rights and obligations provided for in this Agreement and a majority of the States members of that organization are Contracting Parties to this Agreement and to the Treaty on Principles Governing the Activities of States in the Exploration and Use of Outer Space, including the Moon and Other Celestial Bodies.

ARTICLE 7

1. This Agreement shall be open to all States for signature. Any State which does not sign this Agreement before its entry into force in accordance with paragraph 3 of this article may accede to it at any time.

2. This Agreement shall be subject to ratification by signatory States. Instruments of ratification and instruments of accession shall be deposited with the Governments of the United Kingdom of Great Britain and Northern Ireland, the Union of Soviet Socialist Republics and the United States of America, which are hereby designated the Depositary Governments.

3. This Agreement shall enter into force upon the deposit of instruments of ratification by five Governments including the Governments designated as Depositary Governments under this Agreement.

4. For States whose instruments of ratification or accession are deposited subsequent to the entry into force of this Agreement, it shall enter into force on the date of the deposit of their instruments of ratification or accession.

5. The Depositary Governments shall promptly inform all signatory and acceding States of the date of each signature, the date of deposit of each instrument of ratification of and accession to this Agreement, the date of its entry into force and other notices.

6. This Agreement shall be registered by the Depositary Governments pursuant to Article 102 of the Charter of the United Nations.

ARTICLE 8

Any State Party to the Agreement may propose amendments to this Agreement. Amendments shall enter into force for each State Party to the Agreement accepting the amendments upon their acceptance by a majority of the States Parties to the Agreement and thereafter for each remaining State Party to the Agreement on the date of acceptance by it.

ARTICLE 9

Any State Party to the Agreement may give notice of its withdrawal from the Agreement one year after its entry into force by written notification to the Depositary Governments. Such withdrawal shall take effect one year from the date of receipt of this notification.

ARTICLE 10

This Agreement, of which the English, Russian, French, Spanish and Chinese texts are equally authentic, shall be deposited in the archives of the Depositary Governments. Duly certified copies of this Agreement shall be transmitted by the Depositary Governments to the Governments of the signatory and acceding States.

IN WITNESS WHEREOF the undersigned, duly authorized, have signed this Agreement.

DONE in triplicate, at the cities of London, Moscow and Washington, the twenty-second day of April, one thousand nine hundred and sixty-eight.

APPENDIX 3-3

Convention on International Liability for Damage Caused by Space Objects

The States Parties to this Convention,

Recognizing the common interest of all mankind in furthering the exploration and use of outer space for peaceful purposes,

Recalling the Treaty on Principles Governing the Activities of States in the Exploration and Use of Outer Space, including the Moon and Other Celestial Bodies,

Taking into consideration that, notwithstanding the precautionary measures to be taken by States and international intergovernmental organizations involved in the launching of space objects, damage may on occasion be caused by such objects,

Recognizing the need to elaborate effective international rules and procedures concerning liability for damage caused by space objects and to ensure, in particular, the prompt payment under the terms of this Convention of a full and equitable measure of compensation to victims of such damage,

Believing that the establishment of such rules and procedures will contribute to the strengthening of international co-operation in the field of the exploration and use of outer space for peaceful purposes,

Have agreed on the following:

ARTICLE I

For the purposes of this Convention:
  (a) The term "damage" means loss of life, personal injury or other impairment of health; or loss of or damage to property

of States or of persons, natural or juridical, or property of international intergovernmental organizations;
(b) The term "launching" includes attempted launching;
(c) The term "launching State" means:
  (i) A State which launches or procures the launching of a space object;
  (ii) A State from whose territory or facility a space object is launched;
(d) The term "space object" includes component parts of a space object as well as its launch vehicle and parts thereof.

ARTICLE II

A launching State shall be absolutely liable to pay compensation for damage caused by its space object on the surface of the earth or to aircraft flight.

ARTICLE III

In the event of damage being caused elsewhere than on the surface of the earth to a space object of one launching State or to persons or property on board such a space object by a space object of another launching State, the latter shall be liable only if the damage is due to its fault or the fault of persons for whom it is responsible.

ARTICLE IV

1. In the event of damage being caused elsewhere than on the surface of the earth to a space object of one launching State or to persons or property on board such a space object by a space object of another launching State, and of damage thereby being caused to a third State or to its natural or juridical persons, the first two States shall be jointly and severally liable to the third State, to the extent indicated by the following:

  (a) If the damage has been caused to the third State on the surface of the earth or to aircraft in flight, their liability to the third State shall be absolute;
  (b) If the damage has been caused to a space object of the third State or to persons or property on board that space object

elsewhere than on the surface of the earth, their liability to the third State shall be based on the fault of either of the first two States or on the fault of persons for whom either is responsible.

2. In all cases of joint and several liability referred to in paragraph 1 of this article, the burden of compensation for the damage shall be apportioned between the first two States in accordance with the extent to which they were at fault; if the extent of the fault of each of these States cannot be established, the burden of compensation shall be apportioned equally between them. Such apportionment shall be without prejudice to the right of the third State to seek the entire compensation due under this Convention from any or all of the launching States which are jointly and severally liable.

ARTICLE V

1. Whenever two or more States jointly launch a space object, they shall be jointly and severally liable for any damage caused.

2. A launching State which has paid compensation for damage shall have the right to present a claim for indemnification to other participants in the joint launching. The participants in a joint launching may conclude agreements regarding the apportioning among themselves of the financial obligation in respect of which they are jointly and severally liable. Such agreements shall be without prejudice to the right of a State sustaining damage to seek the entire compensation due under this Convention from any or all of the launching States which are jointly and severally liable.

3. A State from whose territory or facility a space object is launched shall be regarded as a participant in a joint launching.

ARTICLE VI

1. Subject to the provisions of paragraph 2 of this Article, exoneration from absolute liability shall be granted to the extent that a launching State establishes that the damage has resulted either wholly or partially from gross negligence or from an act or omission done with intent to cause damage on the part of a claimant State or of natural or juridical persons it represents.

2. No exoneration whatever shall be granted in cases where the damage has resulted from activities conducted by a launching State which are not in conformity with international law including, in particular, the Charter of the United Nations and the Treaty on Principles Governing the Activities of States in the Exploration and Use of Outer Space, including the Moon and Other Celestial Bodies.

ARTICLE VII

The provisions of this Convention shall not apply to damage caused by a space object of a launching State to:

(a) nationals of that launching State;
(b) foreign nationals during such time as they are participating in the operation of that space object from the time of its launching or at any stage thereafter until its descent, or during such time as they are in the immediate vicinity of a planned launching or recovery area as the result of an invitation by that launching State.

ARTICLE VIII

1. A State which suffers damage, or whose natural or juridical persons suffer damage, may present to a launching State a claim for compensation for such damage.

2. If the State of nationality has not presented a claim, another State may, in respect of damage sustained in its territory by any natural or juridical person, present a claim to a launching State.

3. If neither the State of nationality nor the State in whose territory the damage was sustained has presented a claim or notified its intention of presenting a claim, another State may, in respect of damage sustained by its permanent residents, present a claim to a launching State.

ARTICLE IX

A claim for compensation for damage shall be presented to a launching State through diplomatic channels. If a State does not maintain diplomatic relations with the launching State concerned, it may request another State to present its claim to that launching State or otherwise represent its interests under this Convention. It may also present its claim through the Secretary-General of the United Nations, provided the claimant State and the launching State are both Members of the United Nations.

ARTICLE X

1. A claim for compensation for damage may be presented to a launching State not later than one year following the date of the occurrence of the damage or the identification of the launching State which is liable.

2. If, however, a State does not know of the occurrence of the damage or has not been able to identify the launching State which is liable, it may present a claim within one year following the date on which it learned of the aforementioned facts; however, this period shall in no event exceed one year following the date on which the State could reasonably be expected to have learned of the facts through the exercise of due diligence.

3. The time-limits specified in paragraphs 1 and 2 of this Article shall apply even if the full extent of the damage may not be known. In this event, however, the claimant State shall be entitled to revise the claim and submit additional documentation after the expiration of such time-limits until one year after the full extent of the damage is known.

ARTICLE XI

1. Presentation of a claim to a launching State for compensation for damage under this Convention shall not require the prior exhaustion of any local remedies which may be available to a claimant State or to natural or juridical persons it represents.

2. Nothing in this Convention shall prevent a State, or natural or juridical persons it might represent, from pursuing a claim in the courts or administrative tribunals or agencies of a launching State. A State shall not, however, be entitled to present a claim under this Convention in respect of the same damage for which a claim is being pursued in the courts or administrative tribunals or agencies of a launching State or under another international agreement which is binding on the States concerned.

ARTICLE XII

The compensation which the launching State shall be liable to pay for damage under this Convention shall be determined in accordance with international law and the principles of justice and equity, in order to provide such reparation in respect of the damage as will restore the person, natural or juridical, State or international organization on whose behalf the claim is presented to the condition which would have existed if the damage had not occurred.

ARTICLE XIII

Unless the claimant State and the State from which compensation is due under this Convention agree on another form of compensation, the compensation shall be paid in the currency of the claimant State or, if that State so requests, in the currency of the State from which compensation is due.

ARTICLE XIV

If no settlement of a claim is arrived at through diplomatic negotiations as provided for in Article IX, within one year from the date on which the claimant State notifies the launching State that it has submitted the documentation of its claim, the parties concerned shall establish a Claims Commission at the request of either party.

ARTICLE XV

1. The Claims Commission shall be composed of three members: one appointed by the claimant State, one appointed by the launching State and the third member, the Chairman, to be chosen by both parties jointly. Each party shall make its appointment within two months of the request for the establishment of the Claims Commission.

2. If no agreement is reached on the choice of the Chairman within four months of the request for the establishment of the Commission, either party may request the Secretary-General of the United Nations to appoint the Chairman within a further period of two months.

ARTICLE XVI

1. If one of the parties does not make its appointment within the stipulated period, the Chairman shall, at the request of the other party, constitute a single-member Claims Commission.

2. Any vacancy which may arise in the Commission for whatever reason shall be filled by the same procedure adopted for the original appointment.

3. The Commission shall determine its own procedure.

4. The Commission shall determine the place or places where it shall sit and all other administrative matters.

5. Except in the case of decisions and awards by a single-member Commission, all decisions and awards of the Commission shall be by majority vote.

ARTICLE XVII

No increase in the membership of the Claims Commission shall take place by reason of two or more claimant States or launching States being joined in any one proceeding before the Commission. The claimant States so joined shall collectively appoint one member of the Commission in the same manner and subject to the same conditions as would be the case for a single claimant State. When two or more launching States are so joined, they shall collectively appoint one member of the Commission in the same way. If the claimant States or the launching States do not make the appointment within the stipulated period, the Chairman shall constitute a single-member Commission.

ARTICLE XVIII

The Claims Commission shall decide the merits of the claim for compensation and determine the amount of compensation payable, if any.

ARTICLE XIX

1. The Claims Commission shall act in accordance with the provisions of Article XII.

2. The decision of the Commission shall be final and binding if the parties have so agreed; otherwise the Commission shall render a final and recommendatory award, which the parties shall consider in good faith. The Commission shall state the reasons for its decision or award.

3. The Commission shall give its decision or award as promptly as possible and no later than one year from the date of its establishment, unless an extension of this period is found necessary by the Commission.

4. The Commission shall make its decision or award public. It shall deliver a certified copy of its decision or award to each of the parties and to the Secretary-General of the United Nations.

ARTICLE XX

The expenses in regard to the Claims Commission shall be borne equally by the parties, unless otherwise decided by the Commission.

ARTICLE XXI

If the damage caused by a space object presents a large-scale danger to human life or seriously interferes with the living conditions of the population or the functioning of vital centres, the States Parties, and in particular the launching State, shall examine the possibility of rendering appropriate and rapid assistance to the State which has suffered the damage, when it so requests. However, nothing in this article shall affect the rights or obligations of the States Parties under this Convention.

ARTICLE XXII

1. In this Convention, with the exception of Articles XXIV to Articles XXVII, references to States shall be deemed to apply to any international intergovernmental organization which conducts space activities if the organization declares its acceptance of the rights and obligations provided for in this Convention and if a majority of the States members of the organization are States Parties to this Convention and to the Treaty on Principles Governing the Activities of States in the Exploration and Use of Outer Space, including the Moon and Other Celestial Bodies.

2. States members of any such organization which are States Parties to this Convention shall take all appropriate steps to ensure that the organization makes a declaration in accordance with the preceding paragraph.

3. If an international intergovernmental organization is liable for damage by virtue of the provisions of this Convention, that organization and those of its members which are States Parties to this Convention shall be jointly and severally liable; provided, however, that:

   (a) any claim for compensation in respect of such damage shall be first presented to the organization;
   (b) only where the organization has not paid, within a period of six months, any sum agreed or determined to be due as compensation for such damage, may the claimant State invoke the liability of the members which are States Parties to this Convention for the payment of that sum.

4. Any claim, pursuant to the provisions of this Convention, for compensation in respect of damage caused to an organization which has made a declaration in accordance with paragraph 1 of this Article shall be presented by a State member of the organization which is a State Party to this Convention.

ARTICLE XXIII

1. The provisions of this Convention shall not affect other international agreements in force in so far as relations between the States Parties to such agreements are concerned.

2. No provision of this Convention shall prevent States from concluding international agreements reaffirming, supplementing or extending its provisions.

ARTICLE XXIV

1. This Convention shall be open to all States for signature. Any State which does not sign this Convention before its entry into force in accordance with paragraph 3 of this article may accede to it at any time.

2. This Convention shall be subject to ratification by signatory States. Instruments of ratification and instruments of accession shall

be deposited with the Governments of the United Kingdom of Great Britain and Northern Ireland, the Union of Soviet Socialist Republics and the United States of America, which are hereby designated the Depositary Governments.

3. This Convention shall enter into force on the deposit of the fifth instrument of ratification.

4. For States whose instruments of ratification or accession are deposited subsequent to the entry into force of this Convention, it shall enter into force on the date of the deposit of their instruments of ratification or accession.

5. The Depositary Governments shall promptly inform all signatory and acceding States of the date of each signature, the date of deposit of each instrument of ratification of and accession to this Convention, the date of its entry into force and other notices.

6. This Convention shall be registered by the Depositary Governments pursuant to Article 102 of the Charter of the United Nations.

ARTICLE XXV

Any State Party to this Convention may propose amendments to this Convention. Amendments shall enter into force for each State Party to the Convention accepting the amendments upon their acceptance by a majority of the States Parties to the Convention and thereafter for each remaining State Party to the Convention on the date of acceptance by it.

ARTICLE XXVI

Ten years after the entry into force of this Convention, the question of the review of this Convention shall be included in the provisional agenda of the United Nations General Assembly in order to consider, in the light of past application of the Convention, whether it requires revision. However, at any time after the Convention has been in force for five years, and at the request of one third of the States Parties to the Convention, and with the concurrence of the majority of the States Parties, a conference of the States Parties shall be convened to review this Convention.

ARTICLE XXVII

Any State Party to this Convention may give notice of its withdrawal from the Convention one year after its entry into force by written notification to the Depositary Governments. Such withdrawal shall take effect one year from the date of receipt of this notification.

ARTICLE XXVIII

This Convention, of which the English, Russian, French, Spanish and Chinese texts are equally authentic, shall be deposited in the archives of the Depositary Governments. Duly certified copies of this Convention shall be transmitted by the Depositary Governments to the Governments of the signatory and acceding States.

IN WITNESS WHEREOF the undersigned, duly authorized thereto, have signed this Convention.

DONE in triplicate, at the cities of London, Moscow and Washington, this twenty-ninth day of March, one thousand nine hundred and seventy-two.

APPENDIX 3-4

Convention on Registration of Objects Launched into Outer Space

The States Parties to this Convention,

Recognizing the common interest of all mankind in furthering the exploration and use of outer space for peaceful purposes,

Recalling that the Treaty on Principles Governing the Activities of States in the Exploration and Use of Outer Space, including the Moon and Other Celestial Bodies of 27 January 1967 affirms that States shall bear international responsibility for their national activities in outer space and refers to the State on whose registry an object launched into outer space is carried,

Recalling also that the Agreement on the Rescue of Astronauts, the Return of Astronauts and the Return of Objects Launched into Outer Space of 22 April 1968 provides that a launching authority shall, upon request, furnish identifying data prior to the return of an object it has launched into outer space found beyond the territorial limits of the launching authority,

Recalling further that the Convention on International Liability for Damage Caused by Space Objects of 29 March 1972 establishes international rules and procedures concerning the liability of launching States for damage caused by their space objects,

Desiring, in the light of the Treaty on Principles Governing the Activities of States in the Exploration and Use of Outer Space, including the Moon and Other Celestial Bodies, to make provision for the national registration by launching States of space objects launched into outer space,

Desiring further that a central register of objects launched into outer space be established and maintained, on a mandatory basis, by the Secretary-General of the United Nations,

Desiring also to provide for States Parties additional means and procedures to assist in the identification of space objects,

Believing that a mandatory system of registering objects launched into outer space would, in particular, assist in their identification and would contribute to the application and development of international law governing the exploration and use of outer space,

Have agreed on the following:

ARTICLE I

For the purposes of this Convention:

- (a) The term "launching State" means:
    - (i) A State which launches or procures the launching of a space object;
    - (ii) A State from whose territory or facility a space object is launched;
- (b) The term "space object" includes component parts of a space object as well as its launch vehicle and parts thereof;
- (c) The term "State of registry" means a launching State on whose registry a space object is carried in accordance with article II.

ARTICLE II

1. When a space object is launched into earth orbit or beyond, the launching State shall register the space object by means of an entry in an appropriate registry which it shall maintain. Each launching State shall inform the Secretary-General of the United Nations of the establishment of such a registry.

2. Where there are two or more launching States in respect of any such space object, they shall jointly determine which one of them shall register the object in accordance with paragraph 1 of this article, bearing in mind the provisions of article VIII of the Treaty on Principles Governing the Activities of States in the Exploration and Use of Outer Space, including the Moon and Other Celestial Bodies,

and without prejudice to appropriate agreements concluded or to be concluded among the launching States on jurisdiction and control over the space object and over any personnel thereof.

3. The contents of each registry and the conditions under which it is maintained shall be determined by the State of registry concerned.

ARTICLE III

1. The Secretary-General of the United Nations shall maintain a Register in which the information furnished in accordance with article IV shall be recorded.

2. There shall be full and open access to the information in this Register.

ARTICLE IV

1. Each State of registry shall furnish to the Secretary-General of the United Nations, as soon as practicable, the following information concerning each space object carried on its registry:

    (a) name of launching State or States;
    (b) an appropriate designator of the space object or its registration number;
    (c) date and territory or location of launch;
    (d) basic orbital parameters, including:
        (i) nodal period;
        (ii) inclination;
        (iii) apogee;
        (iv) perigee;
    (e) general function of the space object.

2. Each State of registry may, from time to time, provide the Secretary-General of the United Nations with additional information concerning a space object carried on its registry.

3. Each State of registry shall notify the Secretary-General of the United Nations, to the greatest extent feasible and as soon as practicable, of space objects concerning which it has previously transmitted information, and which have been but no longer are in earth orbit.

ARTICLE V

Whenever a space object launched into earth orbit or beyond is marked with the designator or registration number referred to in article IV, paragraph 1 (b), or both, the State of registry shall notify the Secretary-General of this fact when submitting the information regarding the space object in accordance with article IV. In such case, the Secretary-General of the United Nations shall record this notification in the Register.

ARTICLE VI

Where the application of the provisions of this Convention has not enabled a State Party to identify a space object which has caused damage to it or to any of its natural or juridical persons, or which may be of a hazardous or deleterious nature, other States Parties, including in particular States possessing space monitoring and tracking facilities, shall respond to the greatest extent feasible to a request by that State Party, or transmitted through the Secretary-General on its behalf, for assistance under equitable and reasonable conditions in the identification of the object. A State Party making such a request shall, to the greatest extent feasible, submit information as to the time, nature and circumstances of the events giving rise to the request. Arrangements under which such assistance shall be rendered shall be the subject of agreement between the parties concerned.

ARTICLE VII

1. In this Convention, with the exception of articles VIII to XII inclusive, references to States shall be deemed to apply to any international intergovernmental organization which conducts space activities if the organization declares its acceptance of the rights and obligations provided for in this Convention and if a majority of the States members of the organization are States Parties to this Convention and to the Treaty on Principles Governing the Activities of States in the Exploration and Use of Outer Space, including the Moon and Other Celestial Bodies.

2. States members of any such organization which are States Parties to this Convention shall take all appropriate steps to ensure that the

organization makes a declaration in accordance with paragraph 1 of this article.

ARTICLE VIII

1. This Convention shall be open for signature by all States at United Nations Headquarters in New York. Any State which does not sign this Convention before its entry into force in accordance with paragraph 3 of this article may accede to it at any time.

2. This Convention shall be subject to ratification by signatory States. Instruments of ratification and instruments of accession shall be deposited with the Secretary-General of the United Nations.

3. This Convention shall enter into force among the States which have deposited instruments of ratification on the deposit of the fifth such instrument with the Secretary-General of the United Nations.

4. For States whose instruments of ratification or accession are deposited subsequent to the entry into force of this Convention, it shall enter into force on the date of the deposit of their instruments of ratification or accession.

5. The Secretary-General shall promptly inform all signatory and acceding States of the date of each signature, the date of deposit of each instrument of ratification of and accession to this Convention, the date of its entry into force and other notices.

ARTICLE IX

Any State Party to this Convention may propose amendments to the Convention. Amendments shall enter into force for each State Party to the Convention accepting the amendments upon their acceptance by a majority of the States Parties to the Convention and thereafter for each remaining State Party to the Convention on the date of acceptance by it.

ARTICLE X

Ten years after the entry into force of this Convention, the question of the review of the Convention shall be included in the provisional agenda of the United Nations General Assembly in order to consider, in the light of past application of the Convention, whether it requires revision. However, at any time after the Convention has been in force for five years, at the request of one third of the States Parties to the Convention and with the concurrence of the majority of the States Parties, a conference of the States Parties shall be convened to review this Convention. Such review shall take into account in particular any relevant technological developments, including those relating to the identification of space objects.

ARTICLE XI

Any State Party to this Convention may give notice of its withdrawal from the Convention one year after its entry into force by written notification to the Secretary-General of the United Nations. Such withdrawal shall take effect one year from the date of receipt of this notification.

ARTICLE XII

The original of this Convention, of which the Arabic, Chinese, English, French, Russian and Spanish texts are equally authentic, shall be deposited with the Secretary-General of the United Nations, who shall send certified copies thereof to all signatory and acceding States.

IN WITNESS WHEREOF the undersigned, being duly authorized thereto by their respective Governments, have signed this Convention, opened for signature at New York on the fourteenth day of January, one thousand nine hundred and seventy-five.

APPENDIX 3-5

Agreement Governing the Activities of States on the Moon and Other Celestial Bodies

The States Parties to this Agreement,

Noting the achievements of States in the exploration and use of the moon and other celestial bodies,

Recognizing that the moon, as a natural satellite of the earth, has an important role to play in the exploration of outer space,

Determined to promote on the basis of equality the further development of co-operation among States in the exploration and use of the moon and other celestial bodies,

Desiring to prevent the moon from becoming an area of international conflict,

Bearing in mind the benefits which may be derived from the exploitation of the natural resources of the moon and other celestial bodies,

Recalling the Treaty on Principles Governing the Activities of States in the Exploration and Use of Outer Space, including the Moon and Other Celestial Bodies, the Agreement on the Rescue of Astronauts, the Return of Astronauts and the Return of Objects Launched into Outer Space, the Convention on International Liability for Damage Caused by Space Objects, and the Convention on Registration of Objects Launched into Outer Space,

Taking into account the need to define and develop the provisions of these international instruments in relation to the moon and other celestial bodies, having regard to further progress in the exploration and use of outer space,

Have agreed on the following:

ARTICLE 1

1. The provisions of this Agreement relating to the moon shall also apply to other celestial bodies within the solar system, other than the earth, except in so far as specific legal norms enter into force with respect to any of these celestial bodies.

2. For the purposes of this Agreement reference to the moon shall include orbits around or other trajectories to or around it.

3. This Agreement does not apply to extraterrestrial materials which reach the surface of the earth by natural means.

ARTICLE 2

All activities on the moon, including its exploration and use, shall be carried out in accordance with international law, in particular the Charter of the United Nations, and taking into account the Declaration on Principles of International Law concerning Friendly Relations and Co-operation among States in accordance with the Charter of the United Nations, adopted by the General Assembly on 24 October 1970, in the interest of maintaining international peace and security and promoting international co-operation and mutual understanding, and with due regard to the corresponding interests of all other States Parties.

ARTICLE 3

1. The moon shall be used by all States Parties exclusively for peaceful purposes.

2. Any threat or use of force or any other hostile act or threat of hostile act on the moon is prohibited. It is likewise prohibited to use the moon in order to commit any such act or to engage in any such threat in relation to the earth, the moon, spacecraft, the personnel of spacecraft or man-made space objects.

3. States Parties shall not place in orbit around or other trajectory to or around the moon objects carrying nuclear weapons or any other kinds of weapons of mass destruction or place or use such weapons on or in the moon.

4. The establishment of military bases, installations and fortifications, the testing of any type of weapons and the conduct of military manoeuvres on the moon shall be forbidden. The use of military personnel for scientific research or for any other peaceful purposes shall not be prohibited. The use of any equipment or facility necessary for peaceful exploration and use of the moon shall also not be prohibited.

ARTICLE 4

1. The exploration and use of the moon shall be the province of all mankind and shall be carried out for the benefit and in the interests of all countries, irrespective of their degree of economic or scientific development. Due regard shall be paid to the interests of present and future generations as well as to the need to promote higher standards of living and conditions of economic and social progress and development in accordance with the Charter of the United Nations.

2. States Parties shall be guided by the principle of co-operation and mutual assistance in all their activities concerning the exploration and use of the moon. International co-operation in pursuance of this Agreement should be as wide as possible and may take place on a multilateral basis, on a bilateral basis or through international intergovernmental organizations.

ARTICLE 5

1. States Parties shall inform the Secretary-General of the United Nations as well as the public and the international scientific community, to the greatest extent feasible and practicable, of their activities concerned with the exploration and use of the moon. Information on the time, purposes, locations, orbital parameters and duration shall be given in respect of each mission to the moon as soon as possible after launching, while information on the results of each mission, including scientific results, shall be furnished upon completion of the mission. In the case of a mission lasting more than sixty days, information on conduct of the mission, including any scientific results, shall be given periodically, at thirty-day intervals. For missions lasting more than six months, only significant additions to such information need be reported thereafter.

2. If a State Party becomes aware that another State Party plans to operate simultaneously in the same area of or in the same orbit around or trajectory to or around the moon, it shall promptly inform the other State of the timing of and plans for its own operations.

3. In carrying out activities under this Agreement, States Parties shall promptly inform the Secretary-General, as well as the public and the international scientific community, of any phenomena they discover in outer space, including the moon, which could endanger human life or health, as well as of any indication of organic life.

ARTICLE 6

1. There shall be freedom of scientific investigation on the moon by all States Parties without discrimination of any kind, on the basis of equality and in accordance with international law.

2. In carrying out scientific investigations and in furtherance of the provisions of this Agreement, the States Parties shall have the right to collect on and remove from the moon samples of its mineral and other substances. Such samples shall remain at the disposal of those States Parties which caused them to be collected and may be used by them for scientific purposes. States Parties shall have regard to the desirability of making a portion of such samples available to other interested States Parties and the international scientific community for scientific investigation. States Parties may in the course of scientific investigations also use mineral and other substances of the moon in quantities appropriate for the support of their missions.

3. States Parties agree on the desirability of exchanging scientific and other personnel on expeditions to or installations on the moon to the greatest extent feasible and practicable.

ARTICLE 7

1. In exploring and using the moon, States Parties shall take measures to prevent the disruption of the existing balance of its environment, whether by introducing adverse changes in that environment, by its harmful contamination through the introduction of extra-environmental matter or otherwise. States Parties shall also take

measures to avoid harmfully affecting the environment of the earth through the introduction of extraterrestrial matter or otherwise.

2. States Parties shall inform the Secretary-General of the United Nations of the measures being adopted by them in accordance with paragraph 1 of this article and shall also, to the maximum extent feasible, notify him in advance of all placements by them of radio-active materials on the moon and of the purposes of such placements.

3. States Parties shall report to other States Parties and to the Secretary-General concerning areas of the moon having special scientific interest in order that, without prejudice to the rights of other States Parties, consideration may be given to the designation of such areas as international scientific preserves for which special protective arrangements are to be agreed upon in consultation with the competent bodies of the United Nations.

ARTICLE 8

1. States Parties may pursue their activities in the exploration and use of the moon anywhere on or below its surface, subject to the provisions of this Agreement.

2. For these purposes States Parties may, in particular:
   (a) Land their space objects on the moon and launch them from the moon;
   (b) Place their personnel, space vehicles, equipment, facilities, stations and installations anywhere on or below the surface of the moon.

Personnel, space vehicles, equipment, facilities, stations and installations may move or be moved freely over or below the surface of the moon.

3. Activities of States Parties in accordance with paragraphs 1 and 2 of this article shall not interfere with the activities of other States Parties on the moon. Where such interference may occur, the States Parties concerned shall undertake consultations in accordance with article 15, paragraphs 2 and 3, of this Agreement.

ARTICLE 9

1. States Parties may establish manned and unmanned stations on the moon. A State Party establishing a station shall use only that area which is required for the needs of the station and shall immediately inform the Secretary-General of the United Nations of the location and purposes of that station. Subsequently, at annual intervals that State shall likewise inform the Secretary-General whether the station continues in use and whether its purposes have changed.

2. Stations shall be installed in such a manner that they do not impede the free access to all areas of the moon of personnel, vehicles and equipment of other States Parties conducting activities on the moon in accordance with the provisions of this Agreement or of article I of the Treaty on Principles Governing the Activities of States in the Exploration and Use of Outer Space, including the Moon and Other Celestial Bodies.

ARTICLE 10

1. States Parties shall adopt all practicable measures to safeguard the life and health of persons on the moon. For this purpose they shall regard any person on the moon as an astronaut within the meaning of article V of the Treaty on Principles Governing the Activities of States in the Exploration and Use of Outer Space, including the Moon and Other Celestial Bodies and as part of the personnel of a spacecraft within the meaning of the Agreement on the Rescue of Astronauts, the Return of Astronauts and the Return of Objects Launched into Outer Space.

2. States Parties shall offer shelter in their stations, installations, vehicles and other facilities to persons in distress on the moon.

ARTICLE 11

1. The moon and its natural resources are the common heritage of mankind, which finds its expression in the provisions of this Agreement, in particular in paragraph 5 of this article.

2. The moon is not subject to national appropriation by any claim of sovereignty, by means of use or occupation, or by any other means.

3. Neither the surface nor the subsurface of the moon, nor any part thereof or natural resources in place, shall become property of any State, international intergovernmental or non-governmental organization, national organization or non-governmental entity or of any natural person. The placement of personnel, space vehicles, equipment, facilities, stations and installations on or below the surface of the moon, including structures connected with its surface or subsurface, shall not create a right of ownership over the surface or the subsurface of the moon or any areas thereof. The foregoing provisions are without prejudice to the international regime referred to in paragraph 5 of this article.

4. States Parties have the right to exploration and use of the moon without discrimination of any kind, on the basis of equality and in accordance with international law and the terms of this Agreement.

5. States Parties to this Agreement hereby undertake to establish an international regime, including appropriate procedures, to govern the exploitation of the natural resources of the moon as such exploitation is about to become feasible. This provision shall be implemented in accordance with article 18 of this Agreement.

6. In order to facilitate the establishment of the international regime referred to in paragraph 5 of this article, States Parties shall inform the Secretary-General of the United Nations as well as the public and the international scientific community, to the greatest extent feasible and practicable, of any natural resources they may discover on the moon.

7 The main purposes of the international regime to be established shall include:

- (a) The orderly and safe development of the natural resources of the moon;
- (b) The rational management of those resources;
- (c) The expansion of opportunities in the use of those resources;
- (d) An equitable sharing by all States Parties in the benefits derived from those resources, whereby the interests and needs of the developing countries, as well as the efforts of those countries which have contributed either directly or indirectly to the exploration of the moon, shall be given special consideration.

8. All the activities with respect to the natural resources of the moon shall be carried out in a manner compatible with the purposes specified in paragraph 7 of this article and the provisions of article 6, paragraph 2, of this Agreement.

ARTICLE 12

1. States Parties shall retain jurisdiction and control over their personnel, vehicles, equipment, facilities, stations and installations on the moon. The ownership of space vehicles, equipment, facilities, stations and installations shall not be affected by their presence on the moon.

2. Vehicles, installations and equipment or their component parts found in places other than their intended location shall be dealt with in accordance with article 5 of the Agreement on the Rescue of Astronauts, the Return of Astronauts and the Return of Objects Launched into Outer Space.

3. In the event of an emergency involving a threat to human life, States Parties may use the equipment, vehicles, installations, facilities or supplies of other States Parties on the moon. Prompt notification of such use shall be made to the Secretary-General of the United Nations or the State Party concerned.

ARTICLE 13

A State Party which learns of the crash landing, forced landing or other unintended landing on the moon of a space object, or its component parts, that were not launched by it, shall promptly inform the launching State Party and the Secretary-General of the United Nations.

ARTICLE 14

1 States Parties to this Agreement shall bear international responsibility for national activities on the moon, whether such activities are carried on by governmental agencies or by non-governmental entities, and for assuring that national activities are carried out in conformity with the provisions set forth in this Agreement. States Parties shall ensure that non-governmental entities under their jurisdiction shall engage

in activities on the moon only under the authority and continuing supervision of the appropriate State Party.

2. States Parties recognize that detailed arrangements concerning liability for damage caused on the moon, in addition to the provisions of the Treaty on Principles Governing the Activities of States in the Exploration and Use of Outer Space, including the Moon and Other Celestial Bodies and the Convention on International Liability for Damage Caused by Space Objects, may become necessary as a result of more extensive activities on the moon. Any such arrangements shall be elaborated in accordance with the procedure provided for in article 18 of this Agreement.

ARTICLE 15

1. Each State Party may assure itself that the activities of other States Parties in the exploration and use of the moon are compatible with the provisions of this Agreement. To this end, all space vehicles, equipment, facilities, stations and installations on the moon shall be open to other States Parties. Such States Parties shall give reasonable advance notice of a projected visit, in order that appropriate consultations may be held and that maximum precautions may be taken to assure safety and to avoid interference with normal operations in the facility to be visited. In pursuance of this article, any State Party may act on its own behalf or with the full or partial assistance of any other State Party or through appropriate international procedures within the framework of the United Nations and in accordance with the Charter.

2. A State Party which has reason to believe that another State Party is not fulfilling the obligations incumbent upon it pursuant to this Agreement or that another State Party is interfering with the rights which the former State has under this Agreement may request consultations with that State Party. A State Party receiving such a request shall enter into such consultations without delay. Any other State Party which requests to do so shall be entitled to take part in the consultations. Each State Party participating in such consultations shall seek a mutually acceptable resolution of any controversy and shall bear in mind the rights and interests of all States Parties. The Secretary-General of the United Nations shall be informed of the results of the consultations and shall transmit the information received to all States Parties concerned.

3. If the consultations do not lead to a mutually acceptable settlement which has due regard for the rights and interests of all States Parties, the parties concerned shall take all measures to settle the dispute by other peaceful means of their choice appropriate to the circumstances and the nature of the dispute. If difficulties arise in connection with the opening of consultations or if consultations do not lead to a mutually acceptable settlement, any State Party may seek the assistance of the Secretary-General, without seeking the consent of any other State Party concerned, in order to resolve the controversy. A State Party which does not maintain diplomatic relations with another State Party concerned shall participate in such consultations, at its choice, either itself or through another State Party or the Secretary-General as intermediary.

ARTICLE 16

With the exception of articles 17 to 21, references in this Agreement to States shall be deemed to apply to any international intergovernmental organization which conducts space activities if the organization declares its acceptance of the rights and obligations provided for in this Agreement and if a majority of the States members of the organization are States Parties to this Agreement and to the Treaty on Principles Governing the Activities of States in the Exploration and Use of Outer Space, including the Moon and Other Celestial Bodies. States members of any such organization which are States Parties to this Agreement shall take all appropriate steps to ensure that the organization makes a declaration in accordance with the foregoing.

ARTICLE 17

Any State Party to this Agreement may propose amendments to the Agreement. Amendments shall enter into force for each State Party to the Agreement accepting the amendments upon their acceptance by a majority of the States Parties to the Agreement and thereafter for each remaining State Party to the Agreement on the date of acceptance by it.

ARTICLE 18

Ten years after the entry into force of this Agreement, the question of the review of the Agreement shall be included in the provisional agenda of the General Assembly of the United Nations in order to consider, in the light of past application of the Agreement, whether it requires revision. However, at any time after the Agreement has been in force for five years, the Secretary-General of the United Nations, as depository, shall, at the request of one third of the States Parties to the Agreement and with the concurrence of the majority of the States Parties, convene a conference of the States Parties to review this Agreement. A review conference shall also consider the question of the implementation of the provisions of article 11, paragraph 5, on the basis of the principle referred to in paragraph 1 of that article and taking into account in particular any relevant technological developments.

ARTICLE 19

1. This Agreement shall be open for signature by all States at United Nations Headquarters in New York.

2. This Agreement shall be subject to ratification by signatory States. Any State which does not sign this Agreement before its entry into force in accordance with paragraph 3 of this article may accede to it at any time. Instruments of ratification or accession shall be deposited with the Secretary-General of the United Nations.

3. This Agreement shall enter into force on the thirtieth day following the date of deposit of the fifth instrument of ratification.

4. For each State depositing its instrument of ratification or accession after the entry into force of this Agreement, it shall enter into force on the thirtieth day following the date of deposit of any such instrument.

5. The Secretary-General shall promptly inform all signatory and acceding States of the date of each signature, the date of deposit of each instrument of ratification or accession to this Agreement, the date of its entry into force and other notices.

ARTICLE 20

Any State Party to this Agreement may give notice of its withdrawal from the Agreement one year after its entry into force by written notification to the Secretary-General of the United Nations. Such withdrawal shall take effect one year from the date of receipt of this notification.

ARTICLE 21

The original of this Agreement, of which the Arabic, Chinese, English, French, Russian and Spanish texts are equally authentic, shall be deposited with the Secretary-General of the United Nations, who shall send certified copies thereof to all signatory and acceding States.

IN WITNESS WHEREOF the undersigned, being duly authorized thereto by their respective Governments, have signed this Agreement, opened for signature at New York on 18 December 1979.

APPENDIX 3-6

Principles Governing the Use by States of Artificial Earth Satellites for International Direct Television Broadcasting

The General Assembly,

Recalling its resolution 2916 (XXVII) of 9 November 1972, in which it stressed the necessity of elaborating principles governing the use by States of artificial Earth satellites for international direct television broadcasting, and mindful of the importance of concluding an international agreement or agreements,

Recalling further its resolutions 3182 (XXVIII) of 18 December 1973, 3234 (XXIX) of 12 November 1974, 3388 (XXX) of 18 November 1975, 31/8 of 8 November 1976, 32/196 of 20 December 1977, 33/16 of 10 November 1978, 34/66 of 5 December 1979 and 35/14 of 3 November 1980, and its resolution 36/35 of 18 November 1981 in which it decided to consider at its thirty-seventh session the adoption of a draft set of principles governing the use by States of artificial Earth satellites for international direct television broadcasting,

Noting with appreciation the efforts made in the Committee on the Peaceful Uses of Outer Space and its Legal Subcommittee to comply with the directives issued in the above-mentioned resolutions,

Considering that several experiments of direct broadcasting by satellite have been carried out and that a number of direct broadcasting satellite systems are operational in some countries and may be commercialized in the very near future,

Taking into consideration that the operation of international direct broadcasting satellites will have significant international political, economic, social and cultural implications,

Believing that the establishment of principles for international direct television broadcasting will contribute to the strengthening of international cooperation in this field and further the purposes and principles of the Charter of the United Nations,

Adopts the Principles Governing the Use by States of Artificial Earth Satellites for International Direct Television Broadcasting set forth in the annex to the present resolution.

Annex

Principles Governing the Use by States of Artificial Earth Satellites for International Direct Television Broadcasting

A. PURPOSES AND OBJECTIVES

1. Activities in the field of international direct television broadcasting by satellite should be carried out in a manner compatible with the sovereign rights of States, including the principle of non-intervention, as well as with the right of everyone to seek, receive and impart information and ideas as enshrined in the relevant United Nations instruments.

2. Such activities should promote the free dissemination and mutual exchange of information and knowledge in cultural and scientific fields, assist in educational, social and economic development, particularly in the developing countries, enhance the qualities of life of all peoples and provide recreation with due respect to the political and cultural integrity of States.

3. These activities should accordingly be carried out in a manner compatible with the development of mutual understanding and the strengthening of friendly relations and cooperation among all States and peoples in the interest of maintaining international peace and security.

B. APPLICABILITY OF INTERNATIONAL LAW

4. Activities in the field of international direct television broadcasting by satellite should be conducted in accordance with international law, including the Charter of the United Nations, the Treaty on Principles

Governing the Activities of States in the Exploration and Use of Outer Space, including the Moon and Other Celestial Bodies, of 27 January 1967, the relevant provisions of the International 1 Telecommunication Convention and its Radio Regulations and of international instruments relating to friendly relations and cooperation among States and to human rights.

C. RIGHTS AND BENEFITS

5. Every State has an equal right to conduct activities in the field of international direct television broadcasting by satellite and to authorize such activities by persons and entities under its jurisdiction. All States and peoples are entitled to and should enjoy the benefits from such activities. Access to the technology in this field should be available to all States without discrimination on terms mutually agreed by all concerned.

D. INTERNATIONAL COOPERATION

6. Activities in the field of international direct television broadcasting by satellite should be based upon and encourage international cooperation. Such cooperation should be the subject of appropriate arrangements. Special consideration should be given to the needs of the developing countries in the use of international direct television broadcasting by satellite for the purpose of accelerating their national development.

E. PEACEFUL SETTLEMENT OF DISPUTES

7. Any international dispute that may arise from activities covered by these principles should be settled through established procedures for the peaceful settlement of disputes agreed upon by the parties to the dispute in accordance with the provisions of the Charter of the United Nations.

F. STATE RESPONSIBILITY

8. States should bear international responsibility for activities in the field of international direct television broadcasting by satellite carried out by them or under their jurisdiction and for the conformity of any such activities with the principles set forth in this document.

9. When international direct television broadcasting by satellite is carried out by an international intergovernmental organization, the responsibility referred to in paragraph 8 above should be borne both by that organization and by the States participating in it.

G. DUTY AND RIGHT TO CONSULT

10. Any broadcasting or receiving State within an international direct television broadcasting satellite service established between them requested to do so by any other broadcasting or receiving State within the same service should promptly enter into consultations with the requesting State regarding its activities in the field of international direct television broadcasting by satellite, without prejudice to other consultations which these States may undertake with any other State on that subject.

H. COPYRIGHT AND NEIGHBOURING RIGHTS

11. Without prejudice to the relevant provisions of international law, States should cooperate on a bilateral and multilateral basis for protection of copyright and neighbouring rights by means of appropriate agreements between the interested States or the competent legal entities acting under their jurisdiction. In such cooperation they should give special consideration to the interests of developing countries in the use of direct television broadcasting for the purpose of accelerating their national development.

I. NOTIFICATION TO THE UNITED NATIONS

12. In order to promote international cooperation in the peaceful exploration and use of outer space, States conducting or authorizing activities in the field of international direct television broadcasting by satellite should inform the Secretary-General of the United Nations, to the greatest extent possible, of the nature of such activities. On receiving this information, the Secretary-General should disseminate it immediately and effectively to the relevant specialized agencies, as well as to the public and the international scientific community.

J. CONSULTATIONS AND AGREEMENTS BETWEEN STATES

13 A State which intends to establish or authorize the establishment of an international direct television broadcasting satellite service shall without delay notify the proposed receiving State or States of such intention and shall promptly enter into consultation with any of those States which so requests.

14 An international direct television broadcasting satellite service shall only be established after the conditions set forth in paragraph 13 above have been met and on the basis of agreements and/or arrangements in conformity with the relevant instruments of the International Telecommunication Union and in accordance with these principles.

15. With respect to the unavoidable overspill of the radiation of the satellite signal, the relevant instruments of the International Telecommunication Union shall be exclusively applicable.

APPENDIX 3-7

Principles Relating to Remote Sensing of the Earth from Outer Space

The General Assembly,

Recalling its resolution 3234 (XXIX) of 12 November 1974, in which it recommended that the Legal Subcommittee of the Committee on the Peaceful Uses of Outer Space should consider the question of the legal implications of remote sensing of the Earth from space, as well as its resolutions 3388 (XXX) of 18 November 1975, 31/8 of 8 November 1976, 32/196 A of 20 December 1977, 33/16 of 10 November 1978, 34/66 of 5 December 1979, 35/14 of 3 November 1980, 36/35 of 18 November 1981, 37/89 of 10 December 1982, 38/80 of 15 December 1983, 39/96 of 14 December 1984 and 40/162 of 16 December 1985, in which it called for a detailed consideration of the legal implications of remote sensing of the Earth from space, with the aim of formulating draft principles relating to remote sensing,

Having considered the report of the Committee on the Peaceful Uses of Outer Space on the work of its twenty-ninth session (A/41/20) and the text of the draft principles relating to remote sensing of the Earth from space, annexed thereto,

Noting with satisfaction that the Committee on the Peaceful Uses of Outer Space, on the basis of the deliberations of its Legal Subcommittee, has endorsed the text of the draft principles relating to remote sensing of the Earth from space,

Believing that the adoption of the principles relating to remote sensing of the Earth from space will contribute to the strengthening of international cooperation in this field,

Adopts the principles relating to remote sensing of the Earth from space set forth in the annex to the present resolution.

Annex

Principles Relating to Remote Sensing of the Earth from Space

PRINCIPLE I

For the purposes of these principles with respect to remote sensing activities:

(a) The term "remote sensing" means the sensing of the Earth's surface from space by making use of the properties of electromagnetic waves emitted, reflected or diffracted by the sensed objects, for the purpose of improving natural resources management, land use and the protection of the environment;

(b) The term "primary data" means those raw data that are acquired by remote sensors borne by a space object and that are transmitted or delivered to the ground from space by telemetry in the form of electromagnetic signals, by photographic film, magnetic tape or any other means;

(c) The term "processed data" means the products resulting from the processing of the primary data, needed to make such data usable;

(d) The term "analysed information" means the information resulting from the interpretation of processed data, inputs of data and knowledge from other sources;

(e) The term "remote sensing activities" means the operation of remote sensing space systems, primary data collection and storage stations, and activities in processing, interpreting and disseminating the processed data.

PRINCIPLE II

Remote sensing activities shall be carried out for the benefit and in the interests of all countries, irrespective of their degree of economic, social or scientific and technological development, and taking into particular consideration the needs of the developing countries.

PRINCIPLE III

Remote sensing activities shall be conducted in accordance with international law, including the Charter of the United Nations, the Treaty on Principles Governing the Activities of States in the Exploration and Use of Outer Space, including the Moon and Other Celestial Bodies, and the relevant instruments of the International Telecommunication Union.

PRINCIPLE IV

Remote sensing activities shall be conducted in accordance with the principles contained in article I of the Treaty on Principles Governing the Activities of States in the Exploration and Use of Outer Space, including the Moon and Other Celestial Bodies, which, in particular, provides that the exploration and use of outer space shall be carried out for the benefit and in the interests of all countries, irrespective of their degree of economic or scientific development, and stipulates the principle of freedom of exploration and use of outer space on the basis of equality. These activities shall be conducted on the basis of respect for the principle of full and permanent sovereignty of all States and peoples over their own wealth and natural resources, with due regard to the rights and interests, in accordance with international law, of other States and entities under their jurisdiction. Such activities shall not be conducted in a manner detrimental to the legitimate rights and interests of the sensed State.

PRINCIPLE V

States carrying out remote sensing activities shall promote international cooperation in these activities. To this end, they shall make available to other States opportunities for participation therein. Such participation shall be based in each case on equitable and mutually acceptable terms.

PRINCIPLE VI

In order to maximize the availability of benefits from remote sensing activities, States are encouraged, through agreements or other arrangements, to provide for the establishment and operation of data collecting and storage stations and processing and interpretation facilities, in particular within the framework of regional agreements or arrangements wherever feasible.

PRINCIPLE VII

States participating in remote sensing activities shall make available technical assistance to other interested States on mutually agreed terms.

PRINCIPLE VIII

The United Nations and the relevant agencies within the United Nations system shall promote international cooperation, including technical assistance and coordination in the area of remote sensing.

PRINCIPLE IX

In accordance with article IV of the Convention on Registration of Objects Launched into Outer Space and article XI of the Treaty on Principles Governing the Activities of States in the Exploration and Use of Outer Space, including the Moon and Other Celestial Bodies, a State carrying out a programme of remote sensing shall inform the Secretary-General of the United Nations. It shall, moreover, make available any other relevant information to the greatest extent feasible and practicable to any other State, particularly any developing country that is affected by the programme, at its request.

PRINCIPLE X

Remote sensing shall promote the protection of the Earth's natural environment.

To this end, States participating in remote sensing activities that have identified information in their possession that is capable of averting any phenomenon harmful to the Earth's natural environment shall disclose such information to States concerned.

PRINCIPLE XI

Remote sensing shall promote the protection of mankind from natural disasters.

To this end, States participating in remote sensing activities that have identified processed data and analysed information in their possession that may be useful to States affected by natural disasters, or likely to be affected by impending natural disasters, shall transmit such data and information to States concerned as promptly as possible.

PRINCIPLE XII

As soon as the primary data and the processed data concerning the territory under its jurisdiction are produced, the sensed State shall have access to them on a non-discriminatory basis and on reasonable cost terms. The sensed State shall also have access to the available analysed information concerning the territory under its jurisdiction in the possession of any State participating in remote sensing activities on the same basis and terms, taking particularly into account the needs and interests of the developing countries.

PRINCIPLE XIII

To promote and intensify international cooperation, especially with regard to the needs of developing countries, a State carrying out remote sensing of the Earth from space shall, upon request, enter into consultations with a State whose territory is sensed in order to make available opportunities for participation and enhance the mutual benefits to be derived therefrom.

PRINCIPLE XIV

In compliance with article VI of the Treaty on Principles Governing the Activities of States in the Exploration and Use of Outer Space, including the Moon and Other Celestial Bodies, States operating remote sensing satellites shall bear international responsibility for their activities and assure that such activities are conducted in accordance with these principles and the norms of international law, irrespective of whether such activities are carried out by governmental or non-governmental entities or through international organizations to which such States are parties. This principle is without prejudice to the applicability of the norms of international law on State responsibility for remote sensing activities.

PRINCIPLE XV

Any dispute resulting from the application of these principles shall be resolved through the established procedures for the peaceful settlement of disputes.

APPENDIX 3-8

Principles Relevant to the Use of Nuclear Power Sources In Outer Space

The General Assembly,
  Having considered the report of the Committee on the Peaceful Uses of Outer Space on the work of its thirty-fifth session and the text of the Principles Relevant to the Use of Nuclear Power Sources in Outer Space as approved by the Committee and annexed to its report,
  Recognizing that for some missions in outer space nuclear power sources are particularly suited or even essential owing to their compactness, long life and other attributes,
  Recognizing also that the use of nuclear power sources in outer space should focus on those applications which take advantage of the particular properties of nuclear power sources,
  Recognizing further that the use of nuclear power sources in outer space should be based on a thorough safety assessment, including probabilistic risk analysis, with particular emphasis on reducing the risk of accidental exposure of the public to harmful radiation or radioactive material,
  Recognizing the need, in this respect, for a set of principles containing goals and guidelines to ensure the safe use of nuclear power sources in outer space,
  Affirming that this set of Principles applies to nuclear power sources in outer space devoted to the generation of electric power on board space objects for non-propulsive purposes, which have characteristics generally comparable to those of systems used and missions performed at the time of the adoption of the Principles,

Recognizing that this set of Principles will require future revision in view of emerging nuclear power applications and of evolving international recommendations on radiological protection,

Adopts the Principles Relevant to the Use of Nuclear Power Sources in Outer Space as set forth below.

PRINCIPLE 1. APPLICABILITY OF INTERNATIONAL LAW

Activities involving the use of nuclear power sources in outer space shall be carried out in accordance with international law, including in particular the Charter of the United Nations and the Treaty on Principles Governing the Activities of States in the Exploration and Use of Outer Space, including the Moon and Other Celestial Bodies.

PRINCIPLE 2. USE OF TERMS

1. For the purpose of these Principles, the terms "launching State" and "State launching" mean the State which exercises jurisdiction and control over a space object with nuclear power sources on board at a given point in time relevant to the principle concerned.

2. For the purpose of principle 9, the definition of the term "launching State" as contained in that principle is applicable.

3. For the purposes of principle 3, the terms "foreseeable" and "all possible" describe a class of events or circumstances whose overall probability of occurrence is such that it is considered to encompass only credible possibilities for purposes of safety analysis. The term "general concept of defence-in-depth" when applied to nuclear power sources in outer space refers to the use of design features and mission operations in place of or in addition to active systems, to prevent or mitigate the consequences of system malfunctions. Redundant safety systems are not necessarily required for each individual component to achieve this purpose. Given the special requirements of space use and of varied missions, no particular set of systems or features can be specified as essential to achieve this objective. For the purposes of paragraph 2 (d) of principle 3, the term "made critical" does not include actions such as zero-power testing which are fundamental to ensuring system safety.

PRINCIPLE 3. GUIDELINES AND CRITERIA FOR SAFE USE

In order to minimize the quantity of radioactive material in space and the risks involved, the use of nuclear power sources in outer space shall be restricted to those space missions which cannot be operated by non-nuclear energy sources in a reasonable way.

1. General goals for radiation protection and nuclear safety

(a) States launching space objects with nuclear power sources on board shall endeavour to protect individuals, populations and the biosphere against radiological hazards. The design and use of space objects with nuclear power sources on board shall ensure, with a high degree of confidence, that the hazards, in foreseeable operational or accidental circumstances, are kept below acceptable levels as defined in paragraphs 1 (b) and (c).

Such design and use shall also ensure with high reliability that radioactive material does not cause a significant contamination of outer space.

(b) During the normal operation of space objects with nuclear power sources on board, including re-entry from the sufficiently high orbit as defined in paragraph 2 (b), the appropriate radiation protection objective for the public recommended by the International Commission on Radiological Protection shall be observed. During such normal operation there shall be no significant radiation exposure.

(c) To limit exposure in accidents, the design and construction of the nuclear power source systems shall take into account relevant and generally accepted international radiological protection guidelines.

Except in cases of low-probability accidents with potentially serious radiological consequences, the design for the nuclear power source systems shall, with a high degree of confidence, restrict radiation exposure to a limited geographical region and to individuals to the principal limit of 1 mSv in a year. It is permissible to use a subsidiary dose limit of 5 mSv in a year for some years, provided that the average annual effective dose equivalent over a lifetime does not exceed the principal limit of 1 mSv in a year.

The probability of accidents with potentially serious radiological consequences referred to above shall be kept extremely small by virtue of the design of the system.

Future modifications of the guidelines referred to in this paragraph shall be applied as soon as practicable.

(d) Systems important for safety shall be designed, constructed and operated in accordance with the general concept of defence-in-depth. Pursuant to this concept, foreseeable safety-related failures or malfunctions must be capable of being corrected or counteracted by an action or a procedure, possibly automatic.

The reliability of systems important for safety shall be ensured, *inter alia*, by redundancy, physical separation, functional isolation and adequate independence of their components.

Other measures shall also be taken to raise the level of safety.

2. Nuclear reactors

(a) Nuclear reactors may be operated:

- (i) On interplanetary missions;
- (ii) In sufficiently high orbits as defined in paragraph 2 (b);
- (iii) In low-Earth orbits if they are stored in sufficiently high orbits after the operational part of their mission.

(b) The sufficiently high orbit is one in which the orbital lifetime is long enough to allow for a sufficient decay of the fission products to approximately the activity of the actinides. The sufficiently high orbit must be such that the risks to existing and future outer space missions and of collision with other space objects are kept to a minimum. The necessity for the parts of a destroyed reactor also to attain the required decay time before re-entering the Earth's atmosphere shall be considered in determining the sufficiently high orbit altitude.

(c) Nuclear reactors shall use only highly enriched uranium 235 as fuel. The design shall take into account the radioactive decay of the fission and activation products.

(d) Nuclear reactors shall not be made critical before they have reached their operating orbit or interplanetary trajectory.

(e) The design and construction of the nuclear reactor shall ensure that it cannot become critical before reaching the operating orbit during all possible events, including rocket explosion, re-entry, impact on ground or water, submersion in water or water intruding into the core.

(f) In order to reduce significantly the possibility of failures in satellites with nuclear reactors on board during operations in an orbit with a lifetime less than in the sufficiently high orbit (including operations for transfer into the sufficiently high orbit), there shall be a highly reliable operational system to ensure an effective and controlled disposal of the reactor.

3. Radioisotope generators

(a) Radioisotope generators may be used for interplanetary missions and other missions leaving the gravity field of the Earth. They may also be used in Earth orbit if, after conclusion of the operational part of their mission, they are stored in a high orbit. In any case ultimate disposal is necessary.

(b) Radioisotope generators shall be protected by a containment system that is designed and constructed to withstand the heat and aerodynamic forces of re-entry in the upper atmosphere under foreseeable orbital conditions, including highly elliptical or hyperbolic orbits where relevant. Upon impact, the containment system and the physical form of the isotope shall ensure that no radioactive material is scattered into the environment so that the impact area can be completely cleared of radioactivity by a recovery operation.

PRINCIPLE 4. SAFETY ASSESSMENT

1. A launching State as defined in principle 2, paragraph 1, at the time of launch shall, prior to the launch, through cooperative arrangements, where relevant, with those which have designed, constructed or manufactured the nuclear power sources, or will operate the space object, or from whose territory or facility such an object will be launched, ensure that a thorough and comprehensive safety assessment is conducted. This assessment shall cover as well all relevant phases of the mission and shall deal with all systems involved, including the means of launching, the space platform, the nuclear power source and its equipment and the means of control and communication between ground and space.

2. This assessment shall respect the guidelines and criteria for safe use contained in principle 3.

3. Pursuant to article XI of the Treaty on Principles Governing the Activities of States in the Exploration and Use of Outer Space, including the Moon and Other Celestial Bodies, the results of this safety assessment, together with, to the extent feasible, an indication of the approximate intended time-frame of the launch, shall be made publicly available prior to each launch, and the Secretary-General of the United Nations shall be informed on how States may obtain such results of the safety assessment as soon as possible prior to each launch.

PRINCIPLE 5. NOTIFICATION OF RE-ENTRY

1. Any State launching a space object with nuclear power sources on board shall in a timely fashion inform States concerned in the event this space object is malfunctioning with a risk of re-entry of radioactive materials to the Earth. The information shall be in accordance with the following format:

(a) System parameters:
  (i) Name of launching State or States, including the address of the authority which may be contacted for additional information or assistance in case of accident;
  (ii) International designation;
  (iii) Date and territory or location of launch;
  (iv) Information required for best prediction of orbit lifetime, trajectory and impact region;
  (v) General function of spacecraft;
(b) Information on the radiological risk of nuclear power source(s):
  (i) Type of nuclear power source: radioisotopic/reactor;
  (ii) The probable physical form, amount and general radiological characteristics of the fuel and contaminated and/or activated components likely to reach the ground. The term "fuel" refers to the nuclear material used as the source of heat or power.

2. The information, in accordance with the format above, shall be provided by the launching State as soon as the malfunction has become known. It shall be updated as frequently as practicable and the frequency of dissemination of the updated information shall increase as the anticipated time of re-entry into the dense layers of the Earth's

atmosphere approaches so that the international community will be informed of the situation and will have sufficient time to plan for any national response activities deemed necessary.

3. The updated information shall also be transmitted to the Secretary-General of the United Nations with the same frequency.

PRINCIPLE 6. CONSULTATIONS

States providing information in accordance with principle 5 shall, as far as reasonably practicable, respond promptly to requests for further information or consultations sought by other States.

PRINCIPLE 7. ASSISTANCE TO STATES

1. Upon the notification of an expected re-entry into the Earth's atmosphere of a space object containing a nuclear power source on board and its components, all States possessing space monitoring and tracking facilities, in the spirit of international cooperation, shall communicate the relevant information that they may have available on the malfunctioning space object with a nuclear power source on board to the Secretary-General of the United Nations and the State concerned as promptly as possible to allow States that might be affected to assess the situation and take any precautionary measures deemed necessary.

2. After re-entry into the Earth's atmosphere of a space object containing a nuclear power source on board and its components:

   (a) The launching State shall promptly offer and, if requested by the affected State, provide promptly the necessary assistance to eliminate actual and possible harmful effects, including assistance to identify the location of the area of impact of the nuclear power source on the Earth's surface, to detect the re-entered material and to carry out retrieval or clean-up operations;
   (b) All States, other than the launching State, with relevant technical capabilities and international organizations with such technical capabilities shall, to the extent possible, provide necessary assistance upon request by an affected State.

In providing the assistance in accordance with subparagraphs (a) and (b) above, the special needs of developing countries shall be taken into account.

PRINCIPLE 8. RESPONSIBILITY

In accordance with article VI of the Treaty on Principles Governing the Activities of States in the Exploration and Use of Outer Space, including the Moon and Other Celestial Bodies, States shall bear international responsibility for national activities involving the use of nuclear power sources in outer space, whether such activities are carried on by governmental agencies or by non-governmental entities, and for assuring that such national activities are carried out in conformity with that Treaty and the recommendations contained in these Principles. When activities in outer space involving the use of nuclear power sources are carried on by an international organization, responsibility for compliance with the aforesaid Treaty and the recommendations contained in these Principles shall be borne both by the international organization and by the States participating in it.

PRINCIPLE 9. LIABILITY AND COMPENSATION

1. In accordance with article VII of the Treaty on Principles Governing the Activities of States in the Exploration and Use of Outer Space, including the Moon and Other Celestial Bodies, and the provisions of the Convention on International Liability for Damage Caused by Space Objects, each State which launches or procures the launching of a space object and each State from whose territory or facility a space object is launched shall be internationally liable for damage caused by such space objects or their component parts. This fully applies to the case of such a space object carrying a nuclear power source on board. Whenever two or more States jointly launch such a space object, they shall be jointly and severally liable for any damage caused, in accordance with article V of the above-mentioned Convention.

2. The compensation that such States shall be liable to pay under the aforesaid Convention for damage shall be determined in accordance with international law and the principles of justice and equity, in order to provide such reparation in respect of the damage as will restore the person, natural or juridical, State or international organization on

whose behalf a claim is presented to the condition which would have existed if the damage had not occurred.

3. For the purposes of this principle, compensation shall include reimbursement of the duly substantiated expenses for search, recovery and clean-up operations, including expenses for assistance received from third parties.

PRINCIPLE 10. SETTLEMENT OF DISPUTES

Any dispute resulting from the application of these Principles shall be resolved through negotiations or other established procedures for the peaceful settlement of disputes, in accordance with the Charter of the United Nations.

PRINCIPLE 11. REVIEW AND REVISION

These Principles shall be reopened for revision by the Committee on the Peaceful Uses of Outer Space no later than two years after their adoption.

APPENDIX 3-9

Declaration on International Cooperation in the Exploration and Use of Outer Space for the Benefit and in the Interest of All States, Taking into Particular Account the Needs of Developing Countries

The General Assembly,

Having considered the report of the Committee on the Peaceful Uses of Outer Space on the work of its thirty-ninth session and the text of the Declaration on International Cooperation in the Exploration and Use of Outer Space for the Benefit and in the Interest of All States, Taking into Particular Account the Needs of Developing Countries, as approved by the Committee and annexed to its report, [2]

Bearing in mind the relevant provisions of the Charter of the United Nations,

Recalling notably the provisions of the Treaty on the Principles Governing the Activities of States in the Exploration and Use of Outer Space, including the Moon and Other Celestial Bodies, [3]

Recalling also its relevant resolutions relating to activities in outer space,

Bearing in mind the recommendations of the Second United Nations Conference on the Exploration and Peaceful Uses of Outer Space, [4] and of other international conferences relevant in this field,

Recognizing the growing scope and significance of international cooperation among States and between States and international organizations in the exploration and use of outer space for peaceful purposes,

Considering experiences gained in international cooperative ventures,

Convinced of the necessity and the significance of further strengthening international cooperation in order to reach a broad and efficient collaboration in this field for the mutual benefit and in the interest of all parties involved,

Desirous of facilitating the application of the principle that the exploration and use of outer space, including the Moon and other celestial bodies, shall be carried out for the benefit and in the interest of all countries, irrespective of their degree of economic or scientific development, and shall be the province of all mankind,

Adopts the Declaration on International Cooperation in the Exploration and Use of Outer Space for the Benefit and in the Interest of All States, Taking into Particular Account the Needs of Developing Countries, set forth in the annex to the present resolution.

Annex. Declaration on International Cooperation in the Exploration and Use of Outer Space for the Benefit and in the Interest of all States, Taking into Particular Account the Needs of Developing Countries

1. International cooperation in the exploration and use of outer space for peaceful purposes (hereafter "international cooperation") shall be conducted in accordance with the provisions of international law, including the Charter of the United Nations and the Treaty on the Principles Governing the Activities of States in the Exploration and Use of Outer Space, including the Moon and Other Celestial Bodies. It shall be carried out for the benefit and in the interest of all States,

irrespective of their degree of economic, social or scientific and technological development, and shall be the province of all mankind. Particular account should be taken of the needs of developing countries.

2. States are free to determine all aspects of their participation in international cooperation in the exploration and use of outer space on an equitable and mutually acceptable basis. Contractual terms in such cooperative ventures should be fair and reasonable and they should be in full compliance with the legitimate rights and interests of the parties concerned as, for example, with intellectual property rights.

3. All States, particularly those with relevant space capabilities and with programmes for the exploration and use of outer space, should contribute to promoting and fostering international cooperation on an equitable and mutually acceptable basis. In this context, particular attention should be given to the benefit for and the interests of developing countries and countries with incipient space programmes stemming from such international cooperation conducted with countries with more advanced space capabilities.

4. International cooperation should be conducted in the modes that are considered most effective and appropriate by the countries concerned, including, inter alia, governmental and non-governmental; commercial and non-commercial; global, multilateral, regional or bilateral; and international cooperation among countries in all levels of development.

5. International cooperation, while taking into particular account the needs of developing countries, should aim, *inter alia*, at the following goals, considering their need for technical assistance and rational and efficient allocation of financial and technical resources:

   (a) Promoting the development of space science and technology and of its applications;
   (b) Fostering the development of relevant and appropriate space capabilities in interested States;
   (c) Facilitating the exchange of expertise and technology among States on a mutually acceptable basis.

6. National and international agencies, research institutions, organizations for development aid, and developed and developing countries alike should consider the appropriate use of space

applications and the potential of international cooperation for reaching their development goals.

7. The Committee on the Peaceful Uses of Outer Space should be strengthened in its role, among others, as a forum for the exchange of information on national and international activities in the field of international cooperation in the exploration and use of outer space.

8. All States should be encouraged to contribute to the United Nations Programme on Space Applications and to other initiatives in the field of international cooperation in accordance with their space capabilities and their participation in the exploration and use of outer space.

APPENDIX 6

Sample Contract for Launch Services

APPENDIX 6

Sample Contract for Launch Services

The following exhibit contains excerpts from a Contract for Launch Services between Iridium Satellite LLC and Space Exploration Technologies Corp, dated March 19, 2010. This document is Exhibit 10.5 to the Form 10-Q/A Amended Quarterly Report filed by Iridium Communications Inc. with the U.S. Securities and Exchange Commission on March 29, 2011. Portions of the contract not relevant to chapter 6 have been redacted due to space restrictions.

CONTRACT FOR LAUNCH SERVICES

No. IS-10-008

Between

Iridium Satellite LLC

and

Space Exploration Technologies Corp.

The attached Contract and information contained therein is confidential and proprietary to Iridium Satellite LLC, its Affiliates and Space Exploration Technologies Corp. and shall not be published or disclosed to any third party except as permitted by the terms and conditions of this Contract.

CONFIDENTIAL TREATMENT HAS BEEN REQUESTED FOR PORTIONS OF THIS EXHIBIT. THE COPY FILED HEREWITH OMITS THE INFORMATION SUBJECT TO A CONFIDENTIALITY REQUEST. OMISSIONS ARE DESIGNATED [*** . . . ***]. A COMPLETE VERSION OF THIS EXHIBIT HAS BEEN FILED SEPARATELY WITH THE SECURITIES AND EXCHANGE COMMISSION.

TABLE OF CONTENTS

	Page
Article 1 DEFINITIONS	1
Article 2 SERVICES TO BE PROVIDED	8
Article 3 CONTRACT PRICE	11
Article 4 PAYMENT	11
Article 5 LAUNCH SCHEDULE	13
Article 6 LAUNCH SCHEDULE ADJUSTMENTS	15
Article 7 REPRESENTATIONS AND WARRANTIES	20
Article 8 COORDINATION AND COMMUNICATION BETWEEN CUSTOMER AND CONTRACTOR	20
Article 9 ADDITIONAL CONTRACTOR AND CUSTOMER OBLIGATIONS PRIOR TO LAUNCH	22

Article 10 CUSTOMER ACCESS	22
Article 11 LAUNCH VEHICLE QUALIFICATION	23
Article 12 PERMITS AND APPROVALS AND COMPLIANCE WITH UNITED STATES GOVERNMENT REQUIREMENTS	25
Article 13 CHANGES	26
Article 14 INDEMNITY, EXCLUSION OF WARRANTY, WAIVER OF LIABILITY AND ALLOCATION OF CERTAIN RISKS	27
Article 15 INSURANCE	32
Article 16 REFLIGHT	34
Article 17 TERMINATION	35
Article 18 DISPUTE RESOLUTION	37
Article 19 CONFIDENTIALITY	39
Article 20 INTELLECTUAL PROPERTY	41
Article 21 RIGHT OF OWNERSHIP AND CUSTODY	42
Article 22 FORCE MAJEURE	42
Article 23 EFFECTIVE DATE OF CONTRACT	43
Article 24 MISCELLANEOUS	44

CONTRACT FOR LAUNCH SERVICES

This CONTRACT FOR LAUNCH SERVICES (hereinafter "this Contract") is made and entered into as of the 19th Day of March, 2010, by and between **Iridium Satellite LLC**, a limited liability company organized and existing under the laws of Delaware, having its office at 6707 Democracy Boulevard, Suite 300, Bethesda, MA 20817 ("Customer") and **Space Exploration Technologies Corp.**, a Delaware corporation, having its office at 1 Rocket Road, Hawthorne, CA 90250 ("Contractor").

Article 1

Definitions

Article 2

Services to Be Provided

2.1 <u>Launch Services</u>. Contractor shall provide Launch Services for [*** ... ***] dedicated Launches of Satellite Batches ("Firm Launch(es)") and up to an additional [*** ... ***] dedicated Launches of Satellite Batches, if so exercised by Customer ("Additional Launch(es)"), in accordance with this Section 2.1 and the Statement of Work.

    2.1.1 If any Launch Service is not a Launch Success, Customer may exercise Additional Launch(es) (on a per Launch Service basis) at anytime up to the end of the Launch Campaign Period plus [*** ... ***] Days. Such Additional Launch(es) will be performed by Contractor within [*** ... ***] months of Customer's exercise of an Additional Launch, subject to available Launch Opportunities. The Launch Service Price, Milestones and Milestone Payments for such Additional Launch(es) under this Section 2.1.1 shall be the same Launch Service Price, Milestone and Milestone Payments as set forth for the first Firm Launch Service in accordance with Exhibit C.

    2.1.2 Customer may reserve the right to procure up to [*** ... ***] Additional Launches, if any Satellite or Satellite Batch(es) experience a loss or failure following Launch for reasons other than a Launch Failure. In order to reserve such Additional Launches, Customer must pay a reservation fee of [*** ... ***] US Dollars (US $[*** ... ***]) no later than [*** ... ***] months after EDC. The Additional Launches may be exercised by Customer at anytime up to the end of the Launch Campaign Period plus [*** ... ***] Days. Such Additional Launch(es) will be performed by Contractor within [*** ... ***] months (or such longer period as mutually agreed by Customer and Contractor) of Customer's exercise of an Additional Launch, subject to available Launch Opportunities. If exercised, the pricing for such Additional Launch(es) shall be in accordance with Exhibit E. If Customer does not exercise an Additional Launch pursuant to this Section 2.1.2, the reservation fee paid for such Additional Launch to Contractor will be refunded to Customer within [*** ... ***] Days of Customer's notice of such effect.

    2.1.3 Notwithstanding Sections 2.1.1 or 2.1.2, Customer may procure up to [*** ... ***] further Additional Launch(es), that may be exercised up through [*** ... ***], that at the time of such exercise are assigned a Launch Slot ending no later than [*** ... ***] (subject to available Launch Opportunities). The pricing for such Additional Launch(es) shall be determined in accordance with [*** ... ***] associated with [*** ... ***] as of the applicable exercise date thereof. The Milestones and Milestone Payment percentages for such Additional Launch(es) under this Section 2.1.3 shall be the same as set forth in Exhibit E. For the Additional Launch(es) exercised by Customer in accordance with this Section 2.1.3, the Launch Slot for such Launch Service shall be designated so as to occur within [*** ... ***] years of the Launch Services exercise date, provided however that such Launch Service is performed prior to [*** ... ***]. Customer shall pay to Contractor a reservation fee of [*** ... ***] US Dollars (US$[*** ... ***]) for each Additional Launch procured pursuant to this Section 2.1.3 no later than [*** ... ***]. Such reservation fee will be applied to the first Milestone Payment for the applicable Additional Launch. If Customer does not exercise an Additional Launch pursuant to this Section 2.1.3, [*** ... ***] to Customer within [*** ... ***] Days of Customer's notice of such effect.

    2.1.4   Customer and Contractor agree and acknowledge that the total number of Additional Launches that may be exercised by Customer under Sections 2.1.2 and 2.1.3, in the aggregate, is limited to [*** ... ***].

For the avoidance of doubt, any Refight Options exercised by Customer in accordance with Article 16 shall not be considered a Firm Launch or Additional Launch.

2.2 <u>Determination of Launch Success</u>. If, within [*** ... ***] Days following any Launch Service, a Satellite experiences an anomaly, failure, defect or other non-conformance with its performance specifications or operational characteristics, Customer shall promptly inform Contractor in writing and provide reasonable detail regarding such anomaly, failure, defect or non-conformance. Contractor shall, within [*** ... ***] following receipt of Customer's notice pursuant to this Section 2.2 confirm that: (i) all of the Launch Success criteria were achieved for the corresponding Launch Service; or (ii) an independent or intervening event not attributable to Contractor's failure to meet the Launch Success criteria caused the Satellite anomaly, failure, defect or other non-conformance with its performance specifications or operational characteristics, such confirmation in each case, to be based on flight telemetry and other objective data. If Contractor does not provide such confirmation within the stipulated time period, then the corresponding Launch shall be deemed a Launch Failure.

2.3 <u>Satellite Dispenser</u>. Contractor shall design, manufacture, test and qualify the Dispenser, which shall be capable of performing all interface, separation and deployment functions in accordance with the SOW. Contractor shall deliver models, data, software, hardware, and test/support equipment to Customer's Associate Contractor as required by the SOW.

2.4 <u>Separation System</u>.

    2.4.1   Contractor shall itself or through a Third Party design, manufacture, test and qualify the Separation System, which shall be capable of performing all interface, separation and deployment functions in accordance with the SOW. Contractor shall deliver models, data, software, hardware, and test/support equipment to Customer's Associate Contractor as required by the SOW.

    2.4.2   [*** ... ***]. If Contractor obtains [*** ... ***]. Without limiting the foregoing, Contractor agrees that [*** ... ***] if Customer [*** ... ***]. Furthermore, if Contractor obtains [*** ... ***], upon Customer request, Contractor will reasonably assist Customer and provide Customer with [*** . . . ***] as is necessary for Customer [*** . . . ***], including for [*** . . . ***]. Such information includes, but is not limited to, [*** . . . ***].

2.5 <u>Primary and Backup Launch Site</u>. The primary Launch Site for all Launch Services shall be VAFB and KWAJ is designated as the alternate Launch Site in the event of VAFB unavailability as provided for in this Section 2.5.

    2.5.1   Change of Launch Site Not Attributable to Contractor. No later than [*** . . . ***] months (or such shorter period that Customer may reasonably agree to in writing) prior to any scheduled Launch, Contractor shall notify Customer in writing if VAFB is not available for such Launch Services due to Launch Site or Launch Range unavailability for reasons not primarily attributable to Contractor (and notwithstanding Contractor's Reasonable Efforts to maintain or preserve Customer's scheduled Launch Date or Launch Slot) and include in such notification: (i) the reasons for Launch Site or Launch Range unavailability; (ii) the duration of such Launch Site or Launch Range unavailability; and (iii) the next available Launch Opportunity at VAFB (to the best knowledge of Contractor at that time) and at KWAJ. Within [*** . . . ***] Days of receipt of Contractor's notice, Customer shall notify Contractor of its election for the Launch Services to be performed at KWAJ, subject to available Launch Opportunities, or during the next available Launch Opportunity at VAFB. If Customer elects to proceed with the Launch Services at the next available Launch Opportunity at KWAJ or VAFB, then the Adjustment Fee associated with any Launch schedule adjustments as provided for in Article 6 shall not apply to either Contractor or Customer. Notwithstanding the foregoing, in the event of a Launch Site unavailability within [*** . . . ***] months prior to any scheduled Launch Date for any Launch Services under this Contract that results in or is reasonably likely to result in a delay to or displacement of a Customer Launch Slot, Contractor and Customer will abide by the provisions of this Section 2.5.1 in connection with the selection of a Launch Site for the affected Launch Service.

    2.5.2   Change of Launch Site Attributable to Contractor. No later than [*** . . . ***] months (or such shorter period that Customer may reasonably agree to in writing) prior to any scheduled Launch, Contractor shall notify Customer in writing if VAFB is not available for such Launch Services due to Launch Site or Launch Range unavailability for reasons primarily attributable to Contractor and include in such notification: (i) the reasons for Launch Site or Launch Range unavailability; (ii) the duration of such Launch Site or Launch Range unavailability; and (iii) the next available Launch Opportunity at VAFB (to the best knowledge of Contractor at that time) and at KWAJ. Within [*** . . . ***] Days of receipt of Contractor's notice, Customer shall notify Contractor of its election for the Launch Services to be performed at KWAJ, subject to available Launch

Opportunities, or during the next available Launch Opportunity at VAFB. If Customer elects to proceed with the Launch Services at the next available Launch Opportunity at KWAJ, then the Adjustment Fee associated with any Launch schedule adjustments as provided for in Section 6.2 shall apply to Contractor. In the event of a Launch Site unavailability within [*** . . . ***] months prior to any scheduled Launch Date for any Launch Services under this Contract that results in or is reasonably likely to result in a delay to or displacement of a Customer Launch Slot, Contractor and Customer will abide by the provisions of this Section 2.5.2 in connection with the selection of a Launch Site for the affected Launch Service.

Article 3

Contract Price

Article 4

Payment

Article 5

Launch Schedule

Article 6

Launch Schedule Adjustments

Article 7

Representations and Warranties

The Contractor makes the representations and warranties contained in this Article 7. Each such representation and warranty shall be deemed made as of the execution date of this Contract, and if necessary, Contractor shall supplement such representations and warranties as of EDC.

7.1 <u>Contractor's Performance</u>. In connection with Contractor's performance of its obligations under this Contract, Contractor shall maintain its ISO 9001 certification and obtain and maintain AS9100 certification, perform work in a skillful and workmanlike manner and otherwise abide by common standards, practices, methods and procedures in the commercial aerospace industry (and not solely in the commercial launch services industry). For the avoidance of doubt and with the exception of any acts of Contractor Gross Negligence, Contractor's undertaking in this Section 7.1 does not apply to the performance of or liability with respect to any Launch Services following the moment of Intentional Ignition with respect to any Launch Service. With the exception of ISO 9001 and AS9100, Customer represents and warrants that its Satellite manufacturer Associate Contractor is subject to substantially similar contractual obligations as those set forth in this Section 7.1.

7.2 [*** ... ***]. During [*** ... ***] hereunder and until such time as [*** ... ***], Contractor shall [*** ... ***] with [*** ... ***] versions of: (i) [*** ... ***]; (ii) if available, [*** ... ***]; (iii) [*** ... ***]; and (iv) [*** ... ***], in each case as soon as [*** ... ***].

7.3 <u>Litigation</u>. Except as set forth on Exhibit F, there are no facts, actions, suits, litigation, arbitration or administrative proceedings pending or, to Contractor's best knowledge, threatened, against the Contractor which would materially adversely effect the Contractor, its financial condition, results of operations and cash flows or otherwise prevent the Contractor from performing under this Contract.

Article 8

Coordination and Communication between Customer and Contractor

Article 9

Additional Contractor and Customer Obligations Prior to Launch

Article 10

Customer Access

Article 11

Launch Vehicle Qualification

Article 12

Permits and Approvals and Compliance with United States Government Requirements

12.1 <u>Compliance with Requirements</u>. Contractor has executed or, [*** . . . ***] months prior to the first Launch Date, shall have executed, agreements with the required United States Government agencies for use of United States Government-owned property and facilities relating to the Launch Site. Customer and Contractor agree that they shall comply with the United States Government's laws, regulations, policies and directives as they relate to the performance of this Contract. Contractor shall provide to Customer reasonable notice (in writing) of the requirements specific to access and operate at the Launch Site. The Parties shall, before Launch, execute and deliver the Agreement for Waiver of Claims and Assumption of Responsibility, the execution of which is required by the United States Department of Transportation (14 C.F.R. Section 440.17(c)) as a condition of granting Contractor's license to conduct Launch Activities and Launch the Satellites ("Government Cross-Waiver").

    12.1.1   Government Need. Customer and Contractor agree that, in the event of a DO/DX Launch that necessitates the postponement of any of Customer's Launch Services, Contractor shall promptly notify Customer of the delay(s) and reschedule the affected Launch Service within the next available Launch Opportunity and in accordance with Section 5.4. Neither the United States Government nor the Contractor shall be liable to Customer for any costs or damages, including any direct, indirect, special, incidental or consequential damages or any other revenue or business injury or loss, arising out of a delay caused by such priority use of property or personnel.

12.2   <u>Compliance with U.S. Government Export/Import Statutes and Regulations</u>.

12.2.1 Compliance with Statutes and Regulations. Each Party hereby acknowledges that it shall comply with all applicable statutes and regulations relating to the export and import of commodities, services or technical data out of and into the United States of America.

12.2.2 Transfers of Technical Data. Each Party shall be responsible for compliance with applicable United States Government regulations relating to the transfer of technical data to the other Party or to Third Parties and Related Third-Parties.

12.2.3 Notification Regarding Personnel. Customer and Contractor hereby agree to identify and promptly notify the other Party, its Foreign Person employees, Foreign Person employees of its Related Third Parties and Foreign Person consultants of any of them who will participate in, or receive any technical data or defense services in connection with the performance of this Contract.

12.2.4 Refusal to Admit or Transmit Information to Foreign Persons Not Covered. Customer acknowledges that Contractor must refuse to admit to any meeting and refuse to transmit any technical data or provide any defense services, to a Foreign Person participant who is not covered by an applicable license or agreement issued by the United States Government and duly executed by the appropriate parties.

Article 13

Changes

Article 14

Indemnity, Exclusion of Warranty, Waiver of Liability and Allocation of Certain Risks

14.1 <u>NO REPRESENTATIONS OR WARRANTIES</u>. EXCEPT AS SET FORTH IN ARTICLE 7, CONTRACTOR HAS NOT MADE NOR DOES IT MAKE ANY REPRESENTATION OR WARRANTY, WHETHER WRITTEN OR ORAL, EXPRESS OR IMPLIED, INCLUDING, WITHOUT LIMITATION, ANY WARRANTY OF DESIGN, OPERATION, WORKMANSHIP, RESULT, CONDITION, QUALITY, SUITABILITY OR MERCHANTABILITY OR OF FITNESS FOR USE OR FOR A PARTICULAR PURPOSE, ABSENCE OF LATENT OR OTHER DEFECTS, WHETHER OR NOT DISCOVERABLE, WITH RESPECT TO THE LAUNCH VEHICLE, SUCCESS OF ANY LAUNCH OR OTHER PERFORMANCE OF ANY LAUNCH SERVICE HEREUNDER.

14.2 <u>Waiver of Liability</u>.

    14.2.1 Contractor and Customer hereby agree to a reciprocal waiver of liability pursuant to which each Party agrees not to bring a claim or sue the other Party, the United States Government and its contractors and subcontractors at every tier or Related Third Parties of the other Party for any property loss or damage it sustains including, but not limited to, in the case of Customer, loss of or damage to the Satellites, or any other property loss or damage, personal injury or bodily injury, including death, sustained by any of its directors, officers, agents and employees or Related Third Parties, arising in any manner in connection with the performance of or activities carried out pursuant to this Contract, or other activities in or around the Launch Site or Satellite processing area, or the operation or performance of the Launch Vehicle or the Satellites. Such waiver of liability applies to all damages of any sort or nature, including but not limited to any direct, indirect, special, incidental or consequential damages or other loss of revenue or business injury or loss such as costs of effecting cover, lost profits, lost revenues, or costs of recovering the Satellites, from damages to the Satellites before, during or after Launch or from the failure of the Satellites to reach their planned orbit or operate properly.

    14.2.2 Claims of liability are waived and released regardless of whether loss, damage or injury arises from the acts or omissions, negligent or otherwise, of either Party or its Related Third Parties. This waiver of liability shall extend to all theories of recovery, including in contract for property loss or damage, tort, product liability and strict liability. In no event shall this waiver of liability prevent or encumber enforcement of the Parties' contractual rights and obligations to each other as specifically provided in this Contract.

    14.2.3 Contractor and Customer shall each extend the waiver and release of claims of liability as provided in Sections 14.2.1 and 14.2.2 to its Related Third Parties (other than employees, directors and officers) by requiring them to waive and release all claims of liability they may have against the other Party, its Related

Third Parties, and the United States Government and its contractors and subcontractors at every tier and to agree to be responsible for any property loss or damage, personal injury or bodily injury, including death, sustained by them arising in any manner in connection with the performance of or activities carried out pursuant to this Contract, or other related activities in or around the Launch Site or Satellite processing area, or the operation or performance of the Launch Vehicle or the Satellites.

14.2.4 The waiver and release by each Party and its Related Third Parties of claims of liability against the other Party and the Related Third Parties of the other Party extends to the successors and assigns, whether by subrogation or otherwise, of the Party and its Related Third Parties. Each Party shall obtain a waiver of subrogation and release of any right of recovery against the other Party and its Related Third Parties from any insurer providing coverage for the risks of loss for which the Party hereby waives claims of liability against the other Party and its Related Third Parties.

14.2.5 In the event of any inconsistency between the provisions of this Section 14.2 and any other provisions of this Contract, the provisions of this Section 14.2 shall take precedence.

14.3 <u>Indemnification—Property Loss and Damage and Bodily Injury</u>. Contractor and Customer each agree to defend, hold harmless and indemnify the other Party and its Related Third Parties, for any liabilities, costs and expenses (including attorneys' fees, costs and expenses), arising as a result of claims brought by Related Third Parties of the indemnifying Party, for property loss or damage, personal injury or bodily injury, including death, sustained by such Related Third Parties, arising in any manner in connection with the activities carried out pursuant to this Contract, other activities in and around the Launch Site or the Satellite processing area, or the operation or performance of the Launch Vehicle or the Satellites. Such indemnification applies to any claim for direct, indirect, special, incidental or consequential damages or other loss of revenue or business injury or loss, including but not limited to costs of effecting cover, lost profits or lost revenues, resulting from any loss of or damage to the Satellites before, during, or after Launch or from the failure of the Satellites to reach their planned orbit or operate properly.

14.3.1 To the extent that claims of liability by Third Parties are not covered by the third party liability insurance referred to in Section 15.1 or an insurance policy of either Contractor or Customer or are not eligible for payment by the United States Government (as provided in Section 14.4), Contractor will defend, hold harmless and indemnify Customer and its Related Third Parties from any and all claims of Third Parties, for property loss or damage, personal injury or bodily injury, including death arising in any manner from the processing, operation, testing or performance of the Launch Vehicle.

14.3.2 To the extent that claims of liability by Third Parties are not covered by the third party liability insurance referred to in Section 15.1 or an insurance policy of either Contractor or Customer or are not eligible for payment by the United States Government (as provided in Section 14.4), Customer will defend, hold harmless and indemnify Contractor and its Related Third Parties for any and all claims of Third Parties, for property loss or damage, personal injury or bodily injury, including death, arising in any manner from the processing, testing, operation or performance of the Satellites, or loss resulting from any loss of or damage to the Satellites before or after Launch or from the failure of the Satellites to reach their planned orbit or operate properly.

14.3.3 Notwithstanding Sections 14.3.2 and 14.3.3 above, Contractor shall not be obligated to defend, hold harmless or indemnify Customer for any claim brought by a Third Party against Customer resulting from any damage to or loss of the Satellites, whether sustained before or after Launch and whether due to the operation, performance, non-performance or failure of the Launch Vehicle or due to any other causes. Customer shall defend, hold harmless and indemnify Contractor for any claims brought by Third Parties against Contractor for damage to or loss of the Satellites, whether sustained before or after Launch or whether due to the operation, performance, non-performance or failure of the Launch Vehicle or due to other causes.

14.3.4 The indemnification for property loss or damage, personal injury or bodily injury provided by this Section 14.3 shall be available regardless of whether such loss, damage or injury arises from the acts or omissions of the Party entitled to indemnification, or its Related Third Parties, as the case may be, unless if due to willful misconduct.

14.3.5 The right of either Party or Related Third Parties to indemnification under this Article is not subject to subrogation or assignment and either Party's obligation set forth herein to indemnify the other Party or Related Third Parties extends only to that Party or those Related Third Parties and not to others who may claim through them by subrogation, assignment or otherwise.

14.4 Indemnification by United States Government.

14.4.1 The Parties recognize that under the US Commercial Space Launch Act (the "CSLA") and subject thereto, the Secretary of Transportation shall, to the extent provided in advance in appropriations acts or to the extent there is enacted additional legislative authority to provide for the payment of claims, provide for the payment by the United States Government of successful claims (including reasonable expenses of litigation or settlement) of a Third Party against Contractor or its subcontractors, or Customer or its contractors or subcontractors, resulting from activities carried out pursuant to a license issued or transferred under the CSLA for death, bodily injury, or loss of or damage to property resulting from activities carried out under the license, but only to the extent that the aggregate of such successful claims arising out of the Launch:

(A) is in excess of the amount of insurance or demonstration of financial responsibility required of Contractor under its license issued pursuant to the CSLA; and

(B) is not in excess of the level that is $1,500,000,000 (plus any additional sums necessary to reflect inflation occurring after January 1, 1989) above the required amount of insurance or demonstration of financial responsibility required by the CSLA.

14.4.2 Contractor makes no representation or warranty that any payment of claims by the United States Government will be available pursuant to the CSLA. Contractor's obligation is to make commercially Reasonable Efforts to obtain such payment as may be available from the United States Government.

14.5 Indemnification—Intellectual Property Infringement.

14.5.1 Contractor Indemnification Contractor shall indemnify, defend and hold harmless Customer, its Related Third Parties, subsidiaries and Affiliates, its subcontractors (if any), their respective officers, employees, agents, servants and assignees, from and against all losses, damages, liabilities, settlements, penalties, fines, costs and expenses (including reasonable attorneys' fees and expenses) arising out of or resulting from any claim, suit or other action or threat by a Third Party arising out of an allegation that: (i) Contractor's performance under this Contract; (ii) the design, manufacture or operation of the Launch Vehicle, Dispenser or Contractor's provision of Launch Services; or (iii) Customer's Exploitation of the Contractor IP, infringes any Third Party's Intellectual Property Rights ("Intellectual Property Claim").

14.5.2 Contractor Resolution or Mitigation If Contractor's performance under this Contract, the design, manufacture or operation of the Launch Vehicle, Dispenser or Contractor's provision of Launch Services, or any part thereof is enjoined or otherwise prohibited as a result of an Intellectual Property Claim, Contractor shall, at its option and expense, (i) resolve the matter so that the injunction or prohibition no longer pertains, (ii) procure for Customer the right to use the infringing item, and/or (iii) modify the infringing item so that it becomes non-infringing while remaining in compliance with the requirements of this Contract. Customer shall, at Contractor's expense, reasonably cooperate with Contractor to mitigate or remove any infringement. If Contractor is unable to accomplish (i), (ii) or (iii) as stated above, Customer shall have the right to terminate any or all of the unperformed Launch Services under this Contract and receive a refund of all payments associated thereto.

14.5.3 Combinations and Modifications. Contractor shall have no liability under Section 14.5.1 or Section 14.5.2 for any Intellectual Property Claim to the extent arising from (i) use of any technology, service or other deliverable furnished by Contractor to Customer under this Contract in combination with other items not provided, recommended, or approved (in writing) by Contractor, (ii) modifications of any technology, service or other deliverable after delivery by a person or entity other than Contractor unless authorized by written directive or instructions furnished by Contractor to Customer under this Contract or (iii) the compliance of any technology, service or other deliverable with specific designs, specifications or instructions of Customer.

14.5.4 Customer Indemnification Customer shall defend, hold harmless and indemnify Contractor and its Related Third Parties, subsidiaries and Affiliates, its subcontractors (if any), their respective officers, employees, agents, servants and assignees from and against all losses, damages, liabilities, settlements, penalties, fines, costs and expenses (including reasonable attorneys' fees and expenses) arising out of or resulting from any and all Intellectual Property Claims resulting from the infringement, or claims of infringement, of the Intellectual Property Rights of a Third Party, that may arise from the design, manufacture, or operation of the Satellites, ground support equipment, software and related hardware and equipment or an Intellectual Property Claim alleging that the Contractor aided or enabled infringement in the design, manufacture, or operation of the Satellites by the furnishing of Launch Services.

14.6 <u>Rights and Obligations</u>. The rights and obligations specified in Sections 14.3 and 14.5 shall be subject to the following conditions:

14.6.1 The Party seeking indemnification shall promptly advise the other Party in writing of the filing of any suit, or of any written or oral claim alleging an infringement of any Related Third Party's or any Third Party's rights, upon receipt thereof, and shall provide the Party required to indemnify, at such Party's request and expense, with copies of all relevant documentation.

14.6.2 The Party seeking indemnification shall not make any admission nor shall it reach a compromise or settlement without the prior written approval of the other Party, which approval shall not be unreasonably withheld, conditioned or delayed.

14.6.3 The Party required to indemnify, defend and hold the other harmless shall assist in and shall have the right to assume, when not contrary to the governing rules of procedure, the defense of any claim or suit or settlement thereof, and shall pay all reasonable litigation and administrative costs and expenses, including attorneys' fees, incurred in connection with the defense of any such suit, shall satisfy any judgments rendered by a court of competent jurisdiction in such suits, and shall make all settlement payments.

14.6.4 The Party seeking indemnification may participate in any defense at its own expense, using counsel reasonably acceptable to the Party required to indemnify, provided that there is no conflict of interest and that such participation does not otherwise adversely affect the conduct of the proceedings.

14.7 Inconsistency with Government Agreement. In the event of any inconsistency between any provision of this Article 14 and Article 15, this Article 14 shall take precedence as between the Parties.

14.8 Authority to Destroy Launch Vehicle. The range safety officer or equivalent is hereby authorized to destroy, without liability or indemnity to Customer or Customer's Related Third Parties, the Launch Vehicle and the Satellites in the event that such action is determined in such range safety officer's or equivalent's sole discretion to be necessary to avoid damage to persons or property. Any operation of the Launch Vehicle automatic destruct system that causes the destruction of the Launch Vehicle or Satellites shall also be without liability to Customer or Customer's Related Third Parties.

14.9 Limitation of Liability. EXCEPT IN INSTANCES OF WILLFUL MISCONDUCT, IN NO EVENT SHALL EITHER PARTY BE LIABLE TO THE OTHER PARTY UNDER OR IN CONNECTION WITH THIS CONTRACT UNDER ANY LEGAL OR EQUITABLE THEORY FOR DIRECT, INDIRECT, SPECIAL, CONSEQUENTIAL, EXEMPLARY, INCIDENTAL OR PUNITIVE DAMAGES, OR INDEMNITIES OF ANY KIND, FOR THE COST OF PROCUREMENT OF SUBSTITUTE SERVICES OR FOR LOST REVENUE OR PROFIT ARISING OUT OF OR IN CONNECTION WITH THIS CONTRACT HOWSOEVER CAUSED, WHETHER BASED IN CONTRACT, TORT OR OTHERWISE, INCLUDING NEGLIGENCE, PRODUCT LIABILITY OR STRICT LIABILITY. EXCEPT FOR CONTRACTOR'S INDEMNIFICATION OBLIGATIONS PROVIDED FOR IN SECTION 14.5, CONTRACTOR'S TOTAL AND CUMULATIVE LIABILITY ARISING OUT OF OR IN CONNECTION WITH THIS AGREEMENT SHALL NOT EXCEED THE SUM OF:

(i) IF THE FIRST FIRM LAUNCH HAS NOT BEEN PERFORMED BY CONTRACTOR, THE AGGREGATE PAYMENTS MADE BY CUSTOMER FOR THE [*** . . . ***] MILESTONES IDENTIFIED IN EXHIBIT C PLUS ANY [*** . . . ***] MILESTONE PAYMENTS MADE BY CUSTOMER FOR LAUNCH SERVICES THAT HAVE NOT YET BEEN PERFORMED PLUS ANY AND ANY ACCRUED INTEREST AT THE CONTRACTOR INTEREST RATE; OR

(ii) IF THE FIRST FIRM LAUNCH HAS BEEN PERFORMED BY CONTRACTOR, THE [*** . . . ***] MILESTONE PAYMENTS MADE BY CUSTOMER FOR LAUNCH SERVICES THAT HAVE NOT YET BEEN PERFORMED BY CONTRACTOR PLUS ANY ACCRUED INTEREST AT THE CONTRACTOR INTEREST RATE.

14.10 The Parties acknowledge that the amounts payable hereunder are based in part on the limitations set forth in this Article 14 and that such limitations are a bargained-for and essential part of this Contract. EXCEPT AS EXPRESSLY PROVIDED IN THIS CONTRACT, THIS LIMITATION OF LIABILITY DOES NOT APPLY TO CLAIMS BASED ON FRAUD, WILLFUL MISREPRESENTATION OR WILLFUL MISCONDUCT. CONSISTENT WITH THIS LIMITATION OF LIABILITY, EACH

PARTY SHALL USE REASONABLE EFFORTS TO ENSURE THAT ITS INSURER(S) WAIVE ALL RIGHTS OF SUBROGATION AGAINST THE OTHER PARTY.

Article 15

Insurance

15.1 <u>Third Party Liability Insurance</u>. Contractor shall procure and maintain in effect insurance for third party liability to provide for the payment of claims resulting from property loss or damage or bodily injury, including death, sustained by Third Parties caused by an occurrence resulting from Insured Launch Activities. The insurance shall have limits in amounts required by the Office of the Associate Administrator for Commercial Space Transportation by license issued to Contractor pursuant to the CSLA and shall be subject to standard industry exclusions and/or limitations, including, but not limited to, exclusions and/or limitations with regard to terrorism. Coverage for damage, loss or injury sustained by Third Parties arising in any manner in connection with Insured Launch Activities shall attach upon arrival of the Satellites at the Launch Site or the Satellite processing facility (wherever located), whichever occurs first, and will terminate upon the earlier to occur of the return of all parts of the Launch Vehicle to earth or twelve (12) months following the date of Launch, unless the Satellites are removed from the Satellite processing facility other than for the purpose of transportation to the Launch Site or are removed from the Launch Site other than by Launch, in which case, coverage shall extend only until such removal. Such insurance shall not cover loss of or damage to the Satellites even if such claim is brought by any Third Party or Related Third Parties. Such insurance also shall not pay claims made by the United States Government for loss of or damage to United States Government property in the care, custody and control of Customer or Contractor. The cost of such insurance is included within the Launch Price.

15.2 <u>Insurance Required by Launch License</u>. Contractor shall provide such insurance as is required by the launch license issued by the United States Department of Transportation for loss of or damage to United States Government property. The cost of insurance as required by the launch license is included in the Launch Price.

15.3 <u>Miscellaneous Requirements</u>. The third party liability insurance shall name as named insured Contractor and as additional insured Customer and the respective Related Third Parties of the Parties identified by each Party, the United States Government and any of its agencies and such other persons as Contractor may determine. Such insurance shall provide that the insurers shall waive all rights of subrogation that may arise by contract or at law against the named insured and any additional insured.

15.4 <u>Launch and In-Orbit Insurance</u>. Contractor shall provide customary support to assist Customer in obtaining Launch and In-Orbit Insurance, to the extent obtained by Customer, including: (i) supporting Customer with all necessary presentations (oral, written or otherwise), including attendance and participation in such presentations where requested by Customer; (ii) providing on a timely basis all reasonable and appropriate technical information, data and documentation; and (iii) providing documentation and answers to insurer and underwriter inquiries. In addition, Contractor shall provide any other certifications, confirmations or other information with respect to the Launch Vehicle as reasonably required by Customer's Launch and In-Orbit Insurance insurers and underwriters and shall take any other action reasonably requested by Customer or any such insurers or underwriters that is necessary or advisable in order for Customer to obtain and maintain Launch and In-Orbit Insurance on reasonable and customary terms. For the avoidance of doubt, Contractor shall not bind any first party launch and in-orbit risk insurance for any of the Launch Services to be provided for under this Contract without the prior written approval of the Customer, which shall not be unreasonably withheld. Excluding third-party launch liability insurance (which shall be purchased and maintained by Contractor), insurance coverage for the Satellites, Customer property, equipment, and personnel (and the property, equipment, or personnel of Customer's Related Third Parties) and any other insurance contemplated herein, if purchased by Customer, shall include an express waiver of subrogation as to Contractor and its Related Third Parties.

15.5 <u>Cooperation with Regard to Insurance</u>. Each Party agrees to cooperate with the other Party in obtaining relevant reports and other information in connection with the presentation by either Party of any claim under the insurance required by this Article 15.

15.6 <u>Assistance with Claims for Insurance Recovery</u>. Contractor shall cooperate with and provide reasonable support to Customer in making and perfecting claims for insurance recovery and as to any legal proceeding associated with any claim for insurance recovery. Such support shall include: (i) providing on-site inspections as required by Customer's insurers and underwriters; (ii) participating in review sessions with a competent representative selected by the insurers and underwriters to discuss any continuing issue relating to such occurrence, including information conveyed to either Party; (iii) using commercially Reasonable Efforts to secure access for the insurers and underwriters to all information used in or resulting from any investigation or review of the cause or effects of such occurrence; (iv) making available for inspection and copying all information reasonably available to Contractor that is necessary to establish the basis of a claim; and (v) supporting Customer in establishing the basis of any Total Loss, Constructive Total Loss or Partial Loss. The cooperation and support provided for in this Section 15.6 is included in the Launch Price. Customer's rights and Contractor's obligations set forth in this Article 15 are subject to applicable regulatory and confidentiality requirements.

15.7 <u>Evidence of Insurance</u>. For any of the insurance policies or waivers of subrogation required under this Contract, each Party shall provide the other Party with a certificate evidencing such insurance or waiver within [*** . . . ***] Days of a written request by the other Party and require its insurer(s) to provide the other Party written notice no later than [*** . . . ***] Days before cancellation or a material change in policy coverage or waiver.

Article 16

Reflight

Article 17

Termination

Article 18

Dispute Resolution

Article 19

Confidentiality

Article 20

Intellectual Property

Article 21

Right of Ownership and Custody

Article 22

Force Majeure

Article 23

Effective Date of Contract

Article 24

Miscellaneous

IN WITNESS WHEREOF, the Parties hereto have executed this Contract as of the Day and year first above written:

For Customer   For Contractor

IRIDIUM SATELLITE LLC   **SPACE EXPLORATION TECHNOLOGIES CORP.**

Signature:	/s/ [John Brunette]		Signature:	/s/ [Elon Musk]
Name:	John Brunette		Name:	Elon Musk
Title:	Chief Legal & Administrative Officer		Title:	Chairman and CEO

APPENDIX 9

U.S. Munitions List Categories IV and XV, Sample Technical Assistance (or Manufacturing License) Agreement, and Sample Technology Control Plan

Appendix 9-1

U.S. Munitions List Categories IV and XV

(as of January 1, 2012)

Category IV—Launch Vehicles, Guided Missiles, Ballistic Missiles, Rockets, Torpedoes, Bombs and Mines

*(a) Rockets (including but not limited to meteorological and other sounding rockets), bombs, grenades, torpedoes, depth charges, land and naval mines, as well as launchers for such defense articles, and demolition blocks and blasting caps. (See § 121.11.)

*(b) Launch vehicles and missile and antimissile systems including but not limited to guided, tactical and strategic missiles, launchers, and systems.

(c) Apparatus, devices, and materials for the handling, control, activation, monitoring, detection, protection, discharge, or detonation of the articles in paragraphs (a) and (b) of this category. (See § 121.5.)

*(d) Missile and space launch vehicle powerplants.

*(e) Military explosive excavating devices.

*(f) Ablative materials fabricated or semifabricated from advanced composites (e.g., silica, graphite, carbon, carbon/carbon, and boron filaments) for the articles in this category that are derived directly from or specifically developed or modified for defense articles.

*(g) Non/nuclear warheads for rockets and guided missiles.

(h) All specifically designed or modified components, parts, accessories, attachments, and associated equipment for the articles in this category.

(i) Technical data (as defined in § 120.21 of this subchapter) and defense services (as defined in § 120.8 of this subchapter) directly related to the defense articles enumerated in paragraphs (a) through (h) of this category. (See § 125.4 of this subchapter for exemptions.)

Technical data directly related to the manufacture or production of any defense articles enumerated elsewhere in this category that are designated as Significant Military Equipment (SME) shall itself be designated SME.

Author's Note: Defense articles in categories preceded by an asterisk (*) have been designated as SME, to which special export controls apply.

Category XV—Spacecraft Systems and Associated Equipment

*(a) Spacecraft, including communications satellites, remote sensing satellites, scientific satellites, research satellites, navigation satellites, experimental and multi-mission satellites.

*Note to paragraph (a): Commercial communications satellites, scientific satellites, research satellites and experimental satellites are designated as SME only when the equipment is intended for use by the armed forces of any foreign country.

(b) Ground control stations for telemetry, tracking and control of spacecraft or satellites, or employing any of the cryptographic items controlled under category XIII of this subchapter.

(c) Global Positioning System (GPS) receiving equipment specifically designed, modified or configured for military use; or GPS receiving equipment with any of the following characteristics:

- (1) Designed for encryption or decryption (e.g., Y-Code) of GPS precise positioning service (PPS) signals;
- (2) Designed for producing navigation results above 60,000 feet altitude and at 1,000 knots velocity or greater;

(3) Specifically designed or modified for use with a null steering antenna or including a null steering antenna designed to reduce or avoid jamming signals;

(4) Designed or modified for use with unmanned air vehicle systems capable of delivering at least a 500 kg payload to a range of at least 300 km.

Note: GPS receivers designed or modified for use with military unmanned air vehicle systems with less capability are considered to be specifically designed, modified or configured for military use and therefore covered under this paragraph (d)(4).)

Any GPS equipment not meeting this definition is subject to the jurisdiction of the Department of Commerce (DOC). Manufacturers or exporters of equipment under DOC jurisdiction are advised that the U.S. Government does not assure the availability of the GPS P-Code for civil navigation. It is the policy of the Department of Defense (DOD) that GPS receivers using P-Code without clarification as to whether or not those receivers were designed or modified to use Y-Code will be presumed to be Y-Code capable and covered under this paragraph. The DOD policy further requires that a notice be attached to all P-Code receivers presented for export. The notice must state the following: "ADVISORY NOTICE: This receiver uses the GPS P-Code signal, which by U.S. policy, may be switched off without notice."

(d)   Radiation-hardened microelectronic circuits that meet or exceed all five of the following characteristics:

(1) A total dose of 5×105 Rads (SI);

(2) A dose rate upset of 5×108 Rads (SI)/Sec;

(3) A neutron dose of 1×1014 N/cm^2;

(4) A single event upset of 1×10^{-7} or less error/bit/day;

(5) Single event latch-up free and having a dose rate latch-up of 5×108 Rads(SI)/sec or greater.

(e) All specifically designed or modified systems, components, parts, accessories, attachments, and associated equipment for the articles in this category, including the articles identified in Sec. 1516 of Public Law 105-261: satellite fuel, ground support equipment, test equipment, payload adapter or interface hardware, replacement parts, and non-embedded solid propellant orbit transfer engines (see also categories IV and V).

(f) Technical data (as defined in Sec. 120.10 of this subchapter) and defense services (as defined in Sec. 120.9 of this subchapter) directly related to the articles enumerated in paragraphs (a) through (e) of this category, as well as detailed design, development, manufacturing or production data for all spacecraft and specifically designed or modified components for all spacecraft systems. This paragraph includes all technical data, without exception, for all launch support activities (e.g., technical data provided to the launch provider on form, fit, function, mass, electrical, mechanical, dynamic, environmental, telemetry, safety, facility, launch pad access, and launch parameters, as well as interfaces for mating and parameters for launch.) (See Sec. 124.1 for the requirements for technical assistance agreements before defense services may be furnished even when all the information relied upon by the U.S. person in performing the defense service is in the public domain or is otherwise exempt from the licensing requirements of this subchapter.) Technical data directly related to the manufacture or production of any article enumerated elsewhere in this category that is designated as Significant Military Equipment (SME) shall itself be designated SME. Further, technical data directly related to the manufacture or production of all spacecraft, notwithstanding the nature of the intended end use (e.g., even where the hardware is not SME), is designated SME.

Note to paragraph (f):

The special export controls contained in Sec. 124.15 of this subchapter are always required before a U.S. person may participate in a launch failure investigation or analysis and before the export of any article or defense service in this category for launch in, or by nationals of, a country that is not a member of the North Atlantic Treaty Organization or a major non-NATO ally of the United States. Such special export controls also may be imposed with respect to any destination as deemed appropriate in furtherance of the security and foreign policy of the United States.

Appendix 9–2

Sample Technical Assistance (or Manufacturing License) Agreement

This Technical Assistance (or Manufacturing License) Agreement (the "Agreement") is entered into between [U.S. company name], an entity incorporated in the state of [state], with offices at [address], as LICENSOR, and [foreign company name(s)], whose office(s) is/are situated at [foreign company address(es)], as LICENSEE, and is effective upon the date of signature of the last party to sign this Agreement.

WHEREAS, LICENSOR is [describe the program for which the U.S. company is providing technical assistance (or manufacturing for) and the type of assistance it will provide.]

WHEREAS, LICENSOR is [describe the foreign company's role in the TAA]

NOW THEREFORE, the parties desire to enter into this Agreement as follows:

(1) This Agreement is intended to authorize the transfer of unclassified technical data and defense services to be used in [summary of program to be done under the agreement].

(2) It is understood that this Agreement is entered into as required under U.S. Government Regulations and as such, it is an independent agreement between the parties, the terms of which will prevail, notwithstanding any conflict or inconsistency that may be contained in other arrangements between the parties on the subject matter.

(3) The parties agree to comply with all applicable sections of the International Traffic in Arms Regulations (ITAR) of the U.S. Department of State, and in accordance with such regulations the following conditions apply to this Agreement:

I. ITAR Section 124.7

(1) The parties expect that they may, in furtherance of this Agreement, transfer/manufacture [describe the defense article to be manufactured and all defense articles to be exported in furtherance or support of this agreement. If no hardware is being manufactured or exported, then state so but do not leave blank.]

(2) During the term of this Agreement, LICENSOR will [describe the assistance and technical data, to include any design and manufacturing know-how involved. The applicant may address the assistance and technical data in a separate attachment to the request but must reference the attachment under this article.

(3) This Agreement shall be effective through [date].

(4) Territory.

    a. The transfer of technical data, defense articles, and defense services is authorized between the United States and [list countries of foreign licensees] for end-use by [list end-users].

    b. Sub-licensing rights are granted to the foreign licensees. Sub-licensees are identified in Attachment ___.

        Sub-licensees are required to execute a Non-Disclosure Agreement (NDA) prior to provision of, or access to the defense articles, technical data or defense services. The executed NDA, referencing the DTC Case number and incorporating all the provisions of the Agreement that refer to the United States Government and the Department of State (i.e., §124.8) will be maintained on file by the applicant for five years from the expiration of the agreement.

        If Sub-licensing and Retransfer is not requested, the applicant must specifically state that sub-licensing/retransfer is not authorized.

c. *Dual/Third Country National Employees of Foreign Licensees.* Dual/third country national employees of LICENSEE are [not] authorized as follows:

   i. Pursuant to §124.8(5), this Agreement authorizes access to defense articles and/or retransfer of technical data/defense services to individuals who are dual/third country national employees of the foreign licensees [and its approved sub-licensees—if applicable]. The exclusive nationalities authorized are [list all foreign nationalities of the employees who are not eligible for application of §124.16]. Prior to any access or retransfer, the employee must execute a Non-Disclosure Agreement (NDA) referencing this DTC case number. The applicant must maintain copies of the executed NDAs for five years from the expiration of the agreement.

   ii. Pursuant to §124.16, this Agreement authorizes access to unclassified defense articles and/or retransfer of technical data/defense services to individuals who are dual/third country national employees of the foreign licensees (and its approved sub-licensees—if applicable). The exclusive nationalities authorized are limited to NATO, European Union, Australia, Japan, New Zealand, and Switzerland. All access and/or retransfers must take place completely within the physical territories of these countries or the United States.

   NOTE: If requesting dual/third country national employees for access to classified defense articles and/or retransfer of technical data/defense services who otherwise qualify for access pursuant to §124.16, the applicant must specifically identify those exclusive nationalities under the §124.8(5) clause, and NDAs must be executed for these employees.

d. *Dual/Third Country National Employees of U.S. Licensor(s).* LICENSOR employs dual/third country nationals of the following counties who will participate in this program:

e. Contract employees to any party to the agreement hired through a staffing agency or other contract employee provider shall be treated as employees of the party, and that party is legally responsible for the employees' actions with regard to transfer of ITAR controlled defense articles to include technical data, and defense services. Transfers to the parent company by any contract employees are not authorized. The party is further responsible for certifying that each employee is individually aware of their responsibility with regard to the proper handling of ITAR controlled defense articles, technical data, and defense services.

II.   ITAR Section 124.8

[The following statements must be included verbatim as written in the ITAR.]

(1) This Agreement shall not enter into force, and shall not be amended or extended without the prior written approval of the Department of State of the U.S. Government.

(2) This Agreement is subject to all United States laws and regulations relating to exports and to all administrative acts of the U.S. Government pursuant to such laws and regulations.

(3) The parties to this Agreement agree that the obligations contained in this agreement shall not affect the performance of any obligations created by prior contracts or subcontracts which the parties may have individually or collectively with the U.S. Government.

(4) No liability will be incurred by or attributed to the U.S. Government in connection with any possible infringement of privately owned patent or proprietary rights, either domestic or foreign, by reason of the U.S. Government's approval of this agreement.

(5) The technical data or defense service exported from the United States in furtherance of this Agreement and any defense article which may be produced or manufactured from such technical data or defense service may not be transferred to a person in a third country or to a national of a third country except as specifically authorized in this agreement unless the prior written approval of the Department of State has been obtained.

(6) All provisions in this Agreement which refer to the United States Government and the Department of State will remain binding on the parties after the termination of the agreement.

III. ITAR Section 124.9(a)

[All Manufacturing License Agreements must include the clauses verbatim as required by §124.9(a)]

(1) No export, sale, transfer or other disposition of the licensed article is authorized to any country outside the territory wherein manufacture or sale is herein licensed without the prior written approval of the U.S. Government unless otherwise exempted by the U.S. Government. Sales or other transfers of the licensed article shall be limited to governments of countries wherein manufacture or sale is hereby licensed and to private entities seeking to procure the licensed article pursuant to a contract with any such government unless the prior written approval of the U.S. Government is obtained.

(2) It is agreed that sales by licensee or its sub-licensees under contract made through the U.S. Government will not include either charges for patent rights in which the U.S. Government holds a royalty-fee license, or charges for data which the U.S. Government has a right to use and disclose to others, which are in the public domain, or which the U.S. Government has acquired or is entitled to acquire without restrictions upon their use and disclosure to others.

(3) If the U.S. Government is obligated or becomes obligated to pay to the licensor royalties, fees, or other charges for the use of technical data or patents which are involved in the manufacture, use, or sale of any licensed article, any royalties, fees or other charges in connection with purchases of such licensed article from licensee or its sub-licensees with funds derived through the U.S. Government may not exceed the total amount the U.S. Government would have been obligated to pay the licensor directly.

(4) If the U.S. Government has made financial or other contributions to the design and development of any licensed article, any charges for technical assistance or know-how relating to the item in connection with purchases of such articles from licensee or sub-licensees with funds derived through the U.S. Government must be proportionately reduced to reflect the U.S. Government contributions, and subject to the provisions of paragraphs (a)(2) and (3) of this section (be sure you properly reference the paragraph numbering system used in the agreement and not just repeat the ITAR numbering), no other royalties, or fees or other charges may be assessed against U.S. Government funded purchases of such articles. However, charges may be made for reasonable reproduction, handling, mailing, or similar administrative costs incident to the furnishing of such data."

(5) The parties to this agreement agree that an annual report of sales or other transfer pursuant to this agreement of the licensed articles, by quantity, type, U.S. dollar value, and purchaser or recipient, shall be provided by (applicant or licensee) to the Department of State." This clause must specify which party is obligated to provide the annual report. Such reports may be submitted either directly by the licensee or indirectly through the licensor, and may cover calendar or fiscal years. Reports shall be deemed proprietary information by the Department of State and will not be disclosed to unauthorized persons. See §126.10(b) of this subchapter.

U.S. Munitions List

(6) [Licensee] agrees to incorporate the following statement as an integral provision of a contract, invoice, or other appropriate document whenever the licensed articles are sold or otherwise transferred:
These commodities are authorized for export by the U.S. Government only to (state the country of ultimate destination or approved sales territory). They may not be resold, diverted, transferred, transshipped, or otherwise be disposed of in any other country, either in their original form or after being incorporated through an intermediate process into other end-items, without the prior written approval of the U.S. Department of State.

IV. ITAR Section 124.9(b)

[MLAs for the production of SME must include the clauses verbatim required by §124.9(b)]

(1) A completed Non-transfer and Use Certificate (DSP-83) must be executed by the foreign end-user and submitted to the Department of State of the United States before any transfer may take place. Note: No substitute may be made for a DSP-83 (e.g., end user's certificate or a DSP-83 like document modified by the foreign party).

(2) The prior written approval of the U.S. Government must be obtained before entering into a commitment for the transfer of the licensed article by sale or otherwise to any person or government outside of the approved sales territory.

V. Term and Termination

(1) The Term of this Agreement shall commence upon the date it is approved by the U.S. Department of State, or the date the Agreement is signed by all the parties, whichever is later.

(2) Any party may terminate this Agreement as to each party by giving written notice to the other Parties. [U.S. applicant] shall notify the Department of State of such termination.

VI. Miscellaneous

(1) This Agreement is not intended to provide, and does not provide, LICENSEE with any rights or licenses under patents or copyrights on any data received by it from LICENSORS under the terms of this Agreement.

(2) This Agreement may not be assigned by any party thereto. This does not include a merger where the rights and obligations of any party are assumed by another party, or a transfer or assignment between affiliated subsidiaries of a common parent.

(3) This Agreement constitutes the entire agreement between the parties relative to transfer of technical data, and this Agreement may be modified or amended only by an instrument in writing signed by all parties and approved by the U.S. Department of State Directorate of Defense Trade Controls.

(4) This Agreement may be executed in counterparts, which together shall constitute one Agreement.

IN WITNESS WHEREOF, the parties hereto have caused this agreement to be executed effective as of the day and year of the last signature of this agreement (or) upon approval of the Department of State (if a signed agreement was submitted and no modifications are directed by proviso).

LICENSOR: LICENSEE:

_____   _____

(signature block for U.S. entity)   (signature block for foreign entity)

Appendix 9–3

Sample Technology Control Plan

The following sample Technology Control Plan is provided on the website of the Defense Security Services of the U.S. Department of Defense and may be found here: http://www.dss.mil/isp/foci/sample_tech_con_plan.html.

Sample Technology Control Plan (TCP)

I. Scope

The procedures contained in this plan apply to all elements of the _____ (insert company name and address). Disclosure of classified information to foreign persons in a visitor status or in the course of their employment by _____ (insert company name) is considered an export disclosure under the International Traffic in Arms Regulations (ITAR) and requires a Department of State license or DoS approval of either a Technical Assistance Agreement or a Manufacturing License Agreement.

II. Purpose

To delineate and inform employees and visitors of _____ (insert company name) the controls necessary to ensure that no transfer of classified defense information or controlled unclassified information (defined as technical information or data or a defense service as defined in ITAR paragraphs 120.9 & 120.10) occurs unless authorized by DoS' Office of Defense Trade Controls (ODTC), and to ensure compliance with NISPOM 2-307 and 10-509.

III. Background

_____ (insert company name) _____ (explain the products and services the company provide (e.g., designs, manufactures, integrates

...). Reference customers it provides products and/or services to (including foreign customers).

IV. U.S. Person/Foreign Person

The NISPOM defines a U.S. person as any form of business enterprise or entity organized, chartered or incorporated under the laws of the United States or its possessions and trust territories, and any person who is a citizen or national of the United States.

A U.S. National is defined in the NISPOM as a citizen of the U.S., or a person who, though not a citizen of the U.S., owes permanent allegiance to the U.S. Also see 8 USC 1101(a) (22) or 8 USC 1401 (a) para 1 to 7 for further clarification on those who may qualify as nationals of the United States.

A Foreign National is any person who is not a citizen or national of the United States. A Foreign Person is defined as any foreign interest, and any U.S. person effectively controlled by a foreign interest. A Foreign Interest is any foreign government, agency of a foreign government, or representative of a foreign government; any form of business enterprise or legal entity organized, chartered or incorporated under the laws of any country other than the U.S. or its possessions and trust territories, and any person who is not a citizen or national of the United States.

A. Foreign Persons

1) No foreign person will be given access to classified material or controlled unclassified information on any project or program that involves the disclosure of technical data as defined in ITAR paragraph 120.10 until that individual's license authority has been approved by ODTC.

2) _____ (insert company name) employees who have supervisory responsibilities for foreign persons must receive an export control/licensing

briefing that addresses relevant ITAR requirements as they pertain to classified and controlled unclassified information.

B. Foreign Person Indoctrination

Foreign persons employed by, assigned to (extended visit) or visiting _____ (insert name of company), shall receive a briefing that addresses the following items:

a) that prior to the release of classified material or controlled unclassified information to a foreign person an export authorization issued by ODTC needs to be obtained by _____ (insert company name).

b) that they adhered to the _____'s (insert company name) security rules, policies and procedures and in-plant personnel regulations.

c) that outlines the specific information that has been authorized for release to them.

d) that addresses the _____ 's (insert name of the company) in-plant regulations for the use of facsimile, automated information systems and reproduction machines.

e) that any classified information they are authorized to have access and need to forward overseas will be submitted to the _____'s (insert company name) security department for transmission through government-to-government channels.

f) that information received at _____ (insert company name) for the foreign national and information that the foreign national needs to forward from _____ (insert company name) shall be prepared in English.

g) that violations of security procedures and in-plant regulations committed by foreign nationals are subject to _____ (insert company name) sanctions. (List sanctions.)

V. Access Controls for Foreign Nationals

Address how foreign nationals will be controlled within the company's premises, for example:

1) Badges: (if necessary, address procedures, e.g., composition of the badge, identification on badge that conveys that the individual is a foreign national, privileges and so forth).

2) Escorts: (if necessary, address escort procedures). (NOTE: _____ (insert name of company) supervisors of foreign persons shall ensure that foreign nationals are escorted in accordance with U.S. government and _____ (insert name of company) regulations.

3) Establishment of a segregated work area(s).if necessary.

VI. Export Controlled Information

List specific elements of export controlled information, both classified and unclassified, that can be disclosed to foreign nationals and the program(s) the foreign national is supporting

VII. Non-Disclosure Statement and Acknowledgement

All foreign persons shall sign a non-disclosure statement (attachment A) that acknowledges that classified and controlled unclassified information will not be further disclosed, exported or transmitted by the individual to any foreign national or foreign country unless ODTC authorizes such a disclosure and the receiving party is appropriately cleared in accordance with its government's personnel security system. (NOTE: The company may also want to address other controlled information such as company proprietary or unclassified information that does not require an export authorization but which the contract the information pertains calls for specific handling procedures.

VIII. Supervisory Responsibilities

Supervisors of cleared personnel and foreign national employees and foreign national visitors shall ensure that the employees and visitors are informed of and cognizant of the following:

1) that technical data or defense services that require an export authorization is not transmitted, shipped, mailed, handcarried (or any other means of transmission) unless an export authorization has already been obtained by _____ (insert company name) and the transmission procedures follows U. S. Government regulations.

2) that individuals are cognizant of all regulations concerning the handling and safeguarding of classified information and controlled unclassified information. (NOTE: Companies may also want to address company propriety and other types of unclassified information that require mandated controls.

3) that the individuals execute a technology control plan (TCP) briefing form acknowledging that they have received a copy of the TCP and were briefed on the contents of the plan (Attach. B).

4) that U.S. citizen employees are knowledgeable of the information that can be disclosed or accessed by foreign nationals.

_____   _____
Print name and signature of Senior Management Official
Print name and signature of DSS Official

_____

_____
Print name and signature of Facility Security Officer
Print name and signature of Chairman, Government Security Committee

IX. Employee Responsibilities

All _____ (insert name of company) employees who interface with foreign nationals shall receive a copy of the TCP and a briefing that addresses the following:

1) that documents under their jurisdiction that contain technical data are not released to or accessed by any employee, visitor, or subcontractor who is a foreign national unless an export authorization has been obtained by _____ (insert company name) in accordance with the ITAR or the Export Administration Regulations (EAR).

2) If there is any question as to whether or not an export authorization is required, contact the Facility Security Officer promptly.

3) that technical information or defense services cannot be forwarded or provided to a foreign national regardless of the foreign nationals location unless an export authorization has been approved by DTC and issued to _____ (insert company's name).

Attachment A
Non-Disclosure Statement

I, _____ (insert name of individual) acknowledge and understand that any classified information, technical data or defense services related to defense articles on the U.S. Munitions List, to which I have access to or which is disclosed to me in the course of my (insert which ever term is applicable, employment, assignment or visit) by/at _____ (insert name of company) is subject to export control under the International Traffic in Arms Regulations (title 22, code of Federal Regulations, Parts 120-130). I hereby certify that such data or services will not be further disclosed, exported, or transferred in any manner to any foreign national or any foreign country without prior written approval of the

Office of Defense Trade Controls, U.S. Department of State and in accordance with U.S. government security (National Industrial Security Program Operating Manual) and customs regulations.

_____
Print name

_____
Signature

_____
Date

Attachment B

Technology Control Plan Briefing Acknowledgement

I, _____ (insert individual's name) acknowledge that I have received a copy of the Technology Control Plan for _____ (insert name of program) and a briefing outlining the contents of this TCP. Accordingly, I understand the procedures as contained in this TCP and agree to comply with all _____ (insert company's name) and U.S. government regulations as those regulations pertain to classified information and export controlled information.

_____   _____
Print Name of Individual and date   Print Name of Company Briefing Official and date

_____   _____
Signature of Individual   Signature of Company Briefing Official

APPENDIX 11-1

U.S. Government Orbital Debris Mitigation Standard Practices

**U.S. Government
Orbital Debris Mitigation Standard Practices**

OBJECTIVE
1. CONTROL OF DEBRIS RELEASED DURING NORMAL OPERATIONS

Programs and projects will assess and limit the amount of debris released in a planned manner during normal operations.

MITIGATION STANDARD PRACTICES

1-1. *In all operational orbit regimes:* Spacecraft and upper stages should be designed to eliminate or minimize debris released during normal operations. Each instance of planned release of debris larger than 5 mm in any dimension that remains on orbit for more than 25 years should be evaluated and justified on the basis of cost effectiveness and mission requirements.

OBJECTIVE
2. MINIMIZING DEBRIS GENERATED BY ACCIDENTAL EXPLOSIONS

Programs and projects will assess and limit the probability of accidental explosion during and after completion of mission operations.

MITIGATION STANDARD PRACTICES

2-1. *Limiting the risk to other space systems from accidental explosions during mission operations:* In developing the design of a spacecraft or upper stage, each program, via failure mode and effects analyses or equivalent analyses, should demonstrate either that there is no credible failure mode for accidental explosion, or, if such credible failure modes exist, design or operational procedures will limit the probability of the occurrence of such failure modes.

2-2. *Limiting the risk to other space systems from accidental explosions after completion of mission operations:* All on-board sources of stored energy of a spacecraft or upper stage should be depleted or safed when they are no longer required for mission operations or postmission disposal. Depletion should occur as soon as such an operation does not pose an unacceptable risk to the payload. Propellant depletion burns and compressed gas releases should be designed to minimize the probability of subsequent accidental collision and to minimize the impact of a subsequent accidental explosion.

**U.S. Government
Orbital Debris Mitigation Standard Practices**

OBJECTIVE
3. SELECTION OF SAFE FLIGHT PROFILE AND OPERATIONAL CONFIGURATION

Programs and projects will assess and limit the probability of operating space systems becoming a source of debris by collisions with man-made objects or meteoroids.

MITIGATION STANDARD PRACTICES

3-1. *Collision with large objects during orbital lifetime:* In developing the design and mission profile for a spacecraft or upper stage, a program will estimate and limit the probability of collision with known objects during orbital lifetime.

3-2. *Collision with small debris during mission operations:* Spacecraft design will consider and, consistent with cost effectiveness, limit the probability that collisions with debris smaller than 1 cm diameter will cause loss of control to prevent post-mission disposal.

3-3. *Tether systems* will be uniquely analyzed for both intact and severed conditions.

**U.S. Government
Orbital Debris Mitigation Standard Practices**

OBJECTIVE
4. POSTMISSION DISPOSAL OF SPACE STRUCTURES

Programs and projects will plan for, consistent with mission requirements, cost effective disposal procedures for launch vehicle components, upper stages, spacecraft, and other payloads at the end of mission life to minimize impact on future space operations.

MITIGATION STANDARD PRACTICES

4-1. *Disposal for final mission orbits:* A spacecraft or upper stage may be disposed of by one of three methods:

    a. Atmospheric reentry option: Leave the structure in an orbit in which, using conservative projections for solar activity, atmospheric drag will limit the lifetime to no longer than 25 years after completion of mission. If drag enhancement devices are to be used to reduce the orbit lifetime, it should be demonstrated that such devices will significantly reduce the area-time product of the system or will not cause spacecraft or large debris to fragment if a collision occurs while the system is decaying from orbit. If a space structure is to be disposed of by reentry into the Earth's atmosphere, the risk of human casualty will be less than 1 in 10,000.

    b. Maneuvering to a storage orbit: At end of life the structure may be relocated to one of the following storage regimes:
        I. Between LEO and MEO: Maneuver to an orbit with perigee altitude above 2000 km and apogee altitude below 19,700 km (500 km below semi-synchronous altitude
        II. Between MEO and GEO: Maneuver to an orbit with perigee altitude above 20,700 km and apogee altitude below 35,300 km (approximately 500 km above semi-synchronous altitude and 500 km below synchronous altitude.)
        III. Above GEO: Maneuver to an orbit with perigee altitude above 36,100 km (approximately 300 km above synchronous altitude)
        IV. Heliocentric, Earth-escape: Maneuver to remove the structure from Earth orbit, into a heliocentric orbit.

Because of fuel gauging uncertainties near the end of mission, a program should use a maneuver strategy that reduces the risk of leaving the structure near an operational orbit regime.

    c. Direct retrieval: Retrieve the structure and remove it from orbit as soon as practical after completion of mission.

4-2. *Tether systems* will be uniquely analyzed for both intact and severed conditions when performing trade-offs between alternative disposal strategies.

APPENDIX 11-2

IADC Space Debris Mitigation Guidelines

APPENDIX 11-2

IADC-02-01
Revision 1
September 2007

INTER-AGENCY SPACE DEBRIS COORDINATION COMMITTEE

IADC Action Item number 22.4

IADC Space Debris
Mitigation Guidelines

Issued by Steering Group and Working Group 4

IADC Space Debris Mitigation Guidelines

Table of Contents

1	Scope	5
2	Application	5
3	Terms and definitions	5
3.1	Space Debris	5
3.2	Spacecraft, Launch Vehicles, and Orbital Stages	5
3.3	Orbits and Protected Regions	6
3.4	Mitigation Measures and Related Terms	6
3.5	Operational Phases	7
4	General Guidance	7
5	Mitigation Measures	8
5.1	Limit Debris Released during Normal Operations	8
5.2	Minimise the Potential for On-Orbit Break-ups	8
5.2.1	Minimise the potential for post mission break-ups resulting from stored energy	8
5.2.2	Minimise the potential for break-ups during operational phases	8
5.2.3	Avoidance of intentional destruction and other harmful activities	9
5.3	Post Mission Disposal	9
5.3.1	Geosynchronous Region	9
5.3.2	Objects Passing Through the LEO Region	9
5.3.3	Other Orbits	10
5.4	Prevention of On-Orbit Collisions	10
6	Update	10

Foreword

The Inter-Agency Space Debris Coordination Committee (IADC) is an international forum of governmental bodies for the coordination of activities related to the issues of man-made and natural debris in space. The primary purpose of the IADC is to exchange information on space debris research activities between member space agencies, to facilitate opportunities for co-operation in space debris research, to review the progress of ongoing co-operative activities and to identify debris mitigation options.

Members of the IADC are the Italian Space Agency (ASI), British National Space Centre (BNSC), Centre National d'Etudes Spatiales (CNES), China National Space Administration (CNSA), Deutsches Zentrum fuer Luft-und Raumfahrt e.V. (DLR), European Space Agency (ESA), Indian Space Research Organisation (ISRO), Japan, National Aeronautics and Space Administration (NASA), the National Space Agency of Ukraine (NSAU) and Russian Aviation and Space Agency (Rosaviakosmos).

One of its efforts is to recommend debris mitigation guidelines, with an emphasis on cost effectiveness, that can be considered during planning and design of spacecraft and launch vehicles in order to minimise or eliminate generation of debris during operations. This document provides guidelines for debris reduction, developed via consensus within the IADC.

In the process of producing these guidelines, IADC got information from the following documents and study reports.

- *Technical Report on Space Debris,* Text of the report adopted by the Scientific and Technical Subcommittee of the United Nations Committee on the Peaceful Uses of Outer Space, 1999

- *Interagency report on Orbital Debris 1995,* The National Science and Technology Council Committee on Transportation Research and Development, November 1995

- *U.S. Government Orbital Debris Mitigation Standard Practices,* December 2000

- *Space Debris Mitigation Standard,* NASDA-STD-18, March 28, 1996

- *CNES Standards Collection, Method and Procedure Space Debris – Safety Requirements,* RNC-CNES-Q-40-512, Issue 1- Rev. 0, April 19, 1999

- *Policy to Limit Orbital Debris Generation,* NASA Program Directive 8710.3, May 29, 1997

- *Guidelines and Assessment Procedures for Limiting Orbital Debris,* NASA Safety Standard 1740.14, August 1995

- *Space Technology Items. General Requirements. Mitigation of Space Debris Population.* Russian Aviation & Space Agency Standard OCT 134-1023-2000

- *ESA Space Debris Mitigation Handbook,* Release 1.0, April 7 1999

- *IAA Position Paper on Orbital Debris – Edition 2001,* International Academy of Astronautics, 2001

- *European Space Debris Safety and Mitigation Standard,* Issue 1, Revision 0, September 27 2000

Introduction

It has been a common understanding since the United Nations Committee on the Peaceful Uses of Outer Space (UN COPUOS) published its Technical Report on Space Debris in 1999, that man-made space debris today poses little risk to ordinary unmanned spacecraft in Earth orbit, but the population of debris is growing, and the probability of collisions that could lead to potential damage will consequently increase. It has, however, now become common practice to consider the collision risk with orbital debris in planning manned missions. So the implementation of some debris mitigation measures today is a prudent and necessary step towards preserving the space environment for future generations.

Several national and international organisations of the space faring nations have established Space Debris Mitigation Standards or Handbooks to promote efforts to deal with space debris issues. The contents of these Standards and Handbooks may be slightly different from each other but their fundamental principles are the same:

(1) Preventing on-orbit break-ups

(2) Removing spacecraft and orbital stages that have reached the end of their mission operations from the useful densely populated orbit regions

(3) Limiting the objects released during normal operations.

The IADC guidelines are based on these common principles and have been agreed to by consensus among the IADC member agencies.

IADC Space Debris Mitigation Guidelines

1 Scope

The IADC Space Debris Mitigation Guidelines describe existing practices that have been identified and evaluated for limiting the generation of space debris in the environment.

The Guidelines cover the overall environmental impact of the missions with a focus on the following:

(1) Limitation of debris released during normal operations

(2) Minimisation of the potential for on-orbit break-ups

(3) Post-mission disposal

(4) Prevention of on-orbit collisions.

2 Application

The IADC Space Debris Mitigation Guidelines are applicable to mission planning and the design and operation of spacecraft and orbital stages that will be injected into Earth orbit.

Organisations are encouraged to use these Guidelines in identifying the standards that they will apply when establishing the mission requirements for planned spacecraft and orbital stages.

Operators of existing spacecraft and orbital stages are encouraged to apply these guidelines to the greatest extent possible.

3 Terms and definitions

The following terms and definitions are added for the convenience of the readers of this document. They should not necessarily be considered to apply more generally.

3.1 Space Debris

Space debris are all man made objects including fragments and elements thereof, in Earth orbit or re-entering the atmosphere, that are non functional.

3.2 Spacecraft, Launch Vehicles, and Orbital Stages

3.2.1 Spacecraft — an orbiting object designed to perform a specific function or mission (e.g. communications, navigation or Earth observation). A spacecraft that can no longer fulfil its intended mission is considered non-functional. (Spacecraft in reserve or standby modes awaiting possible reactivation are considered functional.)

3.2.2 Launch vehicle – any vehicle constructed for ascent to outer space, and for placing one or more objects in outer space, and any sub-orbital rocket.

3.2.3 Launch vehicle orbital stages — any stage of a launch vehicle left in Earth orbit.

3.3 Orbits and Protected Regions

3.3.1 Equatorial radius of the Earth - the equatorial radius of the Earth is taken as 6,378 km and this radius is used as the reference for the Earth's surface from which the orbit regions are defined.

3.3.2 Protected regions - any activity that takes place in outer space should be performed while recognising the unique nature of the following regions, A and B, of outer space (see Figure 1), to ensure their future safe and sustainable use. These regions should be protected regions with regard to the generation of space debris.

(1) Region A, **Low Earth Orbit** (or LEO) Region – spherical region that extends from the Earth's surface up to an altitude (Z) of 2,000 km

(2) Region B, the **Geosynchronous Region** - a segment of the spherical shell defined by the following:

 lower altitude = geostationary altitude minus 200 km

 upper altitude = geostationary altitude plus 200 km

 -15 degrees ≤ latitude ≤ +15 degrees

 geostationary altitude (Z_{GEO}) = 35,786 km (the altitude of the geostationary Earth orbit)

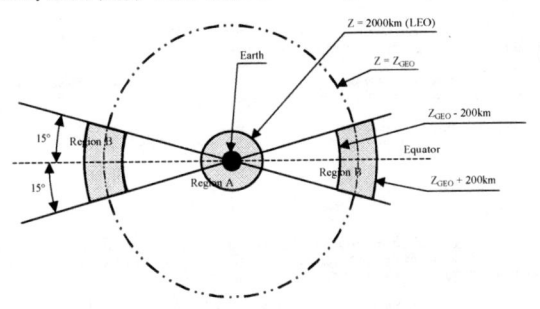

Figure 1 - Protected regions

3.3.3 Geostationary Earth Orbit (GEO) — Earth orbit having zero inclination and zero eccentricity, whose orbital period is equal to the Earth's sidereal period. The altitude of this unique circular orbit is close to 35,786 km.

3.3.4 Geostationary Transfer Orbit (GTO) — an Earth orbit which is or can be used to transfer spacecraft or orbital stages from lower orbits to the geosynchronous region. Such orbits typically have perigees within LEO region and apogees near or above GEO.

3.4 Mitigation Measures and Related Terms

3.4.1 Passivation — the elimination of all stored energy on a spacecraft or orbital stages to reduce the chance of break-up. Typical passivation measures include venting or burning excess propellant, discharging batteries and relieving pressure vessels.

3.4.2 De-orbit — intentional changing of orbit for re-entry of a spacecraft or orbital stage into the Earth's atmosphere to eliminate the hazard it poses to other spacecraft and orbital stages, by applying a retarding force, usually via a propulsion system.

3.4.3 Re-orbit — intentional changing of a spacecraft or orbital stage's orbit

3.4.4 Break-up — any event that generates fragments, which are released into Earth orbit. This includes:

(1) An explosion caused by the chemical or thermal energy from propellants, pyrotechnics and so on

(2) A rupture caused by an increase in internal pressure

(3) A break-up caused by energy from collision with other objects

However, the following events are excluded from this definition:

- A break-up during the re-entry phase caused by aerodynamic forces
- The generation of fragments, such as paint flakes, resulting from the ageing and degradation of a spacecraft or orbital stage.

3.5 Operational Phases

3.5.1 Launch phase - begins when the launch vehicle is no longer in physical contact with equipment and ground installations that made its preparation and ignition possible (or when the launch vehicle is dropped from the carrier-aircraft, if any), and continues up to the end of the mission assigned to the launch vehicle.

3.5.2 Mission phase - the phase where the spacecraft or orbital stage fulfils its mission. Begins at the end of the launch phase and ends at the beginning of the disposal phase.

3.5.3 Disposal phase - begins at the end of the mission phase for a spacecraft or orbital stage and ends when the spacecraft or orbital stage has performed the actions to reduce the hazards it poses to other spacecraft and orbital stages.

4 General Guidance

During an organisation's planning for and operation of a spacecraft and/or orbital stage, it should take systematic actions to reduce adverse effects on the orbital environment by introducing space debris mitigation measures into the spacecraft or orbital stage's lifecycle, from the mission requirement analysis and definition phases.

In order to manage the implementation of space debris mitigation measures, it is recommended that a feasible Space Debris Mitigation Plan be established and documented for each program and project. The Mitigation Plan should include the following items:

(1) A management plan addressing space debris mitigation activities

(2) A plan for the assessment and mitigation of risks related to space debris, including applicable standards

(3) The measures minimising the hazard related to malfunctions that have a potential for generating space debris

(4) A plan for disposal of the spacecraft and/or orbital stages at end of mission

(5) Justification of choice and selection when several possibilities exist

(6) Compliance matrix addressing the recommendations of these Guidelines.

5 Mitigation Measures

5.1 Limit Debris Released during Normal Operations

In all operational orbit regimes, spacecraft and orbital stages should be designed not to release debris during normal operations. Where this is not feasible any release of debris should be minimised in number, area and orbital lifetime.

Any program, project or experiment that will release objects in orbit should not be planned unless an adequate assessment can verify that the effect on the orbital environment, and the hazard to other operating spacecraft and orbital stages, is acceptably low in the long-term.

The potential hazard of tethered systems should be analysed by considering both an intact and severed system.

5.2 Minimise the Potential for On-Orbit Break-ups

On-orbit break-ups caused by the following factors should be prevented using the measures described in 5.2.1 – 5.2.3:

(1) The potential for break-ups during mission should be minimised

(2) All space systems should be designed and operated so as to prevent accidental explosions and ruptures at end-of-mission

(3) Intentional destructions, which will generate long-lived orbital debris, should not be planned or conducted.

5.2.1 Minimise the potential for post mission break-ups resulting from stored energy

In order to limit the risk to other spacecraft and orbital stages from accidental break-ups after the completion of mission operations, all on-board sources of stored energy of a spacecraft or orbital stage, such as residual propellants, batteries, high-pressure vessels, self-destructive devices, flywheels and momentum wheels, should be depleted or safed when they are no longer required for mission operations or post-mission disposal. Depletion should occur as soon as this operation does not pose an unacceptable risk to the payload. Mitigation measures should be carefully designed not to create other risks.

(1) Residual propellants and other fluids, such as pressurant, should be depleted as thoroughly as possible, either by depletion burns or venting, to prevent accidental break-ups by over-pressurisation or chemical reaction.

(2) Batteries should be adequately designed and manufactured, both structurally and electrically, to prevent break-ups. Pressure increase in battery cells and assemblies could be prevented by mechanical measures unless these measures cause an excessive reduction of mission assurance. At the end of operations battery charging lines should be de-activated.

(3) High-pressure vessels should be vented to a level guaranteeing that no break-ups can occur. Leak-before-burst designs are beneficial but are not sufficient to meet all passivation recommendations of propulsion and pressurisation systems. Heat pipes may be left pressurised if the probability of rupture can be demonstrated to be very low.

(4) Self-destruct systems should be designed not to cause unintentional destruction due to inadvertent commands, thermal heating, or radio frequency interference.

(5) Power to flywheels and momentum wheels should be terminated during the disposal phase.

(6) Other forms of stored energy should be assessed and adequate mitigation measures should be applied.

5.2.2 Minimise the potential for break-ups during operational phases

During the design of spacecraft or orbital stages, each program or project should demonstrate, using failure mode and effects analyses or an equivalent analysis, that there is no probable failure mode leading to accidental break-ups. If such failures cannot be excluded, the design or operational procedures should minimise the probability of their occurrence.

During the operational phases, a spacecraft or orbital stage should be periodically monitored to detect malfunctions that could lead to a break-up or loss of control function. In the case that a malfunction is detected, adequate recovery measures should be planned and conducted; otherwise disposal and passivation measures for the spacecraft or orbital stage should be planned and conducted.

5.2.3 Avoidance of intentional destruction and other harmful activities

Intentional destruction of a spacecraft or orbital stage, (self-destruction, intentional collision, etc.), and other harmful activities that may significantly increase collision risks to other spacecraft and orbital stages should be avoided. For instance, intentional break-ups should be conducted at sufficiently low altitudes so that orbital fragments are short lived.

5.3 Post Mission Disposal

5.3.1 Geosynchronous Region

Spacecraft that have terminated their mission should be manoeuvred far enough away from GEO so as not to cause interference with spacecraft or orbital stage still in geostationary orbit. The manoeuvre should place the spacecraft in an orbit that remains above the GEO protected region.

The IADC and other studies have found that fulfilling the two following conditions at the end of the disposal phase would give an orbit that remains above the GEO protected region:

1. A minimum increase in perigee altitude of:

$$235 \text{ km} + (1000 \cdot C_R \cdot A/m)$$

   where C_R is the solar radiation pressure coefficient

   A/m is the aspect area to dry mass ratio ($m^2 kg^{-1}$)

   235 km is the sum of the upper altitude of the GEO protected region (200 km) and the maximum descent of a re-orbited spacecraft due to luni-solar & geopotential perturbations (35 km).

2. An eccentricity less than or equal to 0.003.

Other options enabling spacecraft to fulfil this guideline to remain above the GEO protected region are described in the "Support to the IADC Space Debris Mitigation Guidelines" document.

The propulsion system for a GEO spacecraft should be designed not to be separated from the spacecraft. In the case that there are unavoidable reasons that require separation, the propulsion system should be designed to be left in an orbit that is, and will remain, outside of the protected geosynchronous region. Regardless of whether it is separated or not, a propulsion system should be designed for passivation.

Operators should avoid the long term presence of launch vehicle orbital stages in the geosynchronous region.

5.3.2 Objects Passing Through the LEO Region

Whenever possible spacecraft or orbital stages that are terminating their operational phases in orbits that pass through the LEO region, or have the potential to interfere with the LEO region, should be de-orbited (direct re-entry is preferred) or where appropriate manoeuvred into an orbit with a reduced lifetime. Retrieval is also a disposal option.

A spacecraft or orbital stage should be left in an orbit in which, using an accepted nominal projection for solar activity, atmospheric drag will limit the orbital lifetime after completion of operations. A study on the effect of post-mission orbital lifetime limitation on collision rate and debris population growth has been performed by the IADC. This IADC and some other studies and a number of existing national guidelines have found 25 years to be a reasonable and appropriate

IADC Space Debris Mitigation Guidelines 363

lifetime limit. If a spacecraft or orbital stage is to be disposed of by re-entry into the atmosphere, debris that survives to reach the surface of the Earth should not pose an undue risk to people or property. This may be accomplished by limiting the amount of surviving debris or confining the debris to uninhabited regions, such as broad ocean areas. Also, ground environmental pollution, caused by radioactive substances, toxic substances or any other environmental pollutants resulting from on-board articles, should be prevented or minimised in order to be accepted as permissible.

In the case of a controlled re-entry of a spacecraft or orbital stage, the operator of the system should inform the relevant air traffic and maritime traffic authorities of the re-entry time and trajectory and the associated ground area.

5.3.3 Other Orbits

Spacecraft or orbital stages that are terminating their operational phases in other orbital regions should be manoeuvred to reduce their orbital lifetime, commensurate with LEO lifetime limitations, or relocated if they cause interference with highly utilised orbit regions.

5.4 Prevention of On-Orbit Collisions

In developing the design and mission profile of a spacecraft or orbital stage, a program or project should estimate and limit the probability of accidental collision with known objects during the spacecraft or orbital stage's orbital lifetime. If reliable orbital data is available, avoidance manoeuvres for spacecraft and co-ordination of launch windows may be considered if the collision risk is not considered negligible. Spacecraft design should limit the consequences of collision with small debris which could cause a loss of control, thus preventing post-mission disposal.

6 Update

These guidelines may be updated as new information becomes available regarding space activities and their influence on the space environment.

Table of Authorities

Authority	Citation
Agreement Governing the Activities of States on the Moon and Other Celestial Bodies (Moon Agreement)	G.A. Res. 34/68, U.N. Doc. A/RES/34/68 (Dec. 5, 1979), reprinted in 18 I.L.M. 1434 (1979)
Agreement on the Rescue of Astronauts, the Return of Astronauts and the Return of Objects Launched into Outer Space (Rescue Agreement)	19 U.S.T. 7570, 672 U.N.T.S. 119
Anti-Kickback Act of 1986	41 U.S.C. §§ 51–58
Armed Services Procurement Act of 1947 (ASPA)	10 U.S.C. §§ 2301 et seq.
Arms Export Control Act (AECA)	22 U.S.C. § 2778
Bayh-Dole Act	35 U.S.C. §§ 200–212
Boyle v. United Techs. Corp.	487 U.S. 500 (1988)
Commercial Space Launch Act of 1984, as amended by the Commercial Space Launch Amendments Act of 2004	51 U.S.C. §§ 50901 et seq.
Commercial Space Transportation Regulations	14 C.F.R. pt. 400 et seq.
Communications Satellite Act of 1962	47 U.S.C. §§ 701 et seq.
Competition in Contracting Act of 1984 (CICA)	10 U.S.C. §§ 2301 et seq. & 41 U.S.C. § 403
Contract Disputes Act of 1978 (CDA)	41 U.S.C. §§ 601 et seq.
Convention on International Liability for Damage Caused by Space Objects (Liability Convention)	24 U.S.T. 2389, 961 U.N.T.S. 187

Authority	Citation
Convention on Registration of Objects Launched into Outer Space (Registration Convention)	28 U.S.T. 695, 1023 U.N.T.S. 15
Council on Environmental Quality Regulations for Implementing NEPA	40 C.F.R. pts. 1500–1508
Declaration of the United Nations Conference on the Human Environment (Stockholm Declaration)	Declaration of the United Nations Conference on the Human Environment, June 5–16, 1972, Principle 21; G.A. Res. 2995 (XXVII), Co-operation between States in the Field of the Environment, U.N. GAOR, 2112th Sess. U.N. Doc. A/RES/2995, at 1 (Dec. 15, 1972)
Declaration on International Cooperation in the Exploration and Use of Outer Space for the Benefit and in the Interest of All States, Taking into Particular Account the Needs of Developing Countries	U.N. Res. 51/122 (Dec. 13, 1996)
Defense Federal Acquisition Regulation Supplement (DFARS)	FAR pts. 200–299
Economic Espionage Act of 1996	18 U.S.C. §§ 1831–1839
Executive Order 12,829	Exec. Order No. 12,829, 58 Fed. Reg. 3479 (Jan. 6, 1993), as amended by Exec. Order No. 12,885, 58 Fed. Reg. 65,863 (Dec. 14, 1993)
Export Administration Regulations (EAR)	15 C.F.R. §§ 730–774
False Claims Act	31 U.S.C. §§ 3729–3733
Federal Acquisition Regulations (FAR)	C.F.R. title 48

Table of Authorities

Authority	Citation
Federal Property and Administrative Services Act of 1949 (FPASA)	40 U.S.C. §§ 471 *et seq.* & 41 U.S.C. §§ 251 *et seq.*
G. L. Christian & Assocs. v. United States	312 F.2d 418 (Ct. Cl.), *cert. denied*, 375 U.S. 954 (1963)
International Traffic in Arms Regulations (ITAR)	22 C.F.R. §§ 120–130
ITU Constitution	
ITU Convention	
Land Remote Sensing Policy Act of 1992	15 U.S.C. §§ 5601–5672
NASA FAR Supplement (NFS)	FAR pts. 1800–1899
National Aeronautics and Space Act of 1958	51 U.S.C. §§ 20101 *et seq.*
National Environmental Policy Act (NEPA)	42 U.S.C. §§ 4321 *et seq.*
Office of Federal Procurement Policy Act of 1974	41 U.S.C. §§ 401 *et seq.*
Principles Governing the Use by States of Artificial Earth Satellites for International Direct Television Broadcasting	U.N. Res. 37/92 (Dec. 10, 1982)
Principles Relating to Remote Sensing of Earth from Outer Space	U.N. Res. 41/65 (Dec. 3, 1986)
Principles Relevant to the Use of Nuclear Power Sources in Outer Space	U.N. Res 47/68 (Dec. 14, 1992)
Procurement Integrity Act	41 U.S.C. § 423
The Paquete Habana	175 U.S. 677 (1900)
Treaty on Principles Governing the Activities of States in the Exploration and Use of Outer Space, including the Moon and Other Celestial Bodies (Outer Space Treaty)	18 U.S.T. 2410, T.I.A.S. No. 6347

Authority	Citation
Truth in Negotiations Act (TINA)	10 U.S.C. § 2306A & 41 U.S.C. § 254B
United States Munitions List (USML)	22 C.F.R. § 121.1
Vienna Convention on Law of Treaties	Vienna Convention on the Law of Treaties, May 23, 1969, 1155 U.N.T.S. 331; 8 I.L.M. 679 (1969)

Index

Note: page numbers in *italics* indicate figures and tables.

A-4 rocket, 33
Accounting, 185–186
Ad Hoc Committee on Space, 73
Administrative Procedure Act, 215
Advisory Committee on the Future of the United States Space Program, 78
AECA. *See* Arms Export Control Act
Aerospace flights, 26
AES. *See* Automated Export System
AESDirect, 148, 155
Agreement Governing the Activities of States on the Moon and Other Celestial Bodies (1979), xv, 66
Agreement on the Rescue of Astronauts, the Return of Astronauts, and the Return of Objects Launched into Outer Space (1968), xiv–xv, 63–64
Agreement on Trade-Related Aspects of Intellectual Property (TRIPS), 233
Agreements, ITAR, 149–150
Aldrin, Buzz, 42
Allen, Paul, 48
Allowable Cost and Payment Clause, 185
American Geophysical Union, 210
Ansari, Anousheh, 52
Ansari X Prize, 48, 55
Antares, 53
Anti-Kickback Act of 1986, 193

Anti-satellite missile system, 216
Apollo 1, 103
Apollo 11, *15*, 42, *43*
Apollo 13, 42
Apollo command module, 3
Apollo spaceflight program. *See* Project Apollo
Arbitration, 58–59
Ariane, 47
Ariane 5, 47
Armadillo Aerospace, 49, 51, 56
Arms Export Control Act (AECA), 142
Armstrong, Neil, 42
Article II, U.S. Constitution, 59
Article VI, U.S. Constitution, 59
Artificial satellites. *See* Satellites
Atlantis, 43, *44*
Augustine, Norman, 78
Augustine Report, 78–79
Automated Export System (AES), 148

Bayh-Dole Act of 1980, 195–196
Beidou, 25
Berne Convention for the Protection of Literary and Artistic Works, 233
Bezos, Jeff, 56
Bigelow Aerospace, 52
Bigelow, Robert, 52
Bilateral investment treaties (BITs), 58
BIS. *See* Bureau of Industry and Security
BITs. *See* Bilateral investment treaties
Blue Origin, 49, 51, 53, 56–57

Boeing Company, 53
Boeing Launch Services, 52
Bogotá Declaration, 62
Boycott requests, 160
Branson, Richard, 49
Bureau of Industry and Security (BIS), 141, 154–155
Bush administration, 81
Business travel, international, 163

California Spaceport, 55
Cape Canaveral Spaceport, 55
Carmack, John, 56
CAS. *See* Cost Accounting Standards
Cassini, 16, *37*, 213
CCDev. *See* Crew Development program
CCiCap initiative. *See* Commercial Crew integrated Capability initiative
CCL. *See* Commerce Control List
Cecil Field Spaceport, 55
Center for Space Standards and Innovation (CSSI), 218
Challenger, 23, *43*, 44, 78, 103
Chang Zheng ("Long March"), 46
Charming Betsy doctrine, 60
China, 25, 46–47
CHSF. *See* Commercial human spaceflight
CICA. *See* Competition in Contracting Act of 1984
Cislunar space, 2
Clarke, Arthur C., 24
Classified information, 191–192
Clinton administration, 79
Clinton, Hillary Rodham, 220
Clinton, William J., 36, 81
Coase theorem, 124, 125
Code of Federal Regulations, 80
Cold War, 33, 77
Columbia, 43, 44, 104
Commerce Control List (CCL), 152
Commerce Country Chart, 154

Commercial Crew and Cargo Program (CCCP), 53
Commercial Crew integrated Capability (CCiCap) initiative, 54
Commercial human spaceflight (CHSF), 96
Commercial Orbital Transportation Service (COTS), 53, 93
Commercial remote sensing regulation, 80–81
Commercial Remote Sensing Regulatory Affairs Office (CRSRAO), 133
Commercial Resupply Service (CRS), 53
Commercial software, 198
Commercial Space Launch Act of 1984 (Launch Act)
 federal preemption debate and, 112–113
 liability issues regarding spaceports and, 108–109
 licensing commercial spaceflight and, 83–85
 passage of, 76
 state space tourism laws and, 108–110
 three-tier liability regime under, *106*
Commercial Space Launch Amendments Act of 2004 (CSLAA), 80, 84, 109
Commercial spaceflight, 48–55
 beyond LEO, 54–55
 licensing, 83–101
 orbital, 51–54
 spaceports, 55
 suborbital, 49–51
Committee on Space Research (COSPAR), 69, 211

Committee on the Peaceful Uses
    of Outer Space (COPUOS),
    xiv, xv, 67, 213, 214, 219–220
Communications Act of 1934, 123
Communications Satellite Act of
    1962 (Comsat Act), 75, 122
Communications satellites, 33–34
Communications stations, 17
Competition in Contracting Act of
    1984 (CICA), 179, 182, 184
Competitive negotiation, 183
Computer software, 198
Comsat Act (1962). *See*
    Communications Satellite
    Act of 1962
Constellation program, 45
Consultative Committee for Space
    Data Systems, 69
Contamination
  exhaust products from launch
    vehicle propulsion
    systems, *208*
  launch vehicles, 208–210
  planetary protection, 210–212
  types of, 207
Contract administration
  accounting, 185–186
  allocation of liability risk, 189–191
  changes, 187
  disputes, 188
  protecting classified information,
    191–192
  socioeconomic obligations, 191
  subcontracting, 188–189
  termination, 187
Contract Disputes Act of 1978, 188
Contracting officers (COs), 182
Contractors, ethical obligations of,
    192–194
Contracts
  administration, 185–192
  cooperative research and
    development agreements,
    180–181

FAR-based, 179–180
federal contracting process,
    181–185
intellectual property rights, 194–198
laws and regulations, 176–179
Space Act Agreements, 181
types of government, 179–181
Contractual risk allocation, 115–116
Control stations, 17
Convention on International
    Liability for Damage Caused
    by Space Objects (1972), xv,
    64–65, 86, 104, 213
Convention on Registration of
    Objects Launched into Outer
    Space (1975), xv, 230
Convention on the Recognition
    and Enforcement of Foreign
    Arbitral Awards, 59
Convention on the Registration of
    Objects Launched into Outer
    Space (1976), 65–66
Cooperative research and
    development agreements
    (CRADA), 180–181
Copenhagen Suborbitals, 51
COPUOS. *See* Committee on the
    Peaceful Uses of Outer Space
Corona, 35, *35*
COs. *See* Contracting officers
Cosmos 954, 213, 216
COSPAR. *See* Committee on Space
    Research
COSPAR Planetary Protection
    Policy (CPPP), 211–212, *212*
Cost Accounting Standards
    (CAS), 186
Cost-reimbursement contracts, 180
COTS. *See* Commercial Orbital
    Transportation Service
CPPP. *See* COSPAR Planetary
    Protection Policy
CRADA. *See* Cooperative research
    and development agreement

Crew, 97–98
Crew Development program (CCDev), 53
Cross-waiver requirements, 106–107
CRS. See Commercial Resupply Service
CRSRAO. See Commercial Remote Sensing Regulatory Affairs Office
CSLAA (2004). See Commercial Space Launch Amendments Act of 2004
CSSI. See Center for Space Standards and Innovation
CST-100, 54
Cygnus, 53

D-Trade, 148
Dassault Aviation, 51
Data dissemination, 138–139
Data rights, 196–198
DBS. See Direct broadcasting satellites
DCS. See Destination control statement
DDTC. See Directorate of Defense Trade Controls
Declaration of the United Nations Conference on the Human Environment, 206
Defense article, 144
Defense Authorization Act of 1997, 137
Defense Federal Acquisition Regulation Supplement (DFARS), 179, 196–198, *197–198*
Defense service, 144
Delta-v, 11–12
Denied persons, 159–160
Department of State Circular 175, 59
Destination control statement (DCS), 162
Destination controls, 157–159

DFARS. See Defense Federal Acquisition Regulation Supplement
Diamandis, Peter, 49
DigitalGlobe, 80
Direct broadcasting satellites (DBS), 126–129
Directorate of Defense Trade Controls (DDTC), 141, 145, 147–150
DISCO I Order, 127–129
DISCO II Order, 127–129
Discovery, 43
Dispute resolution, international, 58–59
Disputes, 188
Document marking, 161–162
DOD. See U.S. Department of Defense
DOT. See U.S. Department of Transportation
Dragon, xvi, 3, 53, 93–94

EA. See Environmental Assessment
EADS Astrium, 51
EAR. See Export Administration Regulations
Early Bird, 34
Earth observation satellites, 25
Earth remote sensing satellites, licensing private, 133–140
Earth Resources Technology Satellite, 36
Earth/space boundary, 3
ECCN. See Export control classification number
Echo, 33–34
ECLSS. See Environmental control and life support systems
Economist, 195
EEI. See Electronic export information
Efficient Use policy, 124

EHF. *See* Extremely high frequency
EIS. *See* Environmental impact statements
Eisenhower, Dwight D., xiv, 33, 71
Electromagnetic bands, *119*
Electronic export information (EEI), 148
Elliptical orbits, 7
Embargoed countries, *157*
Empowered official, 156
End use controls, 157–159
Endeavour, 43
Enterprise, 44, 55
Environmental Assessment (EA), 90–91
Environmental control and life support systems (ECLSS), 20
Environmental impact statements (EIS), 209
Environmental law, international, 205–207
Environmental reviews, 90–91
ERTS-1, 138
ESA. *See* European Space Agency
Ethical obligations, 192–194
Europe, 47
European Space Agency (ESA), 38, 47, 68, *69*, 218, 219
European SSA Preparatory Programme, 218
European Union, 25, 68, *69*
Executive agreements, 59–60
Executive Order 12829, 191
Experimental permits, 95
Exploration of Cosmic Space by Means of Reaction Devices, The (Tsiolkovsky), 31
Export Administration Regulations (EAR), 141, 152–155
Export compliance
  boycott requests, 160
  denied persons and restricted parties, 159–160
  destination and end use controls, 157–159
  empowered official, 156
  foreign national employees, 162–163
  international business travel, 163
  overview of process, 155–156
  record keeping, 164–165
  requesting clearance for information release, 164
  safeguarding technical data, 160
  screening for restrictions and diversion risks, 157–160
  verification by suppliers and contractors, 162
Export control classification number (ECCN), 152–153
Export control laws
  Export Administration Regulations, 152–155
  International Traffic in Arms Regulations, 142–152
  reform, 165–167
Extremely high frequency (EHF), 118

FAA. *See* Federal Aviation Administration
Falcon 9, 53
False Claims Act, 193
FAR. *See* Federal Acquisition Regulations
FAR-based contracts, 179–180
Federal Acquisition Regulations (FAR)
  allocation of liability risk, 189–191
  contracts, 179–180
  cost principles in, 186
  data rights provisions, *197*
  disputes and, 188
  intellectual property rights, 194–198

Federal Acquisition Regulations
(FAR), *continued*
  overview of, 176–179
  subcontracting and, 188–189
Federal Aviation Administration
(FAA), 49, 83
  debris mitigation
    requirements, 219
  demonstration of financial
    responsibility, 91
  EIS of launch-related
    activities, 209
  environmental review, 90–91
  experimental permits, 95
  jurisdiction, 83–86
  obtaining FAA license, 86–92
  payload reviews, 92
  policy review, 88
  postlicensing requirements, 91–92
  preliminary consultation, 87–88
  responsibilities of, 175
  safety review, 88–90
  training, medical, and informed
    consent requirements,
    95–98
  waivers, 92–94
Federal Communications
  Commission (FCC)
  debris mitigation
    requirements, 219
  direct broadcasting satellites,
    126–129
  Earth stations licensing, 129–130
  GSO-like satellite systems
    licensing, 124–125
  licensing of telecommunications
    satellites, 122–130
  NGSO-like satellite system
    licensing, 125–126
  ORBIT act auctions exemption,
    126–127
  responsibilities of, 175
  rules for satellite operations, 222

  space station segment of satellite
    systems licensing, 124–129
  spectrum management principles,
    123–124
Federal contracting process,
  181–185
  government contracting
    personnel and agencies,
    181–182
  protests, 185
  solicitation process, 182–185
Federal Property and Administrative
  Services Act of 1949, 179
Federal Supply Schedule (FSS)
  program, 185
Fengyun-1C, 216
Financial responsibility, 91
Finding of No Significant Impact
  (FONSI), 91
Fixed-price contracts, 179
Flight Opportunities
  Program, 51
Florida, 110
FOIA. *See* Freedom of Information
  Act
FONSI. *See* Finding of No
  Significant Impact
Foreign national employees,
  162–163
14 C.F.R. ch. III., 86
Freedom of Information Act
  (FOIA), 231
Frequency allocations, *121*
Friendship 7, *41*
FSS program. *See* Federal Supply
  Schedule program

G. L. Christian & Associates v.
  United States, 178
Gagarin, Yuri, 39
Galileo, 16, 25, 213
Garriott, Richard, 52, 222
Gates, Robert, 166

Gemini spaceflight program. *See* Project Gemini
General Services Administration (GSA), 176, 185
Genesis I, 52
Genesis II, 52
GEO. *See* Geosynchronous Earth orbit
Geo-Eye Inc., 36
GeoEye, 80
Geostationary orbit (GSO), 7, 8, 117
Geosynchronous Earth orbit (GEO), 6–7, 9
Geosynchronous Satellite Launch Vehicle (GSLV), 48
Germany, 33
Glenn, John, 40, *41*
Global navigation satellite systems (GNSS), 25
Global Positioning System (GPS), 18, 25
Globalstar, 18
Glossary of terms, 26–30, 98–101, 130–131, 139–140, 167–172, 223–224
GNSS. *See* Global navigation satellite systems
Goddard, 56
Goddard, Robert, 31–32, *32*, 55
Google Earth, 36
Google Inc., 36, 54
Google Lunar X-Prize competition, 54
Google Maps, 36
Government
 commercial vs. government contracting, *174*
 contracting laws and regulations, 176–179
 federal appropriations for space activities, *174*
 federal contracting process, 181–185
 space activities, 173–176

Government Accountability Office, 185
GPS. *See* Global Positioning System
Grasshopper, 51
Graveyard orbits, 2, 8
GSA. *See* General Services Administration
GSLV. *See* Geosynchronous Satellite Launch Vehicle
GSLV Mark I, 49
GSLV Mark II, 49
GSLV Mark III, 48, 49
GSO. *See* Geostationary orbit
GSO-like satellite systems, 124–125
Guggenheim Foundation, 32

H-II Transfer Vehicle, 49
H-IIA Launch Vehicle, 49
H-IIB Launch Vehicle, 49
Ham (chimpanzee), 40
HEO. *See* Highly elliptical orbits
Highly elliptical orbits (HEO), 7
Hohmann transfer, 11
Hohmann, Walter, 11
Hubble Space Telescope (HST), 26, 38–39, *40*, 78
Hyperion, 56

IAA. *See* International Academy of Astronautics
IAC. *See* International Astronautical Congress
IADC. *See* Inter-Agency Space Debris Coordination Committee
IAEA. *See* International Atomic Energy Agency
IAF. *See* International Astronautical Federation
ICJ. *See* International Court of Justice
IISL. *See* International Institute of Space Law

Ikonos, 36
Immunity legislation, 109–112
Incentive contracts, 180
Indefinite-delivery contracts, 180
India, 47–48
Indian Space Research Organization (ISRO), 47
Information security, 161
Informed consent requirements, 95–98, 109
INSRP. *See* Interagency Nuclear Safety Review Panel
Insurance
 global market, 114–115
 obtaining, 114
 practices, 113–115
 stages of spaceflight to insure, 113–114
Intellectual property rights, 194–198, 228–233
INTELSAT. *See* International Telecommunications Satellite organization
INTELSAT I, 34
INTELSAT III, 34
Intelsat Ltd., 34
INTELSAT V, 34
Inter-Agency Space Debris Coordination Committee (IADC), 69, 219
Interagency Nuclear Safety Review Panel (INSRP), 215
Interagency Operations Advisory Group, 69
Intergovernmental Agreement on Space Station Cooperation (1998), xv
Internal power systems, 15–16
International Academy of Astronautics (IAA), 69
International agreements, 59–60

International Astronautical Congress (IAC), 69, 210
International Astronautical Federation (IAF), 69
International Atomic Energy Agency (IAEA), 214
International Code of Conduct for Outer Space Activities, 220
International Council of Scientific Unions, 69, 211
International Court of Justice (ICJ), 58
International Institute of Space Law (IISL), 69
International law
 customary, 58
 international dispute resolution, 58–59
 lawmaking process, 57–59
 sources of, 57–58
International organizations, 67–70
International Register of Objects Launched into Outer Space, 67
International Scientific Optical Network (ISON), 218
International Space Station (ISS), xvi, 2, *21*, 26, 43
 overview of, 44–45
 patent jurisdiction over activity on, 230–231
 space tourism and, 52
International Telecommunication Union (ITU), xv
 allocation and allotment, 120–122
 coordination of satellite services and, 118–120
 overview of, 68
 supersychronous graveyard orbits requirement, 221–222
International Telecommunications Satellite (INTELSAT) organization, 24, 34, 75

International Traffic in Arms
    Regulations (ITAR), 141,
    142–152
  agreements, 149–150
  checklist for ITAR-controlled
    technical data, *146*
  exemptions, 150–151
  export authorizations,
    147–148
  licenses, 149
  overview of, 142–145
  special export controls for foreign
    satellite launches, 151
  United States Munitions List in,
    145–147
  violations and penalties,
    151–152
International treaties, 232–233
Inventions, 195–196
Iridium 33, 24, 217
Iridium Communications, Inc.,
    18, 217
ISON. *See* International Scientific
    Optical Network
Israel, 137
ISRO. *See* Indian Space Research
    Organization
ISS. *See* International Space Station
ITAR. *See* International Traffic in
    Arms Regulations
ITU. *See* International
    Telecommunications Union
James Webb Space Telescope, 39
Japan, 48
Japan Aerospace Exploration
    Agency (JAXA), 49, 68

JAXA. *See* Japan Aerospace
    Exploration Agency

Kaituozhe ("Pioneer"), 46
Kármán line, 3
Kármán, Theodore von, 3

Keldysh Institute of Applied
    Mathematics, 218
Kennedy, John F., 73, 75
Kessler, Donald, 24, 217
Kessler syndrome, 24, 217
Kibo, 49
Kodiak Launch Complex, 55
Kyl-Bingaman Amendment, 137

Lagrange, Joseph Louis, 8
Lagrange points, 8
Laliberté, Guy, 52
Land Remote-Sensing
    Commercialization Act of
    1984 (Remote Sensing Act),
    76, 77
Land Remote Sensing Policy Act of
    1992, 36, 79, 134
Landsat I, 36, 77, 138
Landsat Program, 35–36
Last in Time rule, 60
Launch Act (1984). *See* Commercial
    Space Launch Act of 1984
Launch vehicles, 208–210
Law of salvage, 222
Law of universal gravitation,
    8, 9
Laws of motion, 8–9
Layka, 33
LEOs. *See* Low Earth orbit
Letter contracts, 180
Lex specialis, 60
Liability
  contracts and allocation of,
    189–191
  at national level, 104–113
  regimes within the United States,
    104–113
  state space tourism laws, 105–107
Liability Convention. *See*
    Convention on International
    Liability for Damage Caused
    by Space Objects

Licenses
    Earth remote sensing satellites,
        133–140
    Earth stations, 129–130
    export, 149
    space station segment of satellite
        systems, 124–129
    telecommunications satellites,
        122–130
Light, speed of, 2
Lindbergh, Charles, 32
Lockheed Martin Commercial
    Launch Services, 52
Low Earth orbits (LEO), 5–6, 9,
    12, 45
Lunokhod-2, 222
Lynx, 56

Manned spaceflight, 39–45
Manufacturing license agreements
    (MLAs), 149
Mars Climate Orbiter, 38
Mars Exploration Rovers, 38, *39*
Mars Global Surveyor, 38
Mars Odyssey, 38
Mars Pathfinder, 38
Mars Rover Curiosity, 38
Mars Science Laboratory (MSL),
    38, 213
Masten Space Systems, 49
Maximum probable loss (MPL), 91,
    105–106
McDonnell Douglas Corp., 218
Medical requirements, 95–98
Memorandum of Understanding
    Concerning Licensing of
    Private Remote Sensing
    Satellite Systems (Remote
    Sensing MOU), 82
MEOs. *See* Middle Earth orbits
Mercury spaceflight program. *See*
    Project Mercury
*Method of Attaining Extreme
    Altitude, A* (Goddard), 31

Microgravity, 20
Micrometeoroids, 22–24
Mid-Atlantic Regional Spaceport,
    55, 108
Middle Earth orbits (MEOs), 6
Mir space station, 44
MLAs. *See* Manufacturing license
    agreements
Mojave Air and Spaceport, 55
Molniya, 7, 34
Molniya orbit, 7
Moon Agreement. *See* Agreement
    Governing the Activities
    of States on the Moon and
    Other Celestial Bodies
Moon rocks, 227
Motion, laws of, 8–9
MPL. *See* Maximum probable loss
MSL. *See* Mars Science Laboratory

NAS Act (1958). *See* National
    Aeronautics and Space Act
    of 1958
NASA. *See* National Aeronautics
    and Space Administration
NASA FAR Supplement
    (NFS), 179
NASA Patent Waiver
    Regulations, 196
National Aeronautics and Space Act
    Amendment, 77
National Aeronautics and Space Act
    of 1958 (NAS Act), 71, 175,
    181, 196
National Aeronautics and Space
    Administration (NASA), 68
    Augustine Report, 78–79
    Challenger accident, 103
    Commercial Crew and Cargo
        Program, 53–54
    Commercial Orbital
        Transportation Service,
        53, 93
    creation of, 33, 71–72, 175

Flight Opportunities Program, 51
government space activities and, 173–174
intellectual property rights and, 196
Landsat Program, 36
long-range plan, 72–73
space-borne observatories, 38
use of nuclear space power sources by, 214–215
National and Commercial Space Programs Act, 133
National Environmental Policy Act (NEPA), 90, 214–215
National Environmental Satellite, Data, and Information Service (NESDIS), 133
National Industrial Security Program (NISP), 191–192
National Industrial Security Program Operating Manual (NISPOM), 192
National Oceanic and Atmospheric Administration (NOAA)
application process, 134–135
civil environmental and weather satellite programs, 175
commercial remote sensing regulation, 80–81
debris mitigation requirements, 219
government space activities and, 174–175
licensing conditions, 135–137
licensing process, 133–137
licensing requirements, 134
monitoring and compliance requirements, 137
prohibition on collection and release of satellite imagery relating to Israel, 137
weather observation satellites, 35
National Science Foundation, 176

National Telecommunications and Information Administration (NTIA), 122
NATO. *See* North Atlantic Treaty Organization
NEPA. *See* National Environmental Policy Act
NESDIS. *See* National Environmental Satellite, Data, and Information Service
New Horizons, 16
New Mexico, 108, 110
New Shepard, 54, 56, 57
Newton, Isaac, 8
Newton's cannonball, 9, *10*
NFS. *See* NASA FAR Supplement
NGSO-like satellite system licensing, 125–126
NISPOM. *See* National Industrial Security Program Operating Manual
Nixon, Richard M., 75–76
NOAA. *See* National Oceanic and Atmospheric Administration (NOAA)
Non-self-executing treaties, 60
North Atlantic Treaty Organization (NATO), 151
NTIA. *See* National Telecommunications and Information Administration
Nuclear power, 15–16
Nuclear power sources, 213–215

OCI. *See* Organizational conflict of interest
OFAC. *See* Office for Foreign Assets Control
Office for Foreign Assets Control (OFAC), 141
Office of Outer Space Affairs (OOSA), 67

Office of Science and Technology Policy, 215
Office of Security Review, 164
Office of Space Commercialization, 175
Oklahoma Air and Space Port, 55
Olsen, Gregory, 52
OOSA. *See* Office of Outer Space Affairs
Open-market Reorganization for the Betterment of International Telecommunications Act of 2000 (ORBIT Act), 126–127
Open Skies policy, 124
Opinio juris, 58
Optional Rules for Arbitration of Disputes Relating to Outer Space Activities., 59
ORBIT Act. *See* Open-market Reorganization for the Betterment of International Telecommunications Act of 2000
Orbit act auctions exemption, 126–127
Orbital Sciences Corporation, 52, 53
Orbital spaceflight, 51–54
Orbits
  delta-v required for various, *12*
  elements of, *4*
  elliptical, 7
  geosynchronous Earth, 6–7
  Lagrange point, 8
  low Earth, 5–6
  middle Earth, 6
  other specialized, 8
  reaching and maintaining, 8–11
  representative earth, *7*
  spacecraft, *5*, 5–8
  terms used to describe, 3–4

Organizational conflict of interest (OCI), 192
Orion Multi-Purpose Crew Vehicle, 45
Outer space
  hazards of, 19–24
  microgravity of, 20
  micrometeoroids and space debris in, 22–24
  practical uses of, 24–26
  radiation in, 20–21
  sustaining human life in, 19–20
  U. N. General Assembly Resolutions, 66–67
Outer Space Treaty (1967)
  development of pace law and, xiv, xv
  law of salvage, 222
  liability in spaceflight and, 104
  licensing commercial spaceflight and, 86
  overview of, 61–63
  patent laws and, 229–230
  protection of outer space environment and, 206–207
  tangible property rights, 225–227

The Paquete Habana 175 U.S. 677, 58
Paragon Space Development Corporation, 53
Paris Convention for the Protection of Industrial Property (1883), 232
Parking orbits, 8
Patent Cooperation Treaty (PCT), 229, 232
Patents, 228–231
Payload reviews, 92
PCT. *See* Patent Cooperation Treaty
Permanent Court of Arbitration, 59
Physical security, technical data, 161

Pioneer, 213
Pioneer 10, 16
Pioneer 11, 16
Planetary protection, 210–212
Plant Variety Protection Act, 195
PM-2, 57
Polar Satellite Launch Vehicle (PSLV), 48
Policy reviews, 88
Postlicensing requirements, 91
Preliminary consultations, 87–88
Presidential Decision Directive 23, 36
Presidential Directive/National Security Council Memorandum No. 25, 215
President's Science Advisory Committee, 73
Principles Relating to Remote Sensing of Earth from Outer Space, 138–139
Principles Relevant to the Use of Nuclear Power Sources in Outer Space (1992), 213–214
Procurement Integrity Act, 193
Project Apollo, 26, 42–43, 75
Project Gemini, 40–42
Project Mercury, 39–40
Property rights
　intellectual property, 228–233
　tangible property, 225–227
Property rights, intellectual, 194–198
Protests, 185
PSLV. *See* Polar Satellite Launch Vehicle
Public domain, 144–145
Public release, 164

Quayle, Dan, 78

Radar Ocean Reconnaissance Satellites (RORSATs), 213
Radiation, 20–21, *22*

Radio Regulations, 120
Radioisotope heater units (RHUs), 15–16, 213
Radioisotope thermoelectric generators (RTGs), 15–16, 213
Reaction engines, *13*, 13, *14*
Record keeping, 164–165
Registration Convention. *See* Convention on the Registration of Objects Launched into Outer Space
Remote sensing, 34–36
Remote Sensing Act (1984). *See* Land Remote-Sensing Commercialization Act of 1984
Remote Sensing MOU. *See* Memorandum of Understanding Concerning Licensing of Private Remote Sensing Satellite Systems
Request for proposal (RFP), 183
Rescue Agreement. *See* Agreement on the Rescue of Astronauts, the Return of Astronauts, and the Return of Objects Launched into Outer Space
Restricted parties, 159–160
Reusable launch vehicle (RLV), 49, 86
RHUs. *See* Radioisotope heater units
Rice University, 75
Risk, 103–104
RLV. *See* Reusable launch vehicle
Rocketplane Global, 51
Rocketplane Kistler, 53
RORSATs. *See* Radar Ocean Reconnaissance Satellites
RTGs. *See* Radioisotope thermoelectric generators
Russia, 25

Russian Academy of Sciences, 218
Rutan, Burt, 48, 49

SAA. *See* Space Act Agreement
Safety Framework for Nuclear Power Source Applications in Outer Space, 214
Safety reviews, 88–90
Salvage laws, 222
Satellite Industry Association (SIA), 165
Satellite navigation systems, 25
Satellite Orbital Conjunction Reports Assessing Threatening Encounters in Space (SOCRATES), 218
Satellite systems
  components of, *17*, 17–18
  constellations of, 18
  GPS constellation of, *19*
  threats and security, 18
Satellites
  direct broadcasting, 126–129
  earth observation, 25–26
  launching of first artificial, xiv
  licensing private Earth remote sensing, 133–140
  licensing private telecommunications, 117–131
  operational, *6*
  telecommunications, 24–25, 33–34
Saturn V rocket, *15*, 42, *43*
Sealed bidding, 183
Self-executing treaties, 60
SFPs. *See* Space flight participants
Shenzhou 5, 46
Shepard, Alan, 40
SHF. *See* Super-high frequency
Shuttleworth, Mark, 52
SIA. *See* Satellite Industry Association

Sierra Nevada Corporation, 51, 53, 54
Simonyi, Charles, 52
Sirius radio, 18
SOCRATES. *See* Satellite Orbital Conjunction Reports Assessing Threatening Encounters in Space
Solar arrays, 15
Solicitation process, 182–185
Soviet Union, xiv, 34, 39, 42, 213
Soyuz, 3, 42–43, 47, 49, 54
Space Act Agreement (SAA), 181
Space Adventures, 54
Space-borne observatories, 38–39
Space debris
  growth mitigation, 218–221
  micrometeoroids and, 22–24
  overview of, 215–217
  removal, 221–222
  tracking, 22, *23*, *216*, 217–218
Space exploration, 26, 37–45
Space Exploration Technologies Corp. (SpaceX), *xvi*, 3, 51, 54, 93–94
Space Flight Liability and Immunity Act of 2007, 109
Space flight participants (SFPs), 80, 84, 95–97
Space Frequency Coordination Group, 69
Space Launch System, 45
Space law
  birth of, xiv
  codifying as independent body of law, 82
  development of, xiv
  development of U.S., 71–82
  international sources of, 60–67
  organizations with connections to, *70*
  in period of transition, xv–xvii

Space race, xiv
　manned spaceflight, 39–45
　unmanned spaceflight, 33–39
Space Shuttle, 2, 3, *23*, 43–44, 75–76
Space situational awareness
　(SSA), 217
Space Surveillance Network
　(SSN), 218
Space tourism
　commercial spaceflight and, 49–51
　International Space Station and, 52
　liability laws, 107–113
　practical uses of outer space
　　and, 26
Space Transportation System (STS),
　43–44, 75–76
Spacecraft propulsion systems,
　13–15
Spaceflight
　commercial, 48–55
　contractual risk allocation,
　　115–116
　history of, 31–55
　insurance practices, 113–115
　liability and insurance issues for
　　private, 103–116
　liability in, 104–113
　manned, 39–45
　operations, 1–3
　other national space programs,
　　46–48
　physics of, 3–11
　power and propulsion, 11–16
　sources of risk in, 103–104
　space race, 33–45
　theory and early practice, 31–33
　unmanned, 33–39
Spaceport America, 55, 108
Spaceports, 55, *108*, 108–109
SpaceShipOne, 48
SpaceShipTwo, 3, 14, 55, 210
SpaceX. *See* Space Exploration
　Technologies Corp.

Special export controls (SECs), 151
Spectrum management, 123–124
Sputnik 1, xiv, 33, 210
Sputnik 2, 33
SRLV. *See* Suborbital RLV
SSA. *See* Space situational awareness
SSA Sharing Program, 218
SSN. *See* Space Surveillance Network
Starfire, 83
State space tourism laws, 105–107
　purpose of legislation, 107
　specific, 107–112
Stockholm Declaration, 206
STS. *See* Space Transportation
　System
Subcontracting, 188–189
Suborbital research platforms, 51
Suborbital RLV (SRLV), 49
Suborbital spaceflight, 49–51
Super-high frequency (SHF), 118
Syncom, 34

T&M contracts. *See* Time-and-
　materials contracts
TAAs. *See* Technical assistance
　agreements
Tangible property, 225–227
TCP. *See* Technology control plan
Technical assistance agreements
　(TAAs), 149
Technical data, 144–145, 196–198
Technical data, safeguarding,
　160–162
　document marking, 161–162
　information security, 161
　physical security, 161
　verification by suppliers and
　　contractors, 162
Technology control plan
　(TCP), 163
Telecommunications satellites
　development of, 33–34
　licensing private, 117–131

Telecommunications satellites, *continued*
   practical uses of outer space and, 24–25
Television Infrared Operational Satellite (TIROS), 35
Telstar, 34
Tenth Amendment, U.S. Constitution, 59
Tereshkova, Valentine, 39
Termination, 187
Texas, 110
Time-and-materials (T&M) contracts, 180
TIROS. *See* Television Infrared Operational Satellite
TIROS-1, 35, *36*
TIROS-N/NOAA satellites, 35
TIROS Operational System (TOS), 35
Title 51, U.S.C., 82
Tito, Dennis, 52
Tokyo Olympic Games, 34
TOS. *See* TIROS Operational System
Tracking, telemetry, and control (TT&C) links, 17
Trade secrets, 231–232
Trail Smelter Arbitration, 205, 206
Training requirements, 95–98
Treaties
   international, 232–233
   interpretation, 60
   major outer space, 61–66
   self-executing/non-self-executing, 60
   as source of international law, 57–58
TRIPS. *See* Agreement on Trade-Related Aspects of Intellectual Property
Truth in Negotiations Act, 193
Tsiolkovsky, Konstantin, 31

TT&C links. *See* Tracking, telemetry, and control links
1248 Report, 166–167

UHF. *See* Ultra-high frequency
Ultra-high frequency (UHF), 118
Uniform Trade Secrets Act, 231
United Launch Alliance, 53
United Nations
   Commission on International Trade Law, 59
   Committee on the Peaceful Uses of Outer Space, xiv, 67, 138, 213, 214, 219–220
   Declaration of the United Nations Conference on the Human Environment, 206
   General Assembly Resolutions, 66–67
   Office of Outer Space Affairs, 67
   Principles Relating to Remote Sensing of Earth from Outer Space, 138–139
   Principles Relevant to the Use of Nuclear Power Sources in Outer Space, 213–214
United States Munitions List, 166
United States Munitions List (USML), 145–147
Unmanned spaceflight
   communications, 33–34
   remote sensing, 34–36
   space exploration, 37–39
UP Aerospace, 49
U.S. Air Force, 25
U.S. Census Bureau, 148
U.S. Commercial Remote Sensing Policy, Fact Sheet, 81
U.S. Constitution
   Article II, 59
   Article VI, 59
   Tenth Amendment, 59

U.S. Court of Federal Claims, 185
U.S. Department of Agriculture, 176
U.S. Department of Commerce
   Bureau of Industry and Security, 141, 154–155
   civil environmental and weather satellite programs, 175
   commercial remote sensing regulation, 80–81
   Export Administration Regulations, 152–155
   export control reform and, 165
   Land Remote Sensing Policy Act, 79
   National Telecommunications and Information Administration, 122
   Office of Space Commercialization, 175
U.S. Department of Defense (DOD), 151, 164, 166, 174–175, 192, 218
U.S. Department of Energy, 176
U.S. Department of State, 141, 165, 166, 176
U.S. Department of the Treasury, 141
U.S. Department of Transportation (DOT), 76, 85–86
U.S. Geological Survey, 36, 175
U.S. NAVSTAR Global Positioning System. *See* Global Positioning System
U.S. Strategic Command, 218
U.S. Supreme Court, 58
U.S.C., Title 51, 82

V-2 rocket, 33, 34
Vega, 47
Vienna Convention on Law of Treaties, 57
Viking, 16, 213
Virgin Galactic
   space tourism and, 26
   SpaceShipTwo, 3, 14, 210
   suborbital commercial spaceflight and, 49, 51, 55
Virginia, 108, 109, 110
Voshkod capsule, 42
Voyager 1, 2, 37, *38*
Voyager 2, 37, *38*
Voyager (NASA), 16, 213
Voyager (Virgin Galactic), 55

Waivers, 92–94
Warehousing and distribution agreements, 149–150
Warning statements, 111–112
WCT. *See* World Intellectual Property Organization Copyright Treaty
Whipple, Fred, 24
Whipple shields, 24
White, Edward H., II, 42
White Sands Missile Range, 34
WhiteKnightTwo, 55–56
Wiesner, Jerome B., 73
Wiesner Report, 73–74
World Intellectual Property Organization Copyright Treaty (WCT), 233
World Radiocommunication Conferences (WRCs), 120
World Trade Organization (WTO), 127
World War II, 33
WTO. *See* World Trade Organization

XCOR Aerospace, 26, 49, 51, 56
XM radio, 18

Yang Liwei, 46